The series Springer Proceedings in Energy covers a broad range of multidisciplinary subjects in those research fields closely related to present and future forms of energy as a resource for human societies. Typically based on material presented at conferences, workshops and similar scientific meetings, volumes published in this series will constitute comprehensive state-of-the-art references on energy-related science and technology studies. The subjects of these conferences will fall typically within these broad categories:

- Energy Efficiency
- Fossil Fuels
- Nuclear Energy
- Policy, Economics, Management & Transport
- Renewable and Green Energy
- Systems, Storage and Harvesting
- Materials for Energy

eBook Volumes in the Springer Proceedings in Energy will be available online in the world's most extensive eBook collection, as part of the Springer Energy eBook Collection. To submit a proposal or for further inquiries, please contact the Springer Editor in your region:

Kamiya Khatter (India)
Email: kamiya.khatter@springer.com
Loyola D'Silva (All other countries)
Email: loyola.dsilva@springer.com

Springer Proceedings in Energy

Akhila Kumar Sahu · Bhim Charan Meikap ·
Vamsi Krishna Kudapa
Editors

Energy Storage and Conservation

Select Proceedings from MESC 2022

Springer

Editors
Akhila Kumar Sahu
Central Electrochemical Research Institute (CECRI)
Chennai, India

Bhim Charan Meikap
Department of Chemical Engineering
Indian Institute of Technology Kharagpur
Kharagpur, India

Vamsi Krishna Kudapa
Department of Chemical Engineering
University of Petroleum and Energy Studies
Dehradun, India

ISSN 2352-2534 ISSN 2352-2542 (electronic)
Springer Proceedings in Energy
ISBN 978-981-99-2869-9 ISBN 978-981-99-2870-5 (eBook)
https://doi.org/10.1007/978-981-99-2870-5

This Springer imprint is published by the registered company Springer Nature Singapore Pte Ltd.
The registered company address is: 152 Beach Road, #21-01/04 Gateway East, Singapore 189721, Singapore

Steering Committee Members

(1) Dr. Sunil Rai, Vice Chancellor, UPES, Dehradun, Uttarakhand, India.
(2) Professor Gurvinder Singh Virk, Dean School of Engineering, UPES, Dehradun, Uttarakhand, India.
(3) Dr. Nilanjana Banerjee, Head, Energy Cluster, School of Engineering, UPES, Dehradun, Uttarakhand, India.
(4) Dr. Uday Bhan, Program Head, Energy Cluster, School of Engineering, UPES, Dehradun, Uttarakhand, India.
(5) Dr. Satyabrata Nayak, Principal, Department of Petroleum Geosciences, PETRONAS, Kuala Lumpur, Malaysia.
(6) Dr. Nagababu Andraju, Postdoc Research Fellow, School of Electrical Engineering & Computer Science, University of North Dakota, Grand Forks, North Dakota 58202, United States.
(7) Dr. Ibrahim Al Rajawy, Professor, Lincoln University College, Malaysia.
(8) Dr. K. J. Sabareesaan, Head of Section, Petroleum Engineering Department, University of Technology and Applied Sciences, Nizwa, Oman.
(9) Dr. Akhila Kumar Sahu Principal Scientist, CSIR—Central Electrochemical Research Institute, Karaikudi, Tamil Nadu, India.
(10) Dr. Vamsi Krishna Kudapa, Assistant Professor (SG), Department of Chemical & Petroleum Engineering, Energy Cluster, UPES, Dehradun-248007, India.
(11) Dr. Suresh Kumar Govindarajan, Professor, Department of Ocean Engineering, IIT Madras, India.
(12) Mr. Debi Prasad Dash, Executive Director, India Energy Storage Alliance (IESA), Pune, Maharashtra, India.
(13) Mr. Sreenivas Jindam, Head of Energy, Racanaa Energy, Bengaluru, Karnataka, India.
(14) Dr. Krunal M. Gangawane, Assistant Professor, Department of Chemical Engineering, NIT Rourkela, India.
(15) Dr. Praveen Kumar G., Assistant Professor, Department of Chemical Engineering, NIT Calicut, India.

(16) Dr. Sreedevi Upadhyayula, Professor, Department of Chemical Engineering, IIT Delhi, India.
(17) Dr. S. V. S. Raja Prasad, Sr. Associate Professor, National Institute of Construction Management and Research, Hyderabad, Telangana, India.
(18) Dr. Anjireddy Bhavanam, Assistant Professor, Department of Chemical Engineering, NIT Jalandhar, Punjab, India.
(19) Dr. Katlakanti Mohan Reddy, Sr. Associate Professor, Department of Chemistry, Basic Science Cluster, UPES, Dehradun-248007, India.
(20) Dr. Annapurna Boruah, Associate Professor, Department of Chemical & Petroleum Engineering, Energy Cluster, UPES, Dehradun-248007, India
(21) T. Subrahmanyam, Managing Director, Elite Natural Oils & Fuels Pvt. Ltd., Rajahmundry, Andhra Pradesh.
(22) Professor Paritosh Mohanty, Professor, Department of Chemistry, IIT Roorkee
(23) Dr. Tapas Kumar Dora, Associate Professor, Department of Chemical Engineering, GMRIT, RAJAM, Andhra Pradesh.
(24) Dr. M. J. A. Prince, Assistant Professor, Department of Petroleum Engineering, AMET University, Chennai, India.
(25) Dr. Venkata Ramana Avula, Associate Professor, Department of Petroleum Engineering, GIET, Rajahmundry, India.

Program Committee Members

(1) Dr. Kumargaurao D. Punase, Assistant Professor (SG), Energy Cluster, School of Engineering, UPES, Dehradun.
(2) Dr. Vamsi Krishna Kudapa, Assistant Professor (SG), Energy Cluster, School of Engineering, UPES, Dehradun.
(3) Ms. Nitu Kumari Thakur, Assistant Professor, Energy Cluster, School of Engineering, UPES, Dehradun.
(4) Dr. Ashoutosh Panday, Professor, Energy Cluster, School of Engineering, UPES, Dehradun.
(5) Dr. Pushpa Sharma, Professor, Energy Cluster, School of Engineering, UPES, Dehradun.
(6) Dr. Vijay Parthasarthy, Professor, Energy Cluster, School of Engineering, UPES, Dehradun.
(7) Mr. Bhalchandra Shingan, Assistant Professor (SG), Energy Cluster, School of Engineering, UPES, Dehradun.
(8) Dr. Seim Timung, Assistant Professor (SG), Energy Cluster, School of Engineering, UPES, Dehradun.
(9) Dr. Tirumala Rao Kotni, Assistant Professor (SG), Energy Cluster, School of Engineering, UPES, Dehradun.
(10) Dr. Upendra Singh Yadav, Assistant Professor (SG), Energy Cluster, School of Engineering, UPES, Dehradun.

Preface

Energy systems are in a constant state of flux and change as a result of the proliferation of alternative energy sources, advances in technology, growing consumer demand, rising prices, and negative effects on the environment. The traditional sources of energy generation are fossil fuels; however, the application of renewable energy, solar, and wind energy is increasing in meeting present energy demand with current innovative technologies. Even though there has been a steady rise in the cost of energy, the demands of customers continue to rise at a rapid pace. This is due to an increase in the number of people, as well as economic growth, consumption on a per capita basis, supply in remote locations, and static forms for machines and portable devices. It is possible that the energy storage will make flexible generation and delivery of stable electricity possible for satisfying customer demands. A group of academics, industry professionals, and researchers work was included in this book, and it provides a concise understanding of the many different technologies that are used for energy storage and conservation, as well as the challenges that are involved in storing energy and the opportunities that are available in energy conservation. In addition to this, it discusses the various approaches that can be taken in the oil and gas industry to maximize energy efficiency. This book offers in-depth information on a variety of opportunities and challenges in the field of energy storage and conservation (Both Renewable and Non-Renewable).

Altogether, the conclave receives 90 papers, including oral and poster presentations. All the 90 papers were reviewed by allotting two reviewers for each paper. The process followed in selecting and forwarding the papers to two reviewers is as follows:

1. Authors have submitted papers through Easy chair to Energy Conclave 2022.
2. After submission, the Editor from UPES had checked the plagiarism using Turnitin.
3. The papers with plagiarism > 15% is sent back to author, and the rest of the papers were allotted with two reviewers (One Reviewer from UPES and another outside UPES).
4. 10 days' time is given for reviewing the paper and sending the comments.

5. Received reviewer comments were forwarded to authors for necessary corrections, depending upon the nature of corrections editor from UPES has decided to accept or reject the paper.
6. The editor decision is communicated to author.
7. After completing the review process, the editor from UPES has selected 43 Papers for the event and submitting to Springer Nature.

Chennai, India — Akhila Kumar Sahu
Kharagpur, India — Bhim Charan Meikap
Dehradun, India — Vamsi Krishna Kudapa

Contents

About the Editors

Akhila Kumar Sahu Senior Scientist, serving Central Electrochemical Research Institute—Madras Unit, CSIR Madras Complex, Chennai since 2008. He obtained his Ph.D. degree in Chemistry from the University of Madras, India in 2009. He was a Senior Research Fellow at National Aerospace Laboratories, Bangalore, during the period 2002–2005. In 2005, he joined Central Electrochemical Research Institute as a Senior Research Fellow and became Research Associate in the same institute during 2006–2008 and devoted his research work to the field of Materials Sciences and Fuel Cell Technologies. He worked as a Post-Doctoral Researcher, at DGIST, Daegu, South Korea, for 1 year during 2012–2013. Presently, he has about 77 research publications in peer-reviewed journals, 3 international patents, and 1 book chapter to his credit. His present research activities include the development of composite membranes for PEMFCs operation at elevated temperature and reduced relative humidity, new membrane materials for mitigating methanol cross-over in DMFCs, Synthesis of nano-porous/mesoporous carbon materials for fuel cell electrodes, Pt-based, non-precious and non-metallic electrocatalysts for oxygen reduction reaction (ORR), etc.

Bhim Charan Meikap Professor & Head Department of Chemical Engineering serving the Indian Institute of Technology, Kharagpur since 2004. He obtained his Ph.D. degree in Chemical Engineering from the Indian Institute of Technology, Kharagpur in 2001. In 2004, he joined the Asian Institute of Technology (AIT), in Bangkok, Thailand as Visiting Assistant Professor. He also worked as an Assistant Professor National Institute of Technology since 1997. He was also the Chairman, of GATE/JAM 2015, Vice-Chairman, of GATE/JAM 2014 & 2013. His recent work on CO_2 adoption has been widely appreciated. He has received Excellence Award from NBCFDC, Ministry of Social Justice, Govt. of India awarded to IIT Kharagpur for demonstrating successfully the CSR project and recipient of the National Design Award in Environmental Engineering by Institution of Engineers (India), NDRF (2015). Presently, he has about 187 research publications in peer-reviewed journals, 3 international patents, and 6 book chapters to his credit. His present research activities include Pollution Control, Wet scrubbing of air pollutants, Mass Cultivation Of

Microalgae for the Production of High-Value Biofuel Fraction, and Production of Microporous Activated Carbon From Palm Shell By Microwaves For Removal Of Hazardous Gases And Liquid Pollutants to name a few.

Vamsi Krishna Kudapa Sr. Assistant Professor, working in the Department of Chemical and Petroleum Engineering, UPES, Dehradun since 2013. Before UPES, he worked as an Assistant Professor and Head of the Department of Petroleum Engineering, at Sri Aditya Engineering College, Andhra Pradesh. He pursued his Ph.D. in Oil & Gas—Modeling and Simulation from UPES, Dehradun in 2018. In his 10 years of experience labs like the Drilling Fluids & Cementation Lab and Petroleum Product, Testing Lab was developed. In addition, a unique prototype model "Digital Oil Field" was developed at UPES in 2015, which is a combination of a gas lift and oil-spill control mechanism apparatus. Presently, he has 16 publications, 2 patents, and 2 book chapters to his credit. He is presently working on solar energy applications in agriculture, nanoparticle applications in drilling fluids and cementation, enhanced oil recovery, modeling, and simulation of unconventional gas reservoirs.

Chapter 1
A Novel Gel Polymer Electrolyte for Flexible Supercapacitor with High Mechanical and Self-Healing Properties

J. Shodmanov and A. Boymirzaev

Abstract In this work, the method to improve mechanical properties of gel polymer electrolytes, testing ionic conductivity, and increasing self-healing ability were investigated. Gellan with polyacrylamide that have been cross-linked via Na^+ ions have good mechanical and self-healing characteristics. The results showed high tensile stress–strain curves compare before and after self-healing mechanical properties at different times—0.3, 2, 4, 8, 16, 24 h and various temperatures—0, 20, 40, 60, 80, 90 °C indicated that gel polymer electrolytes have good self-healing performance and tensile characteristics. As a result of such polymer electrolyte features, supercapacitors have a high specific capacitance.

Keywords Energy storage materials · Supercapacitor · Electrolyte · Self-healing · Double network · Gellan

1.1 Introduction

Nowadays, to save energy use a capacitor, battery and supercapacitors. Among them, supercapacitors (SC) are attractive alternatives to batteries because they can be charged very quickly and can sustain the vastly greater number of recharge cycles than batteries without losing efficiency. SCs are a safe, energy-saving device that combines the characteristics of a battery and a capacitor with a fast charging and long service life [1–3]. They can be used in all areas of industry today: including machinery, robotics, electronic devices, telephones, computers, etc.

SC devices have been developed the storing electrical charge at the interface between electrode materials and electrolyte. Among them, the electrolyte is a crucial part and play an important role to determine electrochemical stable potential window, rate capability and life cycle of SCs [4]. In the past decades, aqueous, organic, ionic liquids and polymer gels types have been widely used for SCs. Liquid electrolytes

J. Shodmanov · A. Boymirzaev (✉)
Namangan Institute of Engineering and Technology, Namangan, Uzbekistan 160115
e-mail: a.boymirzayev@nammti.uz

A. K. Sahu et al. (eds.), *Energy Storage and Conservation*, Springer Proceedings in Energy, https://doi.org/10.1007/978-981-99-2870-5_1

face several drawbacks that can be encountered such as electrolyte leakage, corrosion or packaging difficulties [5].

Gel-polymer electrolytes (GPE) as a kind of important flexible electrolyte has attracted increasing attention because of its minimum leakage compared to liquid electrolyte and much higher ionic conductivity compared to the solid polymer electrolyte [6, 7]. Besides, GPEs have much dominance such as natural, lightweight, ease of fabrication, non-toxic, biodegradable, non-corrosive, variable geometry shape and which may bring new design chances for energy storage devices in the future flexible SCs [8]. Nevertheless, the application of GPEs in specific areas is still hampered because of their feeble mechanical properties and there are few informed to improve them. However, several hydrogels methods with magnificent mechanical performance has been successfully developed. For example, hydrophobically association [13], slide-ring [9], interpenetrating network [10], macromonomer hydrogel [11], nanocomposite [12], and double network (DN) structure [14, 15]. Among them, double network (DN) structure, soft and highly extensible hydrogels with relatively homogeneous network structures are developed or by combining two a rigid and brittle network with a soft and ductile network [16]. It is well demonstrated that DN hydrogel can efficiently disperse the locally applied stress and dissipate the energy through the combination of two networks, thus enhancing the hydrogel's mechanical strength. For this ability, DN hydrogels have selected increasing interests and we focused on improving the mechanical properties of GPE by this method.

1.2 Result and Discussion

To make a DN self-healing GPE with excellent ionic conductivity and mechanical properties, a different amount of gellan gum (GG) was introduced (denoted as PG DN GPE), and the major tensile stress and strain curve characteristics of DN GPE were significantly increased by the GG concentration. Polyacrylamide (PAM) and Na_2SO_4 were constant at this time. In Fig. 1.1a, b, we have given the typical tensile stress–strain curves of pure PAM and PG DN GPE samples. The concentration of GG has slowly increased from 0 to 5 g, and the result has shown the pure PAM electrolyte without GG was soft (elongation break-10 mm/mm, stress-0.6 MPa). The pure GG hydrogel cross-linked via Na^+ with each other without PAM was so weak that it failed to withstand tester clamping. In contrast, when the GC amount reaches 3 g (denoted PG-3 DN GPE), the highest stress reaches 2 MPa, which is almost four times higher, and the corresponding elongation break is almost 400% larger than that of the pure PAM hydrogel. These outcomes demonstrate that the amount of GG greatly influences the mechanical properties of hydrogels. After continuing to increase the GG content, the mechanical abilities started to decline. The reason for this is that the PAM and GG mixture solution becomes so viscous when the GG amount exceeds 3 g, which may cause inhomogeneity and many tiny bubbles so that the mechanical properties decrease. The water contents and ionic conductivity exhibit a similar rising trend, with GG content increasing from 0 to 5 g. The key to self-healing DN GPE was the use of GG doped into the PAM solution and subsequently converted to a hydrogel by adding Na^+ under basic conditions. Na_2SO_4 plays two critical roles in this case: ion mobility and crass-linked agent. Therefore, we determined the most suitable Na_2SO_4 content to keep a balance between the conductivity of the gel and its effect on its physical performance. The ionic conductivity of the PG DN GPE rapidly increases from 0.0005 to 0.29 Sm/cm^2 with the increase in Na_2SO_4 concentration from 0 to 1 mol (Fig. 1.1e, f). Caused by the salts and enhanced ionic conductivity, consequently increasing the number of conductive ions. The equilibrium water contents of the GPE (shown in Fig. 1.1f) decreased with increasing Na_2SO_4 content, possibly because the salts increased the sample density and also because the formation of the double cross-linked network made the hydrogel structure more compact and prevented the penetration of water molecules.

Considering the physical nature of GG and PAM could demonstrate self-recovery property under appropriate conditions. Curiously, the PG 3 GPE could automatically repair itself when suffering physical damage. This kind of ability allows us to use electrolytes for a long time. For this, PG GPE was surveyed by cutting the dumbbell-shape electrolyte into two pieces completely and then the pieces were physically contacted together without applying any external force to determine self-healing ability (Fig. 1.2a–d). When the GPE was cut into two parts, hydrogen bonds were cleaved at the cut interfaces. However, once the two parts contacted together again, the broken hydrogen bonds were prone to link together. The recombination of hydrogen bonds enabled the severed electrolyte to heal itself, leading to the recovery of the electronic conductivity along with the hydrogel. We further quantitatively evaluated the self-healing efficiency by comparing the tensile properties of original and self-healed PG 3 DN GPE at different temperatures and times. In Fig. 1.2, tensile stress-strain curves before and after self-healing mechanical properties at different times 0.3, 2, 4, 8, 16 and 24 h, various temperatures 0, 20, 40, 60, 80 and 90 °C.

Our team continues investigation to improve the mechanical properties of GPE in the future work, and we are planning to use the method of High-Performance Size Exclusion Chromatography [17, 18] to establish the quantitative relationship between molar mass and tensile strain-stress, elongation degree of polymers in GPE. This kind of method allows expressing the determination of molar masses and polydispersity degrees of GPE components and gives reliable data for the optimal choice of polymer type in GPE to reach better performance exploitation characteristics of SC product.

1.3 Conclusion

In summary, a novel GPE was successfully created, exhibits outstanding mechanical and ionic conductivity properties, and is applied to a flexible supercapacitor. The self-healing ability of electrolyte gives a long time use for the supercapacitor under deformation conditions. Our scientific team continues to improve the mechanical properties of GPE with a high electrochemical performance for energy storage materials.

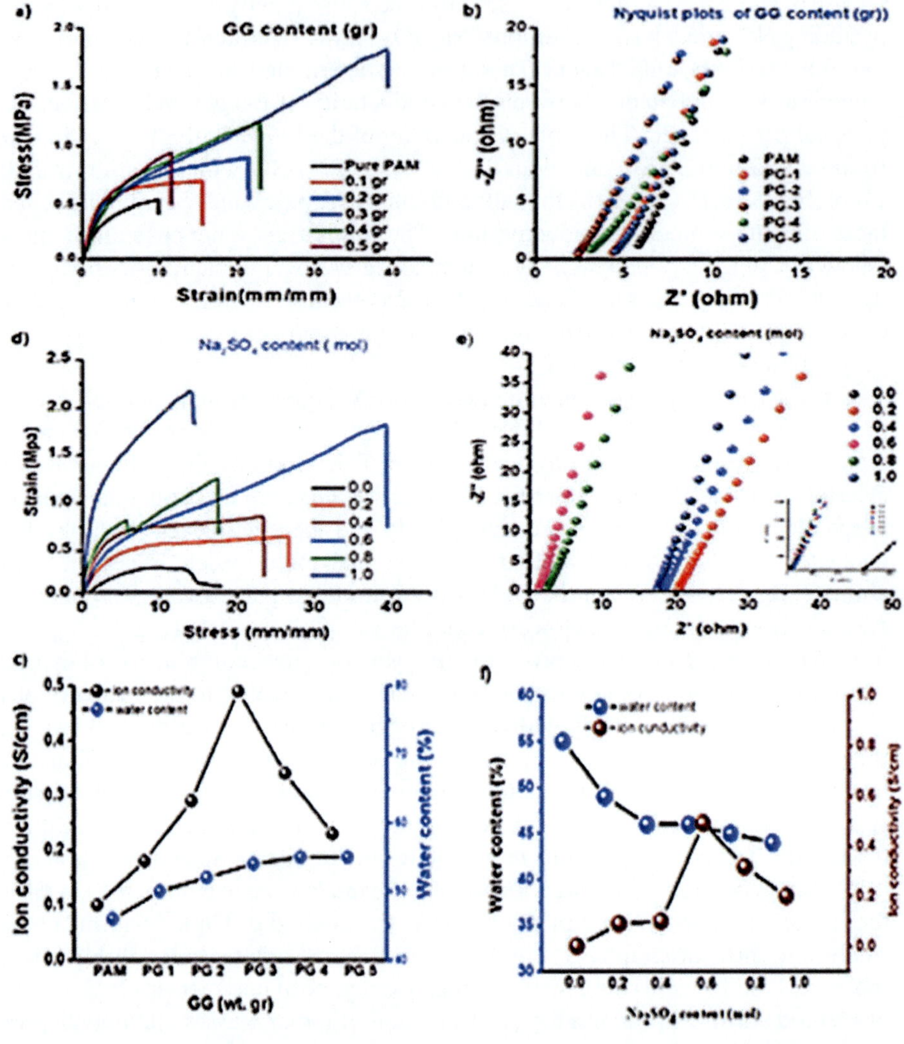

Fig. 1.1 Compare tensile stress–strain curves of various GG content (**a**) Nyquist plots (**b**) effect of GG content on the tensile stress and elongation at break (**c**). Compare tensile stress–strain curves of various Na_2SO_4 content (**d**) Nyquist plots (**e**) effect of GG content on the (**f**) tensile stress and elongation at break

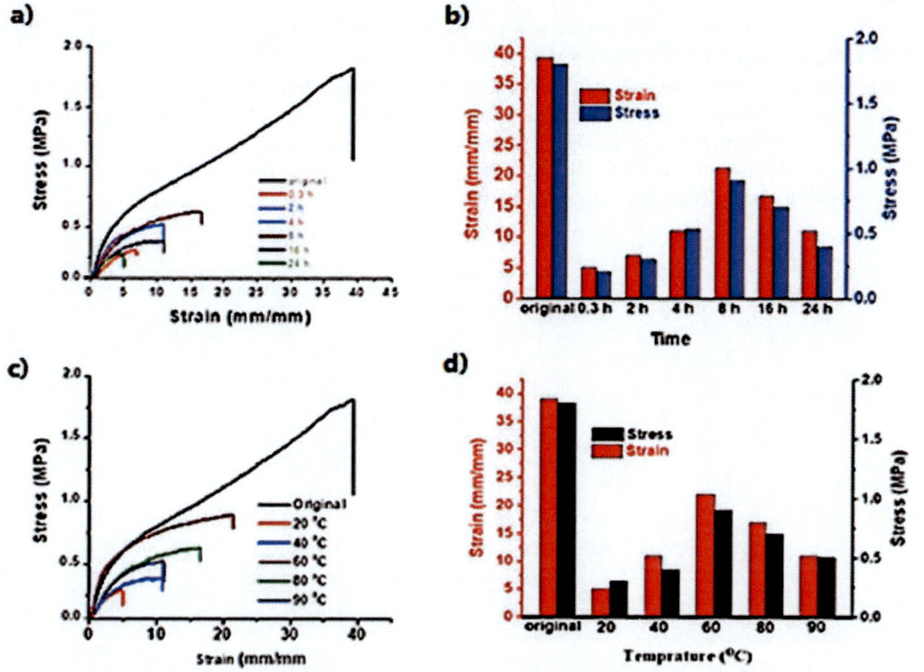

Fig. 1.2 Self-healing test: tensile strain–stress curves after healing different time (**a**) and temperature (**c**), tensile stress and elongation at break respectively (**b**, **d**)

References

1. T. Lin, M. Shi, F. Huang, J. Peng, Q. Bai, J. Li, M. Zhai, ACS Appl. Mater. Interfaces **10**(35), 29684–29693 (2018)
2. J. Wang, F. Liu, F. Tao, Q. Pan, ACS Appl. Mater. Interfaces **9**(33), 27745–27753 (2017)
3. X. Zang, R. Zhang, Z. Zhen, W. Lai, C. Yang, F. Kang, H. Zhu, Nano Energy **40**, 224–232 (2017)
4. W.G. Moon, G.-P. Kim, M. Lee, H.D. Song, J. Yi, ACS Appl. Mater. Interfaces **7**(6), 3503–3511 (2015)
5. M. Hu, J. Wang, J. Liu, P. Wang, Y. Feng, H. Wang, N. Nie, Y. Wang, Y. Huang, Energy Storage Mater. **21**, 174–179 (2019)
6. X. li, L. Liu, X.-Z. Wang, Y.S. Ok, J. Elliott, S. Chang, H.-j. Chung, Sci. Rep. 7 (2017)
7. L. Dagousset, G. Pognon, G.T.M. Nguyen, F. Vidal, S. Jus, P.-H. Aubert, J. Power Sources **391**, 86–93 (2018)
8. H. Wu, Y. Cao, H. Su, C. Wang, Tough. Angew. Chem. **130**(5), 1375–1379 (2018)
9. X. Cheng, J. Pan, Y. Zhao, M. Liao, H. Peng, Adv. Energy Mater. **8**(7), 1702184 (2018)
10. A. Song, Y. Huang, X. Zhong, H. Cao, B. Liu, Y. Lin, M. Wang, X. Li, Electrochim. Acta **245**, 981–992 (2017)
11. A.J. D'Angelo, M.J. Panzer, Design of stretchable and self-healing. Chem. Mater. **31**(8), 2913–2922 (2019)
12. Y. Lin, H. Zhang, H. Liao, Y. Zhao, K. Li, Chem. Eng. J. **367**, 139–148 (2019)
13. Y. Bai, C. Xiong, F. Wei, J. Li, Y. Shu, D. Liu, Energy Fuels **29**(2), 447–458 (2015)
14. Q. Chen, H. Chen, I. Zhu, J. Zheng, J. Mater. Chem. B **3**(18), 3654–3676 (2015)

15. Q. Chen, X. Yan, L. Zhu, H. Chen, B. Jiang, D. Wei, L. Huang, J. Yang, B. Liu, J. Zheng, Chem. Mater. **28**(16), 5710–5720 (2016)
16. J.P. Gong, Y. Katsuyama, T. Kurokawa, Y. Osada, Adv. Mater. **15**(14), 1155–1158 (2003)
17. G.P. Aleksandrova, A.N. Sapozhnikov, A.S. Boymirzaev, B.G. Sukhov, B.A. Trofimov, Russ. J. Gen. Chem. **90**(4), 672–679 (2020)
18. G.P. Aleksandrova, A.S. Boymirzaev, I.V. Klimenkov, B.G. Sukhov, B.A. Trofimov, Nanotechnol. Russ. **14**(1–2), 41–47 (2019)

Chapter 2
Comprehensive Review on the Selection of Materials in City Gas Distribution Value Chain

Bhalchandra Shingan, Mynam Harshita, Nandini Verma, Manish Joshi, and Hrishabh Vishwakarma

Abstract Because of rising energy utilization, natural gas appears to be essential for a country and assumes a critical part in financial development. In Indian market of natural gas, city gas distribution is becoming as a growing area. The pipeline is now becoming a sensible and safe method for transmission of gas from the production field to the downstream clients or petroleum refineries. The issue of guaranteeing the protected activity of gas pipelines turns out to be progressively significant over the long run. In the event that the pipeline is an unforeseen disappointment, it will get an enormous size of annihilation and disturbance to society. The choice of matcrials for pipeline requires thought of plan, development, activities, upkeep, dangers, perils, well-being and financial matters. Codes and standards give obligatory and discretionary prerequisites and rules for the choice of materials. This survey of the course provides the project regarding the material availability of the pipeline for the city gas distribution sector. Additionally, the paper portrays methodology by which a decent way to deal with the usages of guidelines and codes for the vital administrative and obligatory consistence.

Keywords CGD · Steel pipelines · PE pipelines · Distribution system · Materials selection

2.1 Introduction

As worldwide energy request rises, gas presently assumes a major key half in energy provide. It's additional laborious to move and store gas than oil and so it remained behind that ware for a major amount. Throughout the foremost recent few decades, this has modified and gas markets keep it up growing than those of different hydrocarbon fuels. Natural gas is that the cleanest fossil gas and hydrogen-rich of all of

B. Shingan (✉) · M. Harshita · N. Verma · M. Joshi · H. Vishwakarma
Department of Chemical Engineering, School of Engineering, University of Petroleum and Energy Studies, Energy Acres Building, Bidholi, Dehradun 248007, India
e-mail: bshingan@ddn.upes.ac.in

A. K. Sahu et al. (eds.), *Energy Storage and Conservation*, Springer Proceedings in Energy, https://doi.org/10.1007/978-981-99-2870-5_2

the natural electricity sources. Of additional importance is that gas assets found but at this time unexploited keep abundant. The world is prepared for intensive development throughout the subsequent 20 years and a few settles for that it would even overwhelm oil because the nice fuel somewhere within the vary of 2020 and 2030 [1]. The pattern towards gas turning into the highest notch fuel of the planet economy is not presently effectively reversible. We have a tendency besides distinguish the mechanical and enterprise problems to be defeated in taking the planet via the progress. There's an important flip closer to gas that nowadays represents around 20 1/3 of the planet electricity hobby. Many gas-delivering nations have started upon very aggressive styles for unambiguously swollen gas yield. Various new LNG offices square measure being created provide chains differentiating and developing to be continually convertible. There's a developing acknowledgment that uncommon wellsprings of gas, as an example, shale gas, coal bed alkane (CBM) and profound tight gas can contribute a crucial a part of future gas provides as advances evolve. The sector of city gas distribution in India is going to increase around 9223 MMSCM (Million Metric Standard Cubic Meter) in 2020 to 25570 MMSCM by 2030, with a 10% CAGR [1]. Market development is relied upon automobile, industrial, business, and personal end-client users. With a growing exhibit of positive measures occupied by the general public authority of Asian country within the CGD market, a developing parcel of corporations square measure being supported [2]. Assisting with greater growing CGD community inclusion to 228 GA covering 86% of the nation's vicinity. The Indian market for dispersion of city gasoline may well be labeled addicted to the kind, supply, stop-use area, and locale. In 2019, countries' fuel dispersion earning has been oil-fired by the CNG category, prompting extreme ecological guidelines and growing entrance of CNG-fitted automobiles in Asian United States [2]. Upheld by using developing PNG and CNG ventures, a brief increment is anticipated in the India's CGD zone over the estimate time span.

2.2 City Gas Distribution

City gas distribution could be a group of pipelines and appliances for natural gas from the transmission line to a medium-pressure distribution system and consequently to service pipelines as represented in Fig. 2.1. Then, the gas is supplied to residential, commercial, industrial or business sectors and CNG stations set in identified areas. As per PNGRB regulations, CGD network in India includes three types of networks: (i) Primary network, (ii) Secondary network, and (iii) Tertiary network. The first network could be a high-pressure distribution system comprising of distribution mains that are ordinarily steel pipes [3]. The target is to provide gas to distribution network or service lines. It consists of combination of MDPE piping and GI or copper piping components. Subtransmission pipelines link to the central transportation pipelines to the CGS, however, area unit closely held by CGD entity.

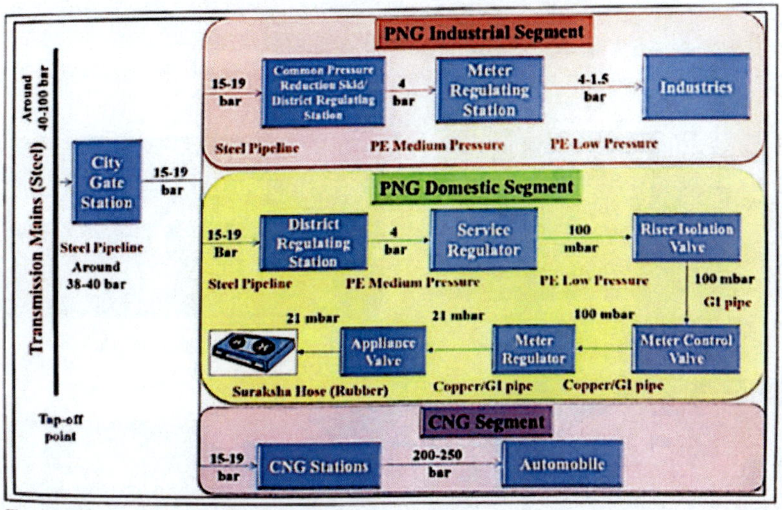

Fig. 2.1 Value chain of CGD [3]

2.3 Selection Criteria of Materials in CGD

The course of choice of pipeline materials is ought to consider ordinary prerequisites like plan, development, pipeline framework activities, support, dangers, risks, security, and financial aspects. It will likewise be fundamental to consent to the relevant guidelines, codes, principles, and particulars, which might incorporate obligatory or discretionary prerequisites and rules [4]. Comprehension, audit, and use of involvement and industry practice are crucial for fruitful materials' execution.

Guidelines, codes, principles, determinations, and experience get updated occasionally. Guidelines and different rules that everyone must follow decide the base prerequisites for safety and ecological consistence.

2.3.1 *Process Conditions for Pipe Material Selection*

The pipeline material selection depends on the following parameters

- Fluid type
- Ambient temperature
- Pressure of operating conditions

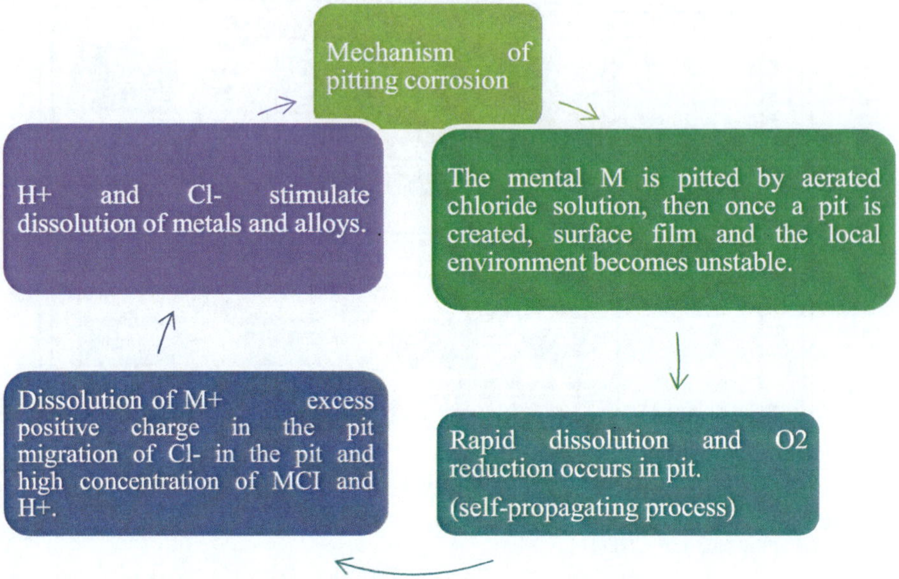

Fig. 2.2 Mechanism of pitting corrosion

First, strategy condition that might influence the decision of texture might be a kind of liquid it transports. For corrosive liquids, you must go for higher erosion obstruction material when contrasted with non-destructive assistance.

- On the contrary hand, conventional steel is enough for non-corrosive liquids like—top off oil, air, nitrogen, and so on.

Second strategy condition that might affect the decision of texture is that the temperature of liquids:

- High temperature
- Low temperature
- Medium temperature
- Cryogenics

Third technique condition that might affect the decision of texture is that the pressure of administration liquids. You wanted high-strength material or higher thickness material for hard-hitting administrations contrast with conventional tension administrations (Fig. 2.2).

2.4 Material Choice

In CGD, material selection depends on pressure level in different types of network [4].

2.4.1 Primary Network

The primary network of pipelines provides the core backbone connecting metric system to varied DRS. The pressure levels for primary network square measure between 26 bar (g) to 19 bar (g). Whereas most of the commercial customers don't seem to be needed to be provided at this pressure, solely a get few units have specific demand for medium pressure delivery that would be connected to through this network. The planning of the first network relies on the demand forecast to be catered [4].

Steel pipelines, square measure trunk pipelines used for transmission of fuel to metric system and downstream of metric system. Gas is transferred from the metric system to CNG stations and DRS via steel pipelines. The pipeline conjointly covers the complete town within the style of ring and used for future iteration and interconnections. These square measure high pipelines operational at a pressure of nineteen bars during a CGD network. Valves square measure put in each two kilometer to regulate the flow. The quality code for steel high pipeline is ASME B 31.8 [5].

2.4.2 Secondary Network

The secondary network system consisting of pipelines operates at pressure between four bars (g) to one bar (g). Pipeline network is planned for cluster of business, commercial, or domestic units at depression. The secondary network is often developed with MDPE pipes.

Polyethylene pipes are widely used everywhere the globe thanks to their best characteristics compared to the traditional ones made of steel and forged iron. The letter of the alphabet pipes among several different blessings is characterized by their high flexibility, lightweight, strain ability, easy pipes affiliation and principally the reduced price of installation.

2.4.3 Tertiary Network

Tertiary network could be an assist with forcing scattering structure containing service regulators administration lines, and client metering systems fabricated using a mix of thermoplastic (MDPE) piping and GI/copper direct parts in a truly tertiary network, the in-action pressure is a more unobtrusive sum than 100 m bar [6]. Circulation main region unit is intended to ensure a continuous proposal to tertiary network or to modern customers through helplines.

2.4.4 GI Pipes and Advantages

Copper pipes are broadly and effectively utilized in petroleum gas benefits and packed air frameworks, basically because of their properties of erosion opposition, effortlessness to join, lightweight and great formability. The course of galvanization is very more straightforward to review as the pipe can be inspected without any problem. Alongside this, the thickness of the lines can be tried utilizing straightforward and non-horrendous techniques. The zinc covering that is given through the course of galvanization gives durability to the lines and makes them rust-free too [7].

2.5 Operations and Maintenance of Pipelines

The primary drivers of pipeline mishaps square measure erosion, gouges plain and crimped scratches, smooth marks on welds, smooth imprints with elective sorts of deformities, delivering deserts inside the line body, circumference and crease weld imperfection, and breaking. Corrosion has constantly been one in everything about premier crucial issues moving the assurance in oil and gas exchange, and its iatrogenic disappointments represent with respect to 25th of all instrumentality disappointments. The water and disastrous medium inside the pipeline influence the speed of utilization. MIC (Microbial disintegration) impediment is sensibly wonderful in oil and gas trade. Particularly, SRB (sulfate diminishing microorganisms) is named MIC story [8].

2.6 Applicable Codes and Standard for CGD [3]

See Tables 2.1, 2.2 and Fig. 2.3.

Table 2.1 Standard code [3]

PE grid	IGTD-2
PE network hydraulics	Synergy software, Gasnet software, pipe phase Simulation software
CNG mother station	OSID-179
Steel grid	ASME B 31.8/OISD-226
CNG RO	OISD-179
Liquid and slurry pipeline	ASME B 31.4
Safety	OISD GDN-192

Table 2.2 For steel pipelines [3]

Fittings	As per ASME A 234
Pipeline execution	ASME B 31.8
Flanges	As per API 105
Steel pipe selection	As Per API 5L
Ball valves	As per API 6D

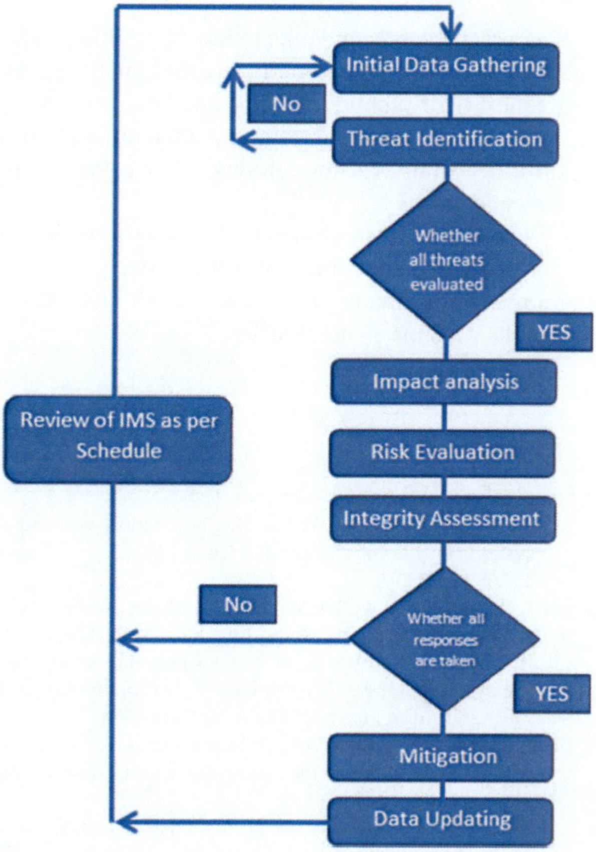

Fig. 2.3 Applicable codes and standards for CGD [3]

2.7 Recent Development

One such improvement is the utilization of Hexane (C6) as a polymer substitution gas (C4) to shape AN olefin Hexane polymer giving better obstruction than mechanical harms and slow break development all through establishment and fix. For PE100 Orange line materials for low/medium strain dispersion framework, the new hexane letter polymer offers a ton of better obstruction than slow break development. In this manner, it's great for Trenchless establishments like HDD or pipe exploding, any place drive the line through the drag inside the ground could well score and scratch the line or covering. Utilizing a Hexane letter administration lifetime of the line isn't impacted notwithstanding the hard establishment methods in view of higher trap of compound chains [9].

2.8 Conclusion

The practice and implementation of a holistic approach to materials selection can result in economic and performance benefits to the pipeline owners and operators. Reliability of pipeline systems and enhancement of the owner's asset value can be augmented through adhering to diligent steps in the material selection. Materials form the vital structures through which the assets, products, revenue, and profits flow and grow.

A systematic assessment of the various factors associated with the materials selection process, concomitant with proper design, construction, operations, maintenance, mitigation of threats, and management of hazards and risks, will help to sustainably operate pipeline systems safely.

References

1. W. Zhuang, S. Zhou, W. Gu, X. Chen, Optimized dispatching of city-scale integrated energy system considering the flexibilities of city gas gate station and line packing. Appl. Energy **290**, 116689 (2021)
2. K. Liao, M. Qin, G. He, N. Yang, S. Zhang, Study on corrosion mechanism and the risk of the shale gas gathering pipelines. Eng. Fail. Anal. **128**, 105622 (2021)
3. R. Narayanaswamy, The process of materials selection for pipeline systems optimization for life cycles, in *ASME India Oil and Gas Pipeline Conference*, vol. 50763 (American Society of Mechanical Engineers, 2017), p. V001T02A001
4. P. Maruschak, I. Danyliuk, O. Prentkovskis, R. Bishchak, A. Pylypenko, A. Sorochak, Degradation of the main gas pipeline material and mechanisms of its fracture. J. Civ. Eng. Manag. **20**(6), 864–872 (2014)
5. M.L. Godec, V.A. Kuuskraa, P. Dipietro, Opportunities for using anthropogenic CO_2 for enhanced oil recovery and CO_2 storage. Energy Fuels **27**(8), 4183–4189 (2013)
6. K. Birkitt, M. Loo-Morrey, C. Sanchez, L. O'Sullivan, Materials aspects associated with the addition of up to 20 mol% hydrogen into an existing natural gas distribution network. Int. J. Hydrogen Energy **46**(23), 12290–12299 (2021)
7. K. Yadav, A. Sircar, Modeling parameters influencing city gas distribution sector based on factor analysis method. Pet. Res. (2021)
8. D. Liu, W. Zhou, X. Pan, Risk evaluation for city gas transmission and distribution system based on information revision. J. Loss Prev. Process Ind. **41**, 194–201 (2016)
9. I. Sannikov, N. Golikov, N. Terentyev, P. Ksenofontov, A. Zhirkov, Investigation of mechanical properties of the main gas pipeline material during a long-term operation in conditions of the North. Procedia Struct. Integr. **30**, 144–148 (2020)

Chapter 3
Scope of Geothermal Energy in Indian Energy Security

Somenath Ganguly and Uday Bhan

Abstract Energy security is a big issue for any nation particularly developing countries like India. With growing Indian economy, energy demand is increasing from industrial to individual level. According to Indian energy scenario, main sources of energy are coal, crude oil, natural gas, solar power, hydropower, nuclear energy and geothermal energy. Due to global warming and climate change, Government of India has imposed regulations/restrictions on energy from fossil fuel and promoting exploring renewable energy sources. Vision of the Indian Government is to achieve "net zero" carbon emission by 2070. Towards this vision, renewable energy like "geothermal energy" will play an important role in Indian energy security in the upcoming days. However, the domain of geothermal energy is very much unexplored so far in India. This work demonstrates the potential of geothermal energy in Indian sub-continent along with its field of applications.

Keywords Geothermal energy · Renewable energy · Global warming · Net zero · Energy security

3.1 Introduction

India is one of the fastest growing economies in the globe. Therefore, there is an ongoing trend of increasing energy demand country is having from industry to individual level. Keeping in mind the total energy consumption of India in terms of compound annual growth rate has increased nearly 5% from the year 2001 to 2011 [1]. Our main source of energy is so far from conventional fossil fuel. Among them coal is the predominant (48%), then comes oil (39%) and natural gas (10%) [1]. Considering the increasing energy demand scenario, a lot of stress has given on the usage of renewable energy like solar energy [2], wind energy, etc. One of them is always "geothermal energy". Geothermal heat is the heat energy stored below the crust. This is an infinite source of heat and can be exploited for many purposes to satisfy

S. Ganguly (✉) · U. Bhan
UPES, Bidholi, via, Prem Nagar, Dehradun, Uttarakhand 248007, India
e-mail: sganguly@ddn.upes.ac.in

A. K. Sahu et al. (eds.), *Energy Storage and Conservation*, Springer Proceedings in Energy, https://doi.org/10.1007/978-981-99-2870-5_3

energy demand of the country. Considering global scenario, total geothermal heat energy utilization is approximately 592,638 TJ/year with a compound growth rate of 6.9% every year. Leading countries utilizing geothermal heat are China, followed by USA and Sweden [3]. Here, an effort has been made to capture the importance of geothermal heat utilization particularly for Indian scenario thus moving towards net zero carbon emission by 2070 as per the standing of Government of India.

3.2 Methodology

In this work, to capture the scope/role of geothermal energy for energy security of India, different government open file reports and published literatures have been consulted. For the year 2010–2011, renewable energy only contributed 1% of the total energy consumption of India. In this domain, contribution of geothermal energy is even small or almost nil. But India has huge potential for geothermal energy.

Geothermal system is an outflow of groundwater having abnormally high temperature [4]. The heat source may be the heat of the earth itself (Hot springs in and around active tectonic plate boundary) or radiogenic heat or heat due to exothermic chemical reactions. There are certain two types of geothermal system [5].

1. Vapor-dominated geothermal system where groundwater temperature is more than 100 °C associated with saturated steam.
2. Water-dominated geothermal system with water temperature range of 60–100 °C.

Hot Dry Rock (HDR) is another type of geothermal system where heat is stored in poorly permeable rock. To extract this heat, "ground source heat pump" is required.

3.2.1 Indian Scenario of Geothermal Energy

Geothermal provinces of India have tremendous potential to contribute towards nation's energy security. In general, geothermal system can be broadly categorized into following classes like hot springs, geysers and fumaroles. Considering Indian geothermal energy scenario, it has been classified into two categories, one is medium enthalpy system (temperature range 100–200 °C), another one is low enthalpy system (temperature < 100 °C) [6]. Heat source for different geothermal provinces is different, which is summarized below [7].

1. For Himalayan region, heat source is younger intrusive granites.
2. West Coast Province includes Son-Narmada-Tapti Lineament Zone, Rajgir-Monghyr in Bihar and Tatapani in Madhya Pradesh it is deep lithospheric heat coming through tectonic fracture/lineament.
3. For Damodar, Godavari and Mahanadi province deep lithospheric heat coming up due to stretching of lithosphere by rifting.

4. For Cambay basin of West Coast, the heat source for tertiary and quaternary sediments are deep lithospheric heat to the subsiding graben due to sediment load.

Geothermal system has been reported from Andaman-Nicobar region in Barren Island also. The geothermal system in this region is manifested by eruption of hot water and occasionally steam (Fumaroles). The reservoir temperature has been estimated >500 °C whereas surface temperature of hot water spring varies from 100 °C and ranges up to 500 °C for fumaroles [8].

3.2.2 Energy Potential of Geothermal System in India

Geothermal energy can be utilized directly as heat source as well as potential to use in thermal power plant for electricity generation. China is the leading nation to use geothermal heat energy with 17,870 MW installed capacity, which accounts approximately 25.2% of global capacity. In the list, the next nation is USA with installed thermal capacity of 17,415.9 MW followed by Sweden with installed thermal capacity of 5600 MW. Considering geothermal power generation till 2021, total installed capacity is approximately 12,729 MW. USA is the leading with installed capacity of 3450 MW, which accounts 28.8% of the global stack. In the list then comes Philippines with installed capacity of 1870 MW and followed by Indonesia with 1340 MW installed capacity [3].

Now considering Indian geothermal provinces, it has potential to generate up to 10,600 MW of power [9]. Potential of geothermal energy as renewable energy is very lately realized in the energy domain of India. Recently, Geological Survey of India (GSI) along with other research collaborates like National Geophysical Research Institute (NGRI) exploring different geothermal provinces of India. Some of the pilot project data are captured in Table 3.1 [10]. Table 3.2 demonstrates thermal data of some major geothermal fields of India [10].

Above discussion and data demonstrate that geothermal energy has huge potential as renewable energy in Indian subcontinent. However, it has been very less

Table 3.1 Pilot project drilling data for some of the geothermal fields of India [10]

Region	Depth (approximate in meters)
Tapoban	728
Manikaran, Jammu & Kashmir, Puga and Ladakh	385
Jammu & Kashmir, Chumathang and Ladakh	220
Chhattisgarh, Tatapani, Surguja	620
Himachal Pradesh and Manikaran	700

Table 3.2 Thermal potential for major Indian geothermal provinces [10]

Geothermal province	Water temperature at the surface (°C)	Reservoir temperature range (°C)	Heat flow (mW/m^2)	Geothermal gradient (°C/km)
Himalaya province	>90	260	468	100
Godavari province	50–60	175–215	93–104	60
Cambay province	40–90	150–175	80–93	70
Sonata province	60–95	105–217	120–290	60–90
West coast province	46–72	102–137	75–129	47–59

explored so far. Gradually more thrust is given in geothermal resource along with other renewable energy sources like wind energy, solar energy, etc.

3.2.3 Indian Energy Security and Scope of Geothermal Energy

Total installed renewable energy capacity of India is around 28 GW, which is approximately 14% of the total nationwide installed capacity. As per TERI [1] "Energy Security Outlook", energy security is defined as "ensuring uninterrupted energy supply to support commercial and economic activities necessary for sustainable growth of economy" for 12th Five-Year Plan (2012–2017). In the 12th Five Year Plan (2012–2017) projected capacity addition for grid-interactive renewable power, the stack of geothermal energy was only 7 MW considering 29,214 MW total renewable energy sector. For the 13th Five-Year Plan (2017–2022), projected growth in renewable energy sector is 34,500 MW. However, in this growth of renewable energy sector what will be the stack of geothermal energy that is not clearly defined in the plan. Recent government policies on promoting renewable energy like Electric Act (2003), National Electricity Policy (2005), National Tariff Policy (2006), Integrated Energy Policy (2006), National Clean Energy Fund, National Action Plan on Climate Change (NAPCC) are giving huge thrust to promote renewable energy considering geothermal energy as one of them where available.

3.3 Discussion and Conclusions

Geothermal energy has huge application in space heating and cooling as well as in thermal power generation particularly in binary cycle power plants. However, thermal power plants run by dry steam and flash steam geothermal heat are also in operation. For Indian scenario, the most favorable configuration is binary cycle power plants since geothermal resources are of medium to low enthalpy range. As per the exploration report of GSI and NGRI, most promising locations for geothermal energy harvesting are Chumathang (Jammu & Kashmir), Puga (Ladakh), Tattapani (Chhattisgarh), Cambay Graben (Gujarat), Manikaran (Himachal Pradesh), Surajkund (Jharkhand), and Bakreshwar (West Bengal). Though having significant geothermal resources, it has been very less explored as a potential energy source by Indian energy industry. Only in 2021, initiative has been taken to establish India's first geothermal power project at Puga in Ladakh. However, significant exploration work in the field of geothermal energy is yet to be done and its proper usage for energy harvesting. This paper is just to capture a small focus towards that.

References

1. TERI (2015) Energy security outlook: defining a secure and sustainable energy future for India. TERI, New Delhi
2. Pandey JK, Aharwal VK (2022) Bypass diodes to improve solar panel efficiency for certain module. J Nano Electron Phys 14(3):03003. https://doi.org/10.21272/jnep.14(3).03003
3. N. Lebbihiat, A. Atia, M. Arıcı, N. Meneceur, Geothermal energy use in Algeria: a review on the current status compared to the worldwide, utilization opportunities and countermeasures. J Clean Prod (2021). https://doi.org/10.1016/j.jclepro.2021.126950
4. E. Barbier, Nature and technology of geothermal energy: a review. Renew Sustain Energy Rev **1**, 1–69 (1997)
5. R.W. Henley, A.J. Ellis, Geothermal systems, ancient and modern: a geochemical review. Earth Sci Rev **19**, 1–50 (1983)
6. H.K. Singh, D. Chandrasekharam, G. Trupti, P. Mohite, B. Singh, C. Varun, S.K. Sinha, Potential geothermal energy resources of India: a review. Curr Sustain/Renew Energy Rep **3**, 80–91 (2016)
7. Sarolkar PB (1994) Subsurface geological studies in Tatapani geothermal field, district Surguja (M.P). Rec GSI 127(6):147–151
8. Manikandan S (2015) Potential of geothermal resources in power generation in India. In: Paper presented as proceedings in world geothermal congress, Melbourne Australia
9. Sircar A, Yadav K (2017) Harnessing geothermal energy: application in India. Technology Publishers, Dehradun, India. ISBN No: 81-901767-5-7
10. K. Yadav, A. Sircar, Geothermal energy provinces in India: a renewable heritage. Int J Geoheritage Park (2020). https://doi.org/10.1016/j.ijgeop.2020.12.002

Chapter 4
Comparative Study of Power Law-Based Rate Models of Ethanol Steam Reforming for Hydrogen Production

Kumargaurao D. Punase

Abstract In the present study, the power law-based kinetic models of ethanol steam reforming from the published literature are evaluated to obtain the higher ethanol conversion and hydrogen yield. The kinetic models developed for Ni(16.1%)–Al_2O_3, Ni(43.6%)–Al–LDH, Ni–Al_2O_3–MgO and Ru–Al_2O_3 catalysts are used based on surface reaction mechanism of ethanol and water. Evaluation of four different is carried out and compared under the operating conditions of temperature, 873 K and the theoretical water/ethanol ratio, 3:1 for a space time of 0–6.97 g_{cat} h/g mol at atmospheric condition. The pseudo-homogeneous model of packed bed reactor is simulated using MATLAB (R2010a) with the assumptions of the steady-state, plug-flow and isothermal conditions. The model-predicted results show the good agreement with the experimental results obtained from the literature.

Keywords Kinetic model · Ethanol steam reforming · Hydrogen · Modeling and simulation

4.1 Introduction

The energy demand has been rapidly increasing over the last decade due to rise in population. The fossil fuel usage has been rapidly increasing over the last few decades at a rate of 30–40% [1]. Various policies and regulations were developed to reduce the emission of carbon. Kyoto Protocol in 1992 provides a solid framework for the European countries to regulate the emission of the CO to about 8% annually. To ease the dependence on the fossil fuels, various alternative sources of energy are being developed. Hydrogen is considered to be future source of energy as more technologies are being developed for the storage and transportation of the hydrogen fuel [1–3]. Hydrogen is produced by various processes especially in the steam reforming

K. D. Punase (✉)
School of Engineering, University of Petroleum and Energy Studies, Dehradun, Uttarakhand 248007, India
e-mail: kgaurav@ddn.upes.ac.in

A. K. Sahu et al. (eds.), *Energy Storage and Conservation*, Springer Proceedings in Energy, https://doi.org/10.1007/978-981-99-2870-5_4

process where the source can be taken from the methanol, natural gas, ethanol, etc. Hydrogen network has led to many interconnected producers, consumers and purification system based on the purity and operating objectives. Ethanol steam reforming (ESR) is considered to be the environmental friendly source of hydrogen because ethanol is renewable, can be produced in biorefinery, easy to transport and distribute. Ethanol produces the higher number of hydrogen molecule per unit weight as compared to water, methane and methanol. Ethanol can be easily converted to hydrogen, which is considered to be thermodynamically feasible in nature. Ethanol reacts with water in the following reaction:

$$C_2H_5OH + 3H_2O \leftrightarrow 6H_2 + CO_2 \qquad \Delta H = +347.4\ kJ/mol$$

ESR has been studied on various catalysts such as Ni, Co, Cu, Fe, Ir, Rh, Ru, Pt and Pd supported over metal oxide, mixed metal oxide, spinel and perovskite structure [1, 2]. Various side reactions and main reactions that occur depend on the catalyst being used and the kind of structure in which metal catalyst is being supported. Researchers have developed an empirical model for the degradation of the ethanol as a primary mechanism in most cases.

Therdthianwong et al. [4] observed the thermal decomposition of ethanol for Ni–Al_2O_3 catalyst from 673 to 923 K with the increase in the coke deposition. At higher temperatures, the hydrogen yield was observed to be increasing and the methane yield was decreasing due to the steam reforming of methane. Mas et al. [5] conducted the experiment on Ni–Al–LDH as a precursor to develop a kinetic rate model. The powers for the model were calculated using the Athena Visual Benchwork. The power law model (PLM) developed was used in the ESR in which the assumptions considered were the isothermal plug flow reactor. The rate at which the ethanol is consumed is independent of the water concentration. The assumption of the heat and mass transfer resistance was neglected, which results in the higher activation energy. Mathure et al. [6] showed that the PLM developed for the Ni/Al_2O_3/MgO where the surface reaction of ethanol consumption is considered to be the rate determining. The kinetic parameters and the power required for the model are calculated by the regression model using the Levenberg–Marquardt process. The PLM is based on the partial pressure of the water and ethanol. The Average Absolute Deviation of the PLM was found to be very low and compared with the mechanistic model for the validation of ESR. Vaidya et al. [7] experimented on the Ru/Al_2O_3 catalyst and developed a PLM based on the experiments conducted. The reaction mechanism is based on the formation of the complex compound between the ethanol and water, which in turn produce intermediates, which will lead to the formation of products. The formation of intermediates between the ethanol and water is considered to be the rate-determining step and the reaction is considered to be the first order with respect to ethanol.

In this comparative study, the power law-based kinetic models [4–7] are used to model a packed bed reactor based on the assumptions: pseudo-homogeneous behavior, steady-state, plug-flow and isothermal conditions. The reactor model is simulated under the operating conditions of temperature, 873 K and the water/ethanol ratio, 3:1 for a space time of 0–6.97 g_{cat} h/g mol at atmospheric condition. The models are evaluated for higher ethanol conversion and hydrogen yield.

4.2 Kinetic Model Based on Power Law

PLM may not be one of the complex models as compared to the various mechanistic approaches such as the Eley–Radial mechanism and the Langmuir–Hinshelwood–Hougen–Watson [8], but it is always taken as the elementary step to understand the mechanism of the ESR reaction and is often used to compare with the experimental data. The mechanism mostly followed by the PLM is the degradation of ethanol and water. The PLM is not as detailed as the mechanistic model but they provide a useful indicator for the validation of the model that can be used in ethanol-based steam reforming. Some authors also ignore the participation of water in the reaction as the concentration of water is excess enough and is considered to be pseudo reaction in nature. The power models used in this study are based on the partial pressure of ethanol and water as it is easy for the operator to control the partial pressure rather than concentration. The PLM was developed by Therdthianwong et al. [4] for Ni/Al_2O_3 at various operating temperatures so as to understand the rate mechanism and the rate parameters. The H_2 and CH_4 yield was being analyzed using the PLM developed by the author. For the estimation of the rate expression model where the power of the rate expression is given by STATISTICA program, it was obtained at a temperature of 673 K under atmospheric pressure with a confidence interval of 98%.

Ethanol steam reforming:

$$C_2H_5OH + H_2O \leftrightarrow CH_4 + CO_2 + 2H_2$$

Methane steam reforming:

$$CH_4 + H_2O \leftrightarrow CO + 3H_2$$

Water gas shift reaction:

$$CO + H_2O \leftrightarrow CO_2 + H_2$$

Mas et al. [5] have developed a model based on the ESR using the Ni/Al/LDH as the catalyst. The PLM is calculated for the temperature range of 823–923 K. The PLM provides the simplistic approach so as to understand the formation of methane during the reaction at a very low temperature and the degradation of methane to form the hydrogen. The CO concentration is said to be the minimum at a very high temperature, which means the presence of the water–gas shift reaction.

Ethanol steam reforming—I:

$$C_2H_5OH + H_2O \rightarrow 2CO + 4H_2$$

Ethanol steam reforming—II:

$$C_2H_5OH + 3H_2O \rightarrow CO_2 + 6H_2$$

Mathure et al. [6] verified that the PLM developed is correlated with the experimental data collected and noted to have the lowest Absolute Average Deviation of 10%, which is considered to be the best fit for the experimental data. The PLM is the empirical model that shows only the consumption of ethanol.

Ethanol steam reforming:

$$C_2H_5OH + H_2O \rightarrow 2CO_2 + 6H_2$$

Water gas shift reaction:

$$CO + H_2O \rightarrow CO_2 + H_2$$

Oxidation reaction—I:

$$CO + 0.5O_2 \rightarrow CO_2$$

Oxidation reaction—II:

$$H_2 + 0.5O_2 \rightarrow H_2O$$

Vaidya et al. [7] developed a first-order PLM that is conducted for a Ru/Al_2O_3 catalyst, which shows the formation of intermediate, which involves the adsorption of water and ethanol on the active site of the catalyst. The PLM is based on the stationary state hypothesis for the formation of complexes, which involves complex formulation for the rate of reaction. Since the reaction is taking place in the atmospheric pressure and the reaction is considered to be the simple first-order reaction with respect to the ethanol.

Ethanol steam reforming:

$$C_2H_5OH + H_2O \leftrightarrow CH_4 + CO_2 + 2H_2$$

Table 4.1 Kinetic models based on power law

Ref.	Catalyst used	Rate equation	Activation energy (kJ/mol)	Temperature (K)
[4]	Ni/Al_2O_3	$r = k * p_{EtOH}^{2.52} * p_{H_2O}^{7}$; $k = 280075$	–	673–923
[5]	Ni–Al–LDH	$r = k * p_{EtOH}^{0.8}$; $k = 1.365 * 10^{-4}$	144	550–923
[6]	$Ni/Al_2O_3/MgO$	$r = k * p_{EtOH}^{0.711} * p_{H_2O}^{2.71}$; $k = 4.39 * 10^2 * e^{\frac{-2.3*10^4}{RT}}$	23	673–873
[7]	Ru/Al_2O_3	$r = k * p_{EtOH}$; $k = 1.5 * 10^5 * \exp^{\frac{-96*10^5}{RT}}$	96	873–973

Methane steam reforming:

$$CH_4 + H_2O \leftrightarrow CO + 3H_2$$

Water gas shift reaction:

$$CO + H_2O \leftrightarrow CO_2 + H_2$$

Table 4.1 shows the rate equations developed for the above studies along with the activation energy and temperature data.

4.3 Analysis of the Power Law Model

The PLM is not as detailed as the mechanistic model but they provide a useful indicator for the validation of the model that can be used in ethanol-based steam reforming. Therdthianwong et al. [4] experimented on dilute solution of ethanol for the production of hydrogen. The PLM was developed for a temperature of 673 K, at a partial pressure lying between $0.0003 < p_{EtOH} < 0.0508$ and $0.7094 < p_{H_2O} < 0.9371$ atm. The reason for the development of PLM at 673 K as it is found that more than 70% conversion of ethanol was achieved, and high-temperature internal steam reformer in the fuel cell was found to be impractical. The reformate gas produces a less coke deposition at a very low temperature and will achieve a higher time on stream process. The smallest hydrogen yield was found out at a temperature of 673 K with relatively low space time. The kinetic model given by Mas et al. [5] provides a substantial overview of the kinetic model based on Ni–Al–LDH where the value of the activation energy and the exponent power is given by the corresponding confidence interval, the correlation between the pre-exponential and activation energy is used to figure out the preexponential factor. The assumptions included in the simplistic PL: the presence of the extraneous effects (diffusion of mass and heat in catalyst particle and mixing inside the reactor), the lower range of temperature and the absence of the water. The maximum yield of 5 hydrogen mol per mol of ethanol was reported at 100% ethanol conversion. Mathure et al. [6] developed a kinetic data using partial pressure to account for the effect of the water–ethanol feed ratio. The power in the model reaffirms the conformity of the non-elementary nature of these reactions. The yield of hydrogen for the $Ni/Al_2O_3/MgO$ catalyst was found out to be 3.0 mol/mol of H_2, the steam to ethanol ratio was 12:1 and the $W/F_{Ao} > 110$ g_{cat} min/mol, which is very low compared to the laboratory made catalyst (4.8–5.2 mol/mol of ethanol). The hydrogen yield remained constant after this temperature. The PLM so obtained is not considered to be the mechanistic model but a rather empirical correlation that can estimate the rate of depletion of ethanol. The reaction order was changed due to a small amount of Mg being added to the catalyst, which can affect the rate at which the ethanol can disappear. The hydrogen yield for the Ni/Al_2O_3 catalyst at a temperature of 673 K is 4.5 mol/mol of ethanol. Vaidya et al. [7] proposed a power law mechanism where the use of Ru/Al_2O_3 catalyst where the reactant A is absorbed on the active site of the catalyst forming a complex which further reacts with another reactant to form the complex. Experimentally, hydrogen yield for Ru/Al_2O_3 was measured from 873 to 973 K. The yield of hydrogen increased from 0.27 to 0.41 mol/mol at a space time between 0.68 and 2.04 g h/mol. Higher the extent of reaction, higher will be the yield of hydrogen. The decomposition of the complex to form the intermediate is considered to be the rate-determining step. These intermediates such as the CH_3CHO, CH_4 and C_2H_4 further react to form the CO_2 and H_2. Since water is considered to be excess in this reaction, the rate is given by

$$r = \frac{k_R p_A}{[1 + b p_A]} \tag{4.1}$$

where the bp_A is considered to be the lumped parameter, where the $bp_A \gg 1$, therefore, the final rate of reaction is given as

$$r = k_R p_A \tag{4.2}$$

The model equation for a pseudo homogeneous packed bed reactor at isothermal and steady state condition is given as:

$$\frac{dy_i}{d\theta} = \gamma * r_i \tag{4.3}$$

where y_i is the mole fraction of ith component, θ. is the space time, r_i is the power law equation of ith model and γ is the stoichiometric coefficient.

The conversion of ethanol is given by the following equation:

$$X_{\text{EtOH}} = \frac{F^0_{\text{EtOH}} - F_{\text{EtOH}}}{F_{\text{EtOH}}} \tag{4.4}$$

F^0_{EtOH} is the initial molar flowrate of ethanol, and F_{EtOH} is the ethanol molar flow rate at space time θ.

The yield of hydrogen is given as:

$$y_{\text{H}_2} = \frac{F_{\text{H}_2}}{6F^0_{\text{EtOH}}} \tag{4.5}$$

F_{H_2} is the molar flow rate of hydrogen at space time θ.

These model equations are solved using MATLAB (2010b) for optimal operating conditions, and the model-predicted results are compared with the experimental results using the SSE (Sum of the squares error) technique. The SSE methods provide a comparison between the empirical data and the experimental data, it can be used to measure the variation within the cluster. If all the clusters are identical, then the SSE is equal to zero.

$$\text{SSE} = \sum \left(x_{\text{exp}} - x_{\text{sim}}\right)^2 \tag{4.6}$$

Equations (4.3), (4.4), (4.5) and (4.6) are solved, and the results are tabulated in Table 4.2 to show the values of ethanol conversion, SSE and hydrogen yield for the selected PLMs.

Figure 4.1 shows the conversion of ethanol *vs.* the space time for the selected PLMs that has been applied for the depletion of the ethanol and are compared with the experimental data. It is found that the highest conversion of ethanol is obtained for Ni/Al_2O_3 catalyst whereas the least SSE value is obtained for Ru/Al_2O_3 catalyst. The lower value of SSE indicates the least deviation of model-predicted values from the experimental values. The least conversion using the power-law model was found using the $Ni–Al_2O_3$. The conversion for the $Ni–MgO–Al_2O_3$ is lower compared to other Ni-based catalysts due to the effect of the side reactions resulting into the deposition of coke. The formation of side-products tends to disappear at a very high temperature of ESR temperature which has resulted in an increase in the conversion

Table 4.2 PLMs with conversion and comparison with the experimental results using SSE

Ref	Rate equation	Conversion (%)	SSE	Hydrogen yield (mol/mol of ethanol)
[4]	$r = k * p^{2.52}_{\text{EtOH}} * p^{7}_{\text{H}_2\text{O}}$	95.368	0.0176	4.57
[5]	$r = k * p^{0.8}_{\text{EtOH}}$	93.09	0.7030	5.06
[6]	$r = k * p^{0.711}_{\text{EtOH}} * p^{2.71}_{\text{H}_2\text{O}}$	82.26	0.0176	2.82
[7]	$r = k * p_{\text{EtOH}}$	93.2	0.0017	0.41

of ethanol for all the kinetic models. At lower values of space time, the ethanol conversion is found out to be very low and limited amount of methane undergoes catalyzed conversion to produce H_2. As the space time is increased, the rate of ethanol conversion increases to give higher yield of hydrogen. The hydrogen yield obtained using the PLM is another selection criterion. The theoretical yield for hydrogen is 6 mol of hydrogen per mol of ethanol. These catalysts are developed to maximize the yield of hydrogen. It can be observed that the yield of hydrogen for the Ni–Al–LDH catalyst is higher as compared to other catalysts (Fig. 4.2). In the Ni–Al–LDH, the ethanol is decomposed to form the methane, which is further reformed to produce H_2.

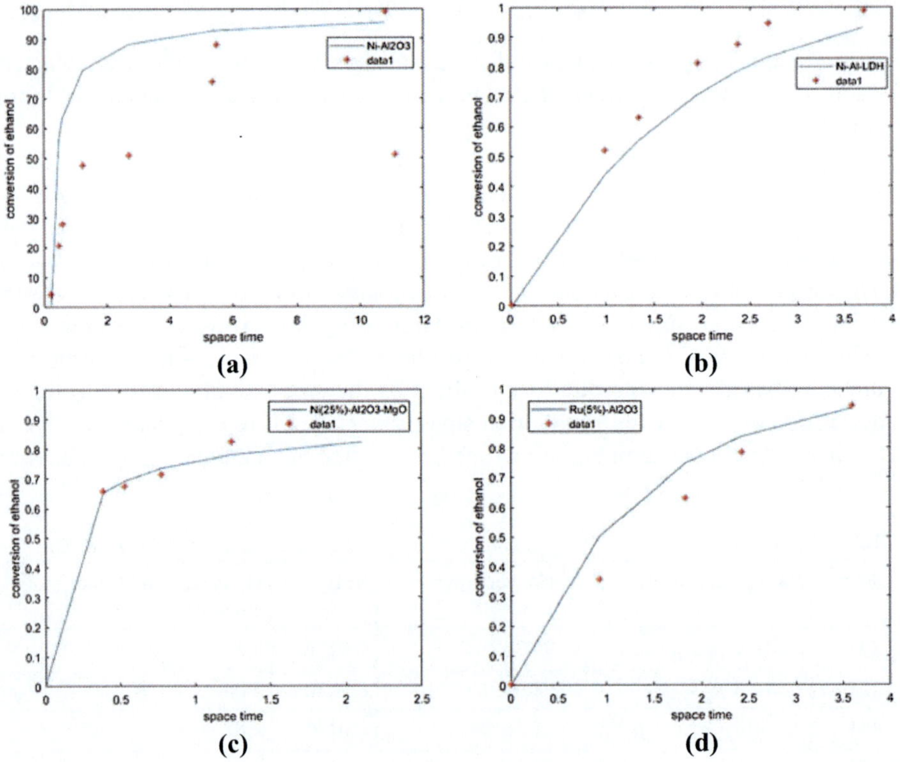

Fig. 4.1 Comparison of all the models based on the space time versus conversion **a** Ni–Al_2O_3 [4] **b** Ni–Al–LDH [5] **c** Ni–MgO–Al_2O_3 [6] **d** Ru–Al_2O_3 [7] at 873 K, water to ethanol ratio 3:1 and 1 atm

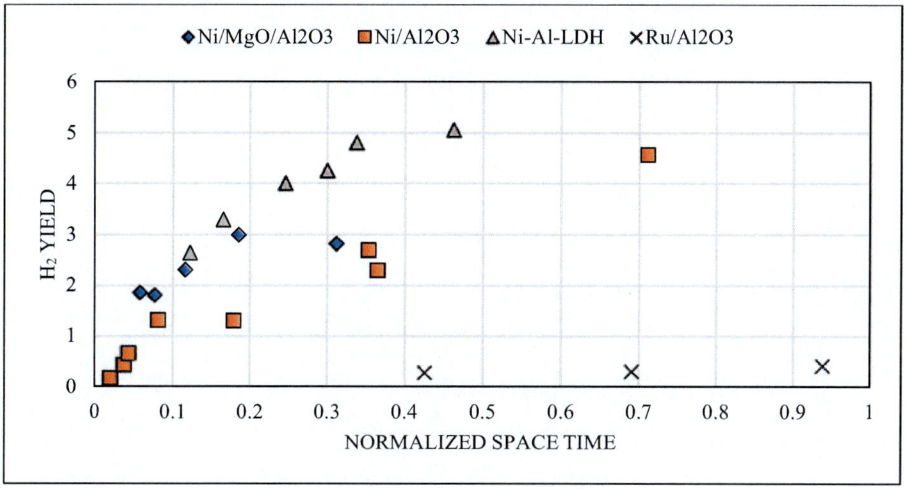

Fig. 4.2 Plot between the normalized space time versus H_2 yield

4.4 Conclusion

The PLM for various catalysts is evaluated based on ethanol conversion and hydrogen yield. These PLMs are formulated based on the partial pressure of ethanol and water as the effect of water to ethanol ratio can be easily studied with respect to ethanol conversion and the yield of hydrogen. The PLM is different for various catalysts based on different operating conditions. This PLM provides the basis for the more complex mechanistic models as it gives a better understanding of how the degradation of the ethanol can take place with respect to space time. These models can also be used to model the pseudo homogeneous packed bed reactor over a range of operating conditions. The model simulated data of a packed bed reactor using the PLMs show the good agreement with the experimental results.

References

1. A. Harayanto, S. Fernando, N. Murali, S. Adhikari, Energy Fuels **19**, 2098 (2005)
2. S. Ogo, Y. Sekine, Fuel Process. Technol. **199**, 106238 (2020)
3. J. Comas, F. Marino, M. Laborde, N. Amadeo, Chem. Eng. J. **98**, 61 (2004)
4. A. Therdthianwong, T. Shakulkoakiet, S. Therdianwong, Science (Asia) 193 (2001)
5. V. Mas, M.L. Bergamini, G. Baronetti, N. Amadeo, M. Lamborde, Top Catal. **51**, 39 (2008)
6. P. Mathure, S. Ganguly, A. Patwardhan, R. Saha, Ind. Eng. Chem. Res. **46**, 8471 (2007)
7. P. Vaidya, A. Rodrigues, Ind. Eng. Chem. Res. **46**, 8471 (2007)
8. A. Akande, A. Aboudheir, R. Idem, A.K. Dalai, Int. J. Hydrogen Energy **31**, 1707 (2006)

Chapter 5
Investigations of Tensile Behaviour of 3D-Printed PLA-GF-PLA Sandwich Composite Structures

Abhishek Sharma, Ranvijay Kumar, and Lochan Sharma

Abstract Three-dimensional (3D) printing is one of the advanced sustainable manufacturing techniques for the production of composite structures for engineering applications. The objects that are printed using 3D printing technique can be produced with lighter weight as compared to traditional manufacturing methods. Improved mechanical properties can be achieved with 3D printing techniques by selecting a suitable combination of fibre along with polymeric material. In this work, PLA (polylactic acid) was reinforced with the glass fibre to prepare sandwich composite as per ASTM D638 type IV standard for tensile testing. The input parameters varied for sample preparation are infill density (50%, 75%, 100%), infill linear sample, number of glass fibre layers (1, 3, 5) and nozzle temperature (210 °C, 215 °C, 220 °C). The effect of variation of these parameters on tensile properties was studied. In sample 3 (single GF layer with 100% infill density and 220 °C infill temperature), the maximum peak and breakpoint strength were observed.

5.1 Introduction

Additive manufacturing is the process in which layer by layer technique of material deposit was used. In this method, firstly, the stereolithography (STL) file is created with the help of the computer aided design (CAD) software. By commanding the 3D printer, the new complex geometry was formed. The additive manufacturing (AM) has many techniques, like stereolithography (SLA), laminated object manufacture (LOM), selective laser sintering (SLS), and fused deposition modelling (FDM), in which plastic filament melts [1–5]. The FDM method is more popular than other techniques due to lesser cost of printer. 3D printers are commonly used in rapid

A. Sharma (✉) · R. Kumar · L. Sharma
Department of Mechanical Engineering, Chandigarh University, Gharaun, Mohali, Punjab, India
e-mail: abhisheksharma23111996@gmail.com

R. Kumar · L. Sharma
University Centre of Research and Development, Chandigarh University, Gharaun, Mohali, Punjab, India

A. K. Sahu et al. (eds.), *Energy Storage and Conservation*, Springer Proceedings in Energy, https://doi.org/10.1007/978-981-99-2870-5_5

prototyping technique, which find their industrial applications such as construction, automotive, aerospace, and medical fields. Fused filament fabrication FFF is a popular 3D printing method currently used in the worldwide because of their lesser cost and high mechanical properties. Previous study revealed that, in fused filament fabrication (FFF), the mechanical properties of the PLA reinforced with carbon fibre (CF) have been significantly improved [6–8]. Materials and procedures must be compatible with each another because of their high demand across all industries for a variety of applications, each with a different set of requirements for the final product in terms of strength, appearance, etc. Available literature suggests that the synthesis of hybrid composite structures has been developed with the use of the PLA composite reinforced with GF. Higher mechanical properties were observed for the composition of PLA reinforced with carbon fibre as compared to the PLA reinforced with GF. In the present study, the number of layers, infill density, infill temperature were used as working parameters. In the previous study, it has been observed that the synthesis of composite structures was also developed with the help of the PLA reinforced with GF by using Fused Deposition Modelling (FDM) technique. In some of the cases, the mechanical properties were not significant while using FDM technique but when the PLA reinforced with the carbon fibre, the mechanical properties have been significantly improved [6, 9, 10]. As per the previous researchers, it was found that when GF is added to the PLA it enhances the mechanical properties of the material. If 20wt% of GF is added to the PLA, the toughness and hardness of the material are improved. This shows that with the increasing of the PLA and GF mixture, there will always be an increase in the mechanical properties like tensile strength of the material, toughness and hardness of the material. It was found that if elastomers are added in the PLA, the toughness and rigidness of the material have been significantly decreased. By comparing both the results, the PLA/GF has higher mechanical properties to that of the PLA/elastomers. It was also found that the GF has a great potential to hold the material to that of the pure PLA. The modulus of the elasticity is high when PLA is reinforced with the GF to that of the pure PLA with the comparison of the PLA/GF and pure PLA, the toughness of the PLA/GF is increased by 31% to that of the pure PLA. This shows that when content of the GF is increased from 1 to 5wt%, the sample strength significantly increased [11–13]. The literature survey reveals that the addition of GF into the polymer matrix increases the mechanical properties and provides the heat resistance properties. Some of the previous studies have been highlighted the use of glass fibre in polymer for the preparation of PLA-GF-based composites for prospective applications. But hitherto, very fewer studies have been reported for the preparation of PLA/GF-based composites by using the materials extrusion-based 3D printing processes (e.g., FFF). In this study, the FFF process was carried out for the preparation of PLA-GF composite structures as per ASTM D638 type IV by varying the number of GF layer, infill percentage and nozzle temperature. Further, the tensile testing was performed to investigate the effect of process parameters on the mechanical properties.

Table 5.1 Properties of materials

Properties	PLA	GF
Tensile strength	42 MPa	4135 MPa
Modulus of elasticity	3.6 GPa	85.5 GPa
Elongation break	7%	4.8%
Glass transition temperature	45–60 °C	–
Density	1.12 g/cc	2.54 g/cc

5.2 Material and Methods

Polylactic acid (PLA) is a very stiff, biocompatible, biodegradable, linear aliphatic thermoplastic polyester. It also has great UV stability. It is generated from entirely renewable sources including corn and wheat. As a result, PLA is currently the polymer that is utilized the most, and it is extensively employed in the biomedical, textile, and food packaging industries. The PLA is mostly used for the preparation of the polymer composite in FDM process. The reinforcement of GF enhances the mechanical properties of polymer composites [14, 15]. The Filament of the Polylactic Acid (size: 1.75 mm diameter) (Orange premium quality Filament) (origin—India) and GF (0.03 mm ultra-thin GF reinforcements' fibreglass cloth 50″ × 39″) was used for the preparation of PLA/GF composite. Table 5.1 shows the properties of the PLA and GF materials. The tensile strength of the PLA is 42 MPa, and GF is 4135 MPa. The modulus of the elasticity of PLA is 3.6 GPa, and for the GF, it is 85.5 GPa. The elongation at break of the PLA is 7% whereas the elongation break of the GF is 4.8%. The glass transition temperature is 45–60 °C of the PLA whereas there is no glass transition temperature in GF. The density of the PLA is 1.12 g/cc, and for the GF, it is 2.54 g/cc. [16]

5.3 Experimentation

5.3.1 3D printing

The FFF-based 3D printer (Creality 3D Ender Pro) is used for making specimen of PLA/GF composite. Taguchi approach of design of experiment (DOE) was used to design the experimental runs. Slicing of a model of tensile specimen was done in the Ultimaker_Cura-4.13.1 software. The process parameter taken into account was the number of layers, infill percentage, and nozzle temperature. Table 5.2 shows the DOE for the preparation of PLA/GF composites.

Table 5.2 DOE for the preparation of PLA/GF composite structures

Serial number	Number of layers	Infill percentage	Nozzle temperature
1	1	50	210
2	1	75	215
3	1	100	220
4	3	50	215
5	3	75	220
6	3	100	210
7	5	50	220
8	5	75	210
9	5	100	215

5.3.2 *Tensile Testing*

The various parameters were taken into an account for the preparation of tensile specimen of PLA reinforced with GF. There were nine samples that were prepared using 3D printer (Creality 3D Ender Pro). The ASTM D638 type IV standards were used for taking dimensions of the tensile specimen. After that, the samples were tested in the universal testing machine (UTM). (Make: Shanta Engineering, maximum capacity: 5000 N, grip separation is 60 mm). The length of the specimen is 115 mm, and width is 12.5 mm. The strain rate was kept 50 mm/min for fracture of the tensile specimen.

5.4 Result and Discussion

5.4.1 *Tensile Testing*

Previous study suggests that when CCF was reinforced with the PLA, the modulus of elasticity observed was 335 MPa [17]. But in the present study, the addition of GF fibres into PLA ensured the Young's modulus to 1165.59 MPa. As per the previous study, when continuous CF reinforced with the PLA, the temperature 230 °C increases, there has been a minimal change in the strength and stiffness of the material. But when the GF was reinforced with the PLA with the rise of temperature, there was a significant increase in the stiffness and tensile strength of PLA/GF specimen. After comparing the two results, it was observed that when single GF layer was added to the test specimen, it shows higher tensile strength to that of the continuous CF [18]. Table 5.3 shows the tensile properties and SN ratios of PLA/GF reinforced composite. The test specimen number 3 (number of layers is 1, infill density 100% and nozzle temperature 220 °C) shows the adequate result having maximum peak

Table 5.3 Tensile properties and SN ratios of PLA/GF reinforced composite

Serial number	Peak Strength (MPa)	SN ratio of Peak Strength (dB	Break Strength (MPa)	SN ratio of Break Strength (dB)	Modulus of Elasticity (MPa)	SN ratio of Modulus of Elasticity (dB)
1	38.51	31.71	34.66	30.79	815.02	58.22
2	42.83	32.63	38.55	31.72	644.06	56.17
3	68.77	36.74	61.89	35.83	1165.59	61.33
4	59.105	35.43	53.2	34.51	1023.46	60.20
5	60.68	35.66	54.61	34.74	517.52	54.27
6	68.49	36.71	61.64	35.79	1151.09	61.22
7	40.82	32.21	36.74	31.30	613.83	55.76
8	38.57	31.72	34.71	30.80	629.71	55.98
9	50.075	33.99	45.07	33.07	820.90	58.28

strength of 68.77 MPa and maximum break strength of 61.89 MPa was observed while test specimen number 1 (number of layers is 1, infill density is 50%, and nozzle temperature is 210 °C) gives minimum peak (38.51 MPa) as well as break (34.66 MPa) strength values (Table 5.3).

5.4.2 Process Parameter Optimization

The mechanical properties are affected by different input parameters. A suitable set of input parameters must be chosen in order to achieve a maximum attribute. In the present study, different input parameters were optimized to get better output responses. Different tensile properties that were measured by SN ratios are used as an input parameter to predict the optimum setting for 3D printing (Fig. 5.1). SN ratio assesses how much response fluctuates in the relation to the nominal value when it breaks at certain point. SN ratio results of PLA/GF composites specimen are shown in Table 5.3.

The SN ratios were calculated using the larger in better-case method because these attributes needed to have larger values. The SN ratio for cases with bigger numbers has improved when calculated using Eq. 5.2.

$$\eta = -10\left[1/n\sum_{k=1}^{n}\frac{1}{y2}\right] \quad (5.1)$$

where, n = number of experiments.

y is an attribute of substance at sample k.

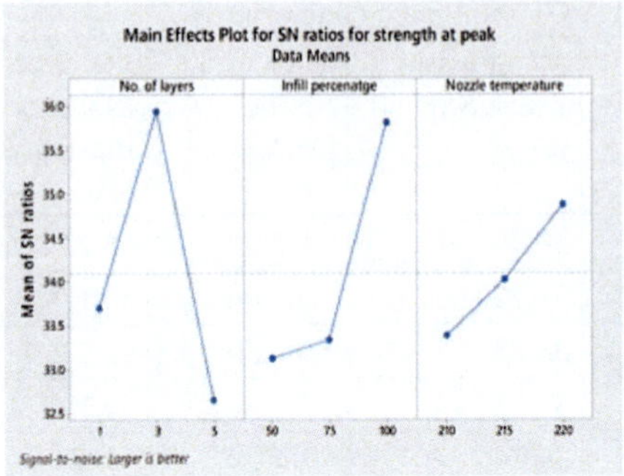

(a) SN ratio at average peak strength

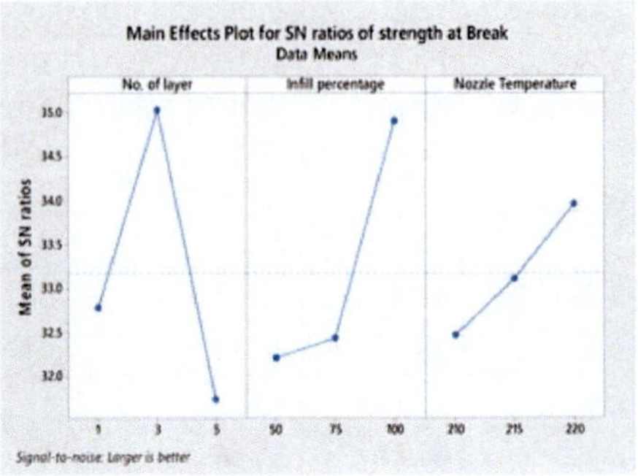

(b) SN ratio value at break strength

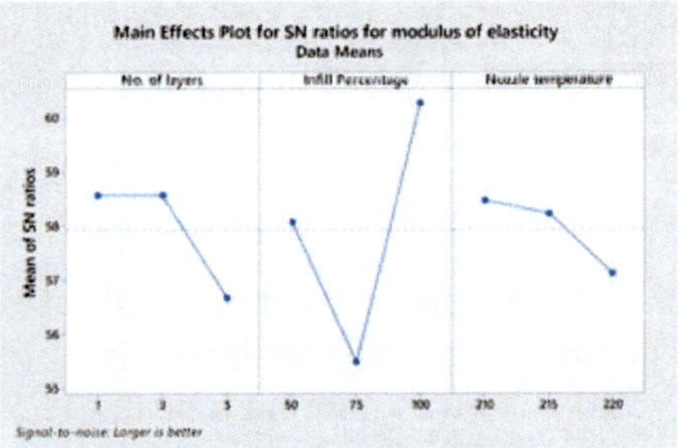

(c) SN ratio value at Modulus of Elasticity

Fig. 5.1 SN main effect plots for **a** peak strength **b** break strength **c** Young's modulus

Figure 5.1a indicates that the SN ratio strength at peak is better having no. of layers three, infill percentage of 100% and the nozzle temperature is 220 °C. Fig. 5.1b indicates that the SN ratio strength at break is better when using three layers, infill percentage of 100%, and the nozzle temperature is 220 °C. In Fig. 5.1c, it indicates that the SN ratio modulus of elasticity is better when using three layers having infill percentage is 100% and nozzle temperature is 210 °C. Table 5.4 can be used to predict that the process is statistically well-regulated because the residual error calculated for the process is 3.78%, and the P is value for the input parameter entitled the number of layers is 0.073, which is near to the significant value, i.e., 0.05. The procedure is, therefore, inferred to be monitored with a higher level of confidence, i.e., 95%.

Table 5.4 ANOVA for the SN ratios for strength at peak

Sources	DF	Seq SS	Adj SS	Adj MS	F	P	% of contribution
Number of layers	2	16.941	16.941	8.4703	12.73	0.073	48.27
Infill percentage	2	13.460	13.460	6.7302	10.12	0.090	38.35
Nozzle temperature	2	3.365	3.365	1.6825	2.53	0.283	9.58
Residual error	2	1.330	1.330	0.6652			3.78
Total	8	35.096					

Table 5.5 SN ratios value for peak strength

Level	Number of layers	Infill percentage	Nozzle temperature
1	33.70	33.12	33.38
2	35.94	33.34	34.02
3	32.64	35.82	34.88
Delta	3.29	2.70	1.49
Rank	1	2	3

Table 5.5 shows the response table for SN ratio of peak strength of PLA/GF composite structures. The number of layers has been ranked 1 followed by infill percentage and nozzle temperature.

To find the optimum responses for tensile behaviour, SN ratio formula can be used, which is given in Eq. 5.2.

$$\gamma_{opt} = T + (T_{A2} - T) + (T_{B3} - T) + (T_{C3} - T) \quad (5.2)$$

where γ_{opt} is the optimum value of the SN ratio of the tensile strength peak, T is the mean ratio of the strength at peak of the nine different samples, T_{A2} is the SN ratio for the number of layers at level 2, T_{B3} is the SN ration for the infill percentage at level 3 and T_{C3} is the SN ration for the nozzle temperature at leave 3

$$T = 34.09\,dB,\ T_{A2} = 35.94\,dB,\ T_{B3} = 35.82,\ T_{C3} = 34.88\,dB$$

$$\gamma_{opt} = 34.09 + (35.94 - 34.09) + (35.82 - 34.09) + (34.88 - 34.09)$$

$$\gamma_{opt} = 38.46\,dB$$

The optimum value for the SN ratio for the tensile strength at peak (Z_{opt}) can be calculated from Eq. 5.2

$$Z^2_{opt} = (10)^{\alpha opt/10}(\text{larger is better}) \quad (5.3)$$

$$Z^2_{opt} = (10)^{38.46/10} = 83.74\,MPa$$

The optimum value has been calculated at 83.74 MPa for strength at peak. Similarly, the optimum values of the has been calculated 75.33 MPa for strength at break and 1183.04 MPa Young's modulus on the predicted setting.

5.5 Conclusion

- The maximum value for tensile modulus was observed for sample 3 (1 number of layers, infill density is 100%, nozzle temperature is 220 °C and infill pattern is linear).
- Among all the parameters, the minimum strength at the peak and the break were observed at sample 1 (1 number of layers, 50% infill density, and nozzle temperature is 210 °C and infill pattern is linear). Due to minimum number of layers, the strength of observed minimum in case of sample 1.
- For sample 3, the maximum value of the modulus of the elasticity observed was 1165.59 MPa, maximum value of the strength at peak was 68.77 MPa, maximum value of the strength at the break was 61.89 MPa
- Overall, the maximum contributing factor for peak strength and break strength was changing number of GF layers which is ranked 1, followed by infill percentage (ranked 2) and nozzle temperature (ranked 3).
- The composite's structure prepared by the proposed process may be used for the preparation of the insulation purposes.

References

1. ASTM F2792–12a, Standard Terminology for Additive Manufacturing Technologies, ASTM International, West Conshohocken, PA (2012)
2. J. Wang, A. Goyanes, S. Gaisford, A.W. Basit, Stereo lithographic (SLA) 3D printing of oral modified-release dosage forms. Int. J. Pharm. (2016)
3. D. Ahn, J.H. Kweon, J. Choi, S. Lee, Quantification of surface roughness of parts processed by laminated object manufacturing. J. Mater Process Technol (2012)
4. S. Greiner, K. Wudy, L. Lanzl, D. Drummer, Selective laser sintering of polymer blends: bulk properties and process behavior. Polym Test (2017)
5. B.N. Turner, R. Strong, S.A. Gold, Review of melt extrusion additive manufacturing processes: I. Process design and modeling. Rapid Prototyp J. (2014)
6. M. Kamaal, M. Anas, H. Rastogi, N. Bhardwaj, A. Rahaman, Effect of FDM process parameters on mechanical properties of 3D-printed carbon fibre–PLA composite. Progress Additive Manuf. **6**(1), 63–69 (2021)
7. V.D.P. Rao, P. Rajiv, V.N. Geethika, Effect of fused deposition modelling (FDM) process parameters on tensile strength of carbon fibre PLA. Mater. Today Proc. **18**, 2012–2018 (2019)
8. J.M. Reverte, M.Á. Caminero, J.M. Chacón, E. García-Plaza, P.J. Núñez, J.P. Becar, Mechanical and geometric performance of PLA-based polymer composites processed by the fused filament fabrication additive manufacturing technique. Materials **13**(8), 1924 (2020)
9. N. Vinoth Babu, N. Venkateshwaran, N. Rajini, S.O. Ismail, F. Mohammad, H.A. Al-Lohedan, S. Suchart, Influence of slicing parameters on surface quality and mechanical properties of 3D-printed CF/PLA composites fabricated by FDM technique. Mater. Technol. **37**(9), 1008–1025 (2022)
10. A.E. Magri, K. El Mabrouk, S. Vaudreuil, M.E. Touhami, Mechanical properties of CF-reinforced PLA parts manufactured by fused deposition modeling. J. Thermoplast. Compos. Mater. **34**(5), 581–595 (2021)

11. G. Wang, D. Zhang, G. Wan, B. Li, G. Zhao, Glass fiber reinforced PLA composite with enhanced mechanical properties, thermal behavior, and foaming ability. Polymer **181**, 121803 (2019)
12. G. Wang, D. Zhang, B. Li, G. Wan, G. Zhao, A. Zhang, Strong and thermal-resistance glass fiber-reinforced polylactic acid (PLA) composites enabled by heat treatment. Int. J. Biol. Macromol. **129**, 448–459 (2019)
13. L. Lin, C. Deng, G.P. Lin, Y.H. Wang, Mechanical properties, heat resistance and flame retardancy of glass fiber-reinforced PLA-PC alloys based on aluminum hypophosphite. Polym.-Plast. Technol. Eng. **53**(6), 613–625 (2014)
14. A.N. Dickson, J.N. Barry, K.A. McDonnell, D.P. Dowling, Fabrication of continuous carbon, glass and Kevlar fibre reinforced polymer composites using additive manufacturing. Addit. Manuf. **16**, 146–152 (2017)
15. J. Bochnia, M. Blasiak, T. Kozior, Tensile strength analysis of thin-walled polymer glass fiber reinforced samples manufactured by 3D printing technology. Polymers **12**(12), 2783 (2020)
16. http://www.matweb.com/search/DataSheet.aspx?MatGUID=ab96a4c0655c4018a8785ac4031b9278
17. K. Chen, L. Yu, Y. Cui, M. Jia, K. Pan, Optimization of printing parameters of 3D-printed continuous glass fiber reinforced polylactic acid composites. Thin-Walled Struct. **164**, 107717 (2021)
18. H. Dou, Y. Cheng, W. Ye, D. Zhang, J. Li, Z. Miao, S. Rudykh, Effect of process parameters on tensile mechanical properties of 3D printing continuous carbon fiber-reinforced PLA composites. Materials **13**(17), 3850 (2020)

Chapter 6
Recent Advances in Pharmaceutical Degradation Using Fenton Oxidation

Anurag Kulabhi, Pranav Shukla, Purusharth Sharma, Amit K. Thakur, and Rahul Kumar

Abstract Fenton oxidation is a promising approach for the removal of pharmaceutical compounds present in an aqueous solution. The homogeneous Fenton-oxidation requires a high concentration of catalyst for efficient degradation. The heterogeneous Fenton-oxidation process is explored to address this limitation. This article provides an overview of the recent findings in the Fenton-based advanced oxidation processes (AOPs) for pharmaceutical removal from wastewater. Different Fenton AOPs and their application for pharmaceutical wastewater treatment have been reviewed. The cost estimation and electrical energy consumption of some Fenton-based AOPs are tabulated. This article has clearly established that Fenton oxidation when combined with other procedures resulted in a high degree of mineralization.

Keywords Fenton oxidation · Pharmaceutical degradation · Cost estimation · Energy analysis

6.1 Introduction

Water plays a vital role in the survival of any ecosystem and is one of the essential pillars of sustainable development. More than three-fourths of the earth is covered with water; still, freshwater is scarce. The WHO stated that over 2 billion people live in areas where the demand for drinkable water exceeds the water supply. The primary reasons for the deterioration of the water quality are improper management of industrial waste and the lack of efficient techniques to completely degrade pollutants present in various aqueous matrices. Over the last few decades, the concentration of pharmaceutical compounds (PCs) in wastewater has increased significantly, and their presence leads to many adverse effects on the environment and ecosystems. The physicochemical technologies, e.g., adsorption have the potential to remove various contaminants from wastewater, they are primarily separation techniques,

A. Kulabhi · P. Shukla · P. Sharma · A. K. Thakur (✉) · R. Kumar
Department of Chemical Engineering, University of Petroleum and Energy Studies, Dehradun 248007, Uttarakhand, India
e-mail: akthakur@ddn.upes.ac.in

A. K. Sahu et al. (eds.), *Energy Storage and Conservation*, Springer Proceedings in Energy, https://doi.org/10.1007/978-981-99-2870-5_6

necessitating further treatment. Advanced Oxidation Processes (AOPs) are one of the emerging techniques for the degradation of PCs from various water matrices [1, 2]. These processes mainly generate the highly reactive hydroxyl ion (OH), which oxidizes the organic compounds into nontoxic compounds [3]. Fenton oxidation is an AOP for the efficient treatment of PCs. The Fenton oxidation can be classified as homogeneous and heterogeneous. In the treatment of wastewater, homogeneous Fenton reactions are commonly used. However, the main drawback of homogeneous Fenton-oxidation primarily stems from the high concentration of catalyst required for efficient contaminant removal (e.g., >50 ppm), which is considerably higher than the acceptable concentration in the water treatment process' effluent (~2 ppm) to be released into the aqueous system [4]. Heterogeneous Fenton oxidation is explored to address the limitations of the homogeneous process. A general classification of Fenton oxidation-based AOP is shown in Fig. 6.1. This paper presents an overview of the Fenton-based AOPs for the degradation of PCs. The recent applications of Fenton oxidation for pharmaceutical degradation are summarized. The cost estimation and energy analysis of the Fenton-based AOPs are also discussed.

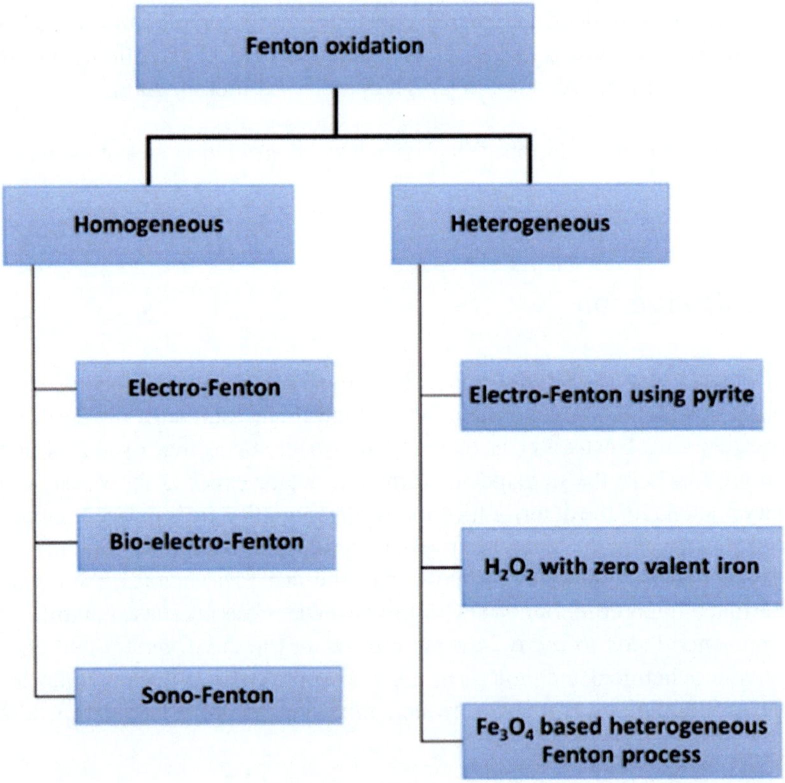

Fig. 6.1 A general classification of Fenton oxidation methods

6.2 Homogeneous Fenton Oxidation

The Fenton AOPs are one of the most effective AOP for the mineralization of PCs. The degradation mechanism comprises of reaction between peroxides (H_2O_2) and iron ions used as a catalyst for the oxidation of organic as well as an inorganic compound [5].

The traditional Fenton oxidation mechanism is as follows.

$$H_2O_2 + Fe^{2+} \rightarrow Fe^{3+} + HO^- + HO^\bullet \tag{6.1}$$

$$HO^\bullet + RH \rightarrow H_2O + R^\bullet \tag{6.2}$$

$$R^\bullet + Fe^{3+} \rightarrow R^+ + Fe^{2+} \tag{6.3}$$

$$R^+ + H_2O \rightarrow ROH + H^+ \tag{6.4}$$

where RH is an organic compound that is recalcitrant. This process is highly efficient and operates under mild operating conditions [6]. Moreover, no toxic gases evolved during the process as H_2O_2 can easily be broken into H_2O and O_2 [7]. Figure 6.2 shows the advantages and disadvantages of Fenton oxidation.

Shetty and Verma [8] investigated the efficiency of Fenton's reagent on pharmaceutical degradation and reported that 900 mg/L concentration of H_2O_2 provided efficient COD removal for subsequent dosage of ferrous sulfate. Simultaneous degradation of eight pharmaceuticals drugs (gemfibrozil, nimesulide, furosemide, paracetamol, propranolol, dipyrone, fluoxetine, and diazepam) for different water matrixes was investigated using the Fenton process. In the case of distilled water, increased and multiple iron (II) additions demonstrated effective mineralization and degradation of selected PCs [9].

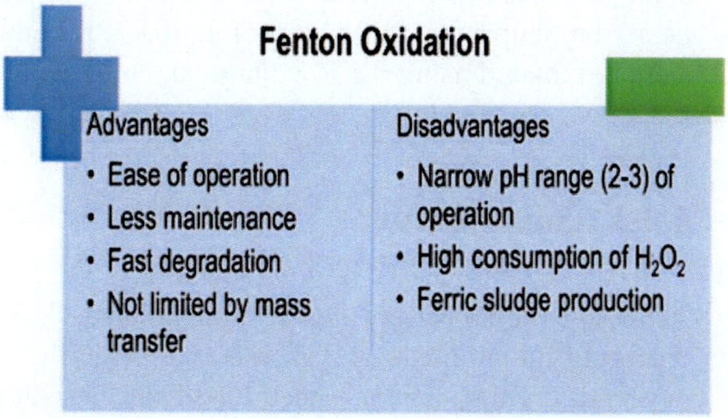

Fig. 6.2 Advantages and disadvantages of Fenton oxidation

6.2.1 *Electro-fenton Process*

The electro-Fenton (EF) process is a combination of electrochemical and Fenton AOPs. The Fenton reaction generates hydroxyl ions for the degeneration of contaminants. One of the drawbacks of the Fenton process is that it requires a higher concentration of reagents for effective mineralization [10]. Thus, Fenton oxidation is combined with the electrochemical process to address this limitation. The advantage of this combination is the cost reduction in the storage and transportation of H_2O_2 as it is electrochemically generated at the cathode [11].

This process is recently used for the degradation of ibuprofen [12], sulfamethazine[13], and % amoxicillin [14]. Kadji et al. [14] showed a 95% degradation of amoxicillin under optimal conditions. Davarnejad et al. [15] used this technique to remove ibuprofen from synthetic water and reduced its concentration from 400 ppm to 6.840 ppm under optimum conditions.

6.2.2 *Bio-electro Fenton*

Bio-electro Fenton (Bio-EF) process is a hybrid process of electro-Fenton and biological treatment. Although biological processes are pretty efficient, some pharmaceutical contaminants are non-biodegradable, and they can not be degraded using biological treatments. Although the EF can degrade many toxic organic contaminants, the process is costly to achieve high degradation efficiency. Therefore, this process is combined with biological treatments for the cost-effective degradation of PCs. The Bio-EF was recently employed for the degeneration of metoprolol [16] diclofenac [17], naproxen [17], carbamazepine [18], and sulfamethoxazole [19]. The biological treatment is employed as pre-treatment or post-treatment for the EF process [20]. If the contaminants can be degraded into bio-degradable forms, EF pretreatment is used otherwise biological methods are used as post-treatment to EF. Zou et al. [17] investigated the degradation of clofibric acid, diclofenac, carbamazepine, naproxen, ibuprofen, and ketoprofen in a 20L continuous Bio-EF system. The PCs were eliminated using 0.1 V voltage, 0.3 mM Fe^{2+}, and a 26-h HRT.

6.2.3 *Sono-fenton*

The conventional wastewater treatment plants (WWTPs) are not designed to degrade emerging contaminants (ECs), and as a result, the aquatic environment is degrading day by day. Thus, there is a need for advanced treatment methods with high mineralization performance, especially for EC [21]. The ultrasonic irradiation (US) method has received substantial attention for the removal of organic contaminants due to its high pollutant degradation efficiency [22]. The method of combining ultrasonic and Fenton oxidation is called Sono-Fenton. The bio-treated hospital wastewater subjected to sonochemical treatment resulted in 58.82% degradation and sono-photo-Fenton resulted in 82.86% degradation of fifteen pharmaceutical contaminants [23]. The degradation of pharmaceutical wastewater loaded with cephalexin was carried out using the sono-Fenton process and resulted in degradation efficiency of 90% at pH 3, initial cephalexin concentration = 50 mg/L, reaction time = 60 min, ferrous concentration = 8 mg/L, and H_2O_2 concentration = 60 mg/L [24].

6.3 Heterogeneous Fenton-Based AOPS

Homogeneous Fenton catalysis has certain limitations, such as sludge formation and the requirement of a narrow pH range (2.8–3.5), which subsequently requires the addition of a large amount of acid. The heterogeneous Fenton oxidation was explored to address these limitations of homogenous Fenton.

6.3.1 Heterogeneous Electro-fenton Using Pyrite

Researchers have used several iron minerals as Fenton catalysts due to their abundance low cost and availability [25]. The commonly used iron minerals in AOPs are pyrite, magnetite, hematite, goethite, etc. Pyrite (FeS_2) may be found in various sediments and hydrothermal deposits and is one of the most prevalent sulfide minerals and nontoxic semiconductors found on earth [26]. In solutions with varying pH levels, pyrite does have hydrophobicity and hydrophobicity on its surface [27]. As a result of their catalytic potential, both natural and synthetic pyrites were employed as a catalyst for pollutant degradation in Fenton processes [28]. Barhoumi et al. [29] employed a pyrite mineral catalyst in heterogeneous electro-Fenton oxidation operating at 300 mA for degradation of levofloxacin and achieved complete mineralization for a 0.23 mM concentration in 8 h. Ye et al. [30] synthesized pyrite core–shell nanoparticles as heterogeneous Fenton catalysts for the degradation of fluoxetine present in urban wastewater and reported complete removal within 1 h at near-neutral pH.

6.3.2 H_2O_2 with Zero Valent Iron

Recently, many researchers have explored the application of zero-valent iron (ZVI) as a catalyst for heterogeneous Fenton oxidation. The mechanism of Fenton oxidation using ZVI is discussed in [31]. The ZVI has been used to remove carbamazepine [32], norfloxacin [33], 4-chlorophenol [34], ciprofloxacin [35], and pentachlorophenol [36]. The nanoscale zero-valent iron has been utilized as a catalyst in the Fenton process in the presence of H_2O_2 for the enhanced degradation of ciprofloxacin antibiotic [37]. Recently, graphene oxide supported for nanoscale zero-valent iron composite has been utilized for the enhanced degradation of atrazine (a herbicide) from an aqueous medium [38].

6.3.3 Fe_3O_4-Based Heterogeneous Fenton Process

Many naturally occurring iron minerals are used in catalysis because of their unique structure and properties. Out of many iron-based catalysts used in the heterogeneous Fenton process, Fe_3O_4 is one of them. It is used extensively because of its high chemical-thermal stability and magnetism [39]. Juanli et al. studied the hydrogel-coated Fe_3O_4 for the degradation of phenol. After three cycles, the results showed catalytic stability and procurement with more than 55% COD removal efficiency [40].

6.4 Energy Consumption and Cost Estimation

The cost analysis is very important to understand the economic feasibility of any process. In general, operating expenses for processes include various critical components such as energy consumption, chemical, and electrode as well as additional costs such as maintenance, labor, sludge dewatering and disposal, and fixed costs. The cost can be expressed in terms of the total cost per kg of COD removed ($/kg COD) or energy consumption per kg COD (kWh/kg COD) removed or energy consumption per kg TOC (kWh/kg TOC) removed. Table 6.1 summarizes the cost analysis of some recent Fenton-based AOPs for pharmaceutical degradation. The electrical energy consumption (EEC) is expressed as,

$$EEC = \frac{1000U \times I \times t}{C \times V} kWh(or\, kg\, COD(or\, TOC))$$

where, U = Voltage (V)
I = Current (A)
t = electrolysis time (h)
V = volume of wastewater (m^3)
C = COD or TOC removed (mg/m^3).

Table 6.1 Summary of the cost estimation of some Fenton-based AOPs for pharmaceutical degradation

AOP	PCs removed	Removal efficiency	Cost/energy estimation	References
Electro-Fenton using graphene-modified cathode	Sulfadiazines	99%	0.21 kW h/m^3	[41]
Heterogeneous electro-Fenton	Diclofenac (DCF)	95%	175.46 kW h/(kg DCF)	[42]
Bio-electro-Fenton	DCF, ibuprofen (IBU), ketoprofen (KTF) and naproxen (NPX)	KTF: 59–61% DCF: 87–97% IBU:80–86% NPX:75–81%	2.39 × 10^{-3} kWh for 80 mL solution	[43]
Sono-electro-Fenton	Cefixime	97.5%	2 kW h/m^3 of solution	[44]
Electro-fenton	Imatinib (IMA)	100%	2 kW h/g TOC	[45]

6.5 Conclusions and Future Research

Fenton oxidation is an efficient method for pharmaceutical degradation from aqueous matrices. The method has several advantages such as less maintenance, ease of operation, and fast degradation. However, large operating costs and sludge formation are some major disadvantages of this method. Researchers have explored homogeneous as well as heterogeneous Fenton oxidation for pharmaceutical degradation. A high degree of mineralization has been obtained by combining Fenton oxidation with other methods. The electrical energy consumption and cost analysis are reviewed. Future work should be focused on the development of an integrated/combined Fenton oxidation-based approach with the objectives of cost minimization and efficient degradation. The scaling up of the EF processes via examination of their technical, economic, and environmental aspects could be another research area for prospective researchers.

References

1. V. Boucher, M. Beaudon, P. Ramirez, P. Lemoine, K. Volk, V. Yargeau, P.A. Segura, Environ. Sci. Water Res. Technol. **7**, 1301–1314 (2021)
2. M. Sharma, A. Yadav, M.K. Mandal, K.K. Dubey, Int. J. Environ. Sci. Technol. **2022**, 1–16 (2022)
3. J.A. Garrido-Cardenas, B. Esteban-García, A. Agüera, J.A. Sánchez-Pérez, F. Manzano-Agugliaro. Int. J. Environ. Res. Public Heal. **17**, 170 (2020), **17**, 170 (2019)
4. R. Guo, X. Xie, J. Chen **36**, 844–851 (2014), https://doi.org/10.1080/09593330.2014.963696
5. M. Hui Zhang, H. Dong, L. Zhao, D. Xi Wang, D. Meng, Sci. Total Environ. **670**, 110–121 (2019)
6. R. B.-C. E. Progress and undefined 1995, osti.gov

7. N. Wang, T. Zheng, G. Zhang, P. Wang, J. Environ. Chem. Eng. **4**, 762–787 (2016)
8. R. Shetty, S. Verma, Int. J. Sci. Res. **4**, 2319–7064 (2013)
9. E. Cuervo Lumbaque, R.M. Cardoso, A. Dallegrave, L.O. Dos Santos, M. Ibáñez, F. Hernández, C. Sirtori, J. Environ. Chem. Eng. **6**, 3951–3961 (2018)
10. N. Oturan, M.A. Oturan, *Electro-fenton process: Background, new developments, and applications*. Elsevier Inc. (2018)
11. O. Ganzenko, C. Trellu, N. Oturan, D. Huguenot, Y. Péchaud, E.D. van Hullebusch, M.A. Oturan, Chemosphere, https://doi.org/10.1016/j.chemosphere.2020.126659
12. D. Liu, H. Zhang, Y. Wei, B. Liu, Y. Lin, G. Li, F. Zhang, Chemosphere **209**, 998–1006 (2018)
13. F. Sopaj, N. Oturan, J. Pinson, F.I. Podvorica, M.A. Oturan, Chem. Eng. J. **384**, 123249 (2020)
14. H. Kadji, I. Yahiaoui, Z. Garti, A. Amrane, F. Aissani-Benissad, Chinese J. Chem. Eng. **32**, 183–190 (2021)
15. R. Davarnejad, K. Zangene, A.R. Fazlali, R. Behfar, Int. J. Eng. Trans. B Appl. **30**, 1639–1646 (2017)
16. X. Yang, R. Zou, K. Tang, H.R. Andersen, I. Angelidaki, Y. Zhang, Sci. Total Environ. **771**, 145385 (2021)
17. R. Zou, I. Angelidaki, X. Yang, K. Tang, H.R. Andersen, Y. Zhang, Sci. Total Environ. **727**, 138684 (2020)
18. W. Wang, Y. Lu, H. Luo, G. Liu, R. Zhang, S. Jin, Water Res. **139**, 58–65 (2018)
19. S. Li, T. Hua, C.S. Yuan, B. Li, X. Zhu, F. Li, Bioresour. Technol. **298**, 122501 (2020)
20. O. Ganzenko, C. Trellu, S. Papirio, N. Oturan, D. Huguenot, E. D. van Hullebusch, G. Esposito, M.A. Oturan, Environ. Sci. Pollut. Res. 2521, **25**, 20283–20292 (2017)
21. L. Xu, X. Zhang, J. Han, H. Gong, L. Meng, X. Mei, Y. Sun, L. Qi, L. Gan, J. Hazard. Mater. **391**, 122229 (2020)
22. L. Qi, W. Lu, G. Tian, Y. Sun, J. Han, L. Xu, Catal. **10**, 1297 (2020)
23. E.A. Serna-Galvis, J. Silva-Agredo, A.M. Botero-Coy, A. Moncayo-Lasso, F. Hernández, R.A. Torres-Palma, Sci. Total Environ. **670**, 623–632 (2019)
24. T.J. Al-Musawi, H. Kamani, E. Bazrafshan, A.H. Panahi, M.F. Silva, G. Abi, Catal. Lett. 1495, **149**, 1186–1196 (2019)
25. M. Munoz, P. Domínguez, Z.M. de Pedro, J.A. Casas, J.J. Rodriguez, Appl. Catal. B Environ. **203**, 166–173 (2017)
26. S. Mozia, D. Darowna, J. Przepiórski, A.W. Morawski, Polish J. Chem. Technol. **15**, 51–60 (2013)
27. M. Khabbaz, M.H. Entezari, J. Environ. Manage. **187**, 416–423 (2017)
28. A. Khataee, P. Gholami, M. Sheydaei, J. Taiwan Inst. Chem. Eng. **58**, 366–373 (2016)
29. N. Barhoumi, L. Labiadh, M.A. Oturan, N. Oturan, A. Gadri, S. Ammar, E. Brillas, Chemosphere **141**, 250–257 (2015)
30. Z. Ye, J.A. Padilla, E. Xuriguera, J.L. Beltran, F. Alcaide, E. Brillas, I. Sirés, Environ. Sci. Technol. **54**, 4664–4674 (2020)
31. Z. Wang, M. Liu, F. Xiao, G. Postole, H. Zhao, G. Zhao, Chinese Chem. Lett. **33**, 653–662 (2022)
32. E. Shirazi, A. Torabian, G. Nabi-Bidhendi, Clean Soil. Air Water **41**, 1062–1072 (2013)
33. W. Zhang, H. Gao, J. He, P. Yang, D. Wang, T. Ma, H. Xia, X. Xu, Sep. Purif. Technol. **172**, 158–167 (2017)
34. L. Wang, J. Yang, Y. Li, J. Lv, J. Zou, Chem. Eng. J. **284**, 1058–1067 (2016)
35. M. Pirsaheb, S. Moradi, M. Shahlaei, X. Wang, N. Farhadian, Environ. Process. **7**, 227–241 (2020)
36. R. Cheng, C. Cheng, G. Hua Liu, X. Zheng, G. Li, J. Li, Chemosphere **141**, 138–143 (2015)
37. S.K. Mondal, A.K. Saha, A. Sinha, J. Clean. Prod. **171**, 1203–1214 (2018)
38. R. Xing, J. He, P. Hao, W. Zhou, Colloids Surfaces A Physicochem. Eng. Asp. **589**, 124466 (2020)
39. P.K. Boruah, B. Sharma, I. Karbhal, M.V. Shelke, M.R. Das, J. Hazard. Mater. **325**, 90–100 (2017)
40. J. Shen, Y. Zhou, S. Li, P. Gu, G. Xue, J. Mater. Sci. **54**, 10684–10694 (2019)

41. G. Ren, M. Zhou, P. Su, W. Yang, X. Lu, Y. Zhang, J. Hazard. Mater. **368**, 830–839 (2019)
42. E. Rosales, S. Diaz, M. Pazos, M.A. Sanromán, Sep. Purif. Technol. **208**, 130–141 (2019)
43. H. Nadais, X. Li, N. Alves, C. Couras, H.R. Andersen, I. Angelidaki, Y. Zhang, Chem. Eng. J. **338**, 401–410 (2018)
44. K. Hasani, A. Peyghami, A. Moharrami, M. Vosoughi, A. Dargahi, Arab. J. Chem. **13**, 6122–6139 (2020)
45. W. Yang, M. Zhou, N. Oturan, Y. Li, M.A. Oturan, Electrochim. Acta **305**, 285–294 (2019)

Chapter 7
Shoreline Change Assessment and Forecasting Using GIS and Digital Shoreline Analysis System Along the Coast of Machilipatnam, Andhra Pradesh

Ashish Aggarwal, Muskan, Mansi Gupta, and Shravya Attri

Abstract Coastlines or shorelines are sites of dynamic activities, and phenomenon such as wave and tidal actions, rate of sediment supply, sea level changes and morphological properties of the area play a critical role in shaping the coastal ecosystems. The present study analyzes the long-term shoreline change detection along the Machilipatnam coast in Andhra Pradesh using Geographic information system (GIS) and Digital shoreline analysis system (DSAS) for the past two decades (2000–2019); and then predicts the position of the shoreline in the upcoming 10 and 20 years. Multi-temporal LANDSAT images were used for shoreline extraction, and the erosion patterns and the shoreline change rates were estimated using four statistically accepted models, End Point Rate (EPR), Net Shoreline Movement (NSM), Linear Regression Rate (LRR) and Weighted Linear Regression (WLR). The results from NSM show 83.13% of negative distance along the study transects. The high percentage of negative distance indicates a very high rate of erosion along the study transects. The average EPR was estimated to be -13.52 with ± 0.74 rate of uncertainty. The statistical significance of erosion was estimated as 82.27%. The average LRR was -13.55 ± 3.1. The high negative results of LRR again suggest an increased erosional environment along the shoreline. In our study, the values for LRR and WLR are same, both of which show landward movement of the shoreline (erosional). The analysis of NSR, EPR and LPR rates, thus, reveals a pronounced shoreline retreat in the Machilipatnam coastal zone during the study period. Beta forecasting was carried out to create Shoreline change envelope (SCE), and prediction maps of shoreline were produced for the next 10 and 20 years. The study has, thus, helped us manifest into a quantification of the trends and changes in the shoreline geometry and is a good precursor to efficient coastal classification and management.

A. Aggarwal (✉) · Muskan · M. Gupta · S. Attri
Petroleum and Earth Sciences Department, University of Petroleum and Energy Studies (UPES), Energy Acres Building, Bidholi, Dehradun 248007, Uttarakhand, India
e-mail: ashi_aggarwal@hotmail.com

A. K. Sahu et al. (eds.), *Energy Storage and Conservation*, Springer Proceedings in Energy, https://doi.org/10.1007/978-981-99-2870-5_7

Keywords Coastal erosion · Digital shoreline assessment system · Net shoreline movement · Endpoint rate · Linear regression rate · Forecasting

7.1 Introduction

The Indian state of Andhra Pradesh constitutes 4.87% of India's landmass. It contributes roughly about 975 km out of the 5422.6 km of the Indian (mainland) coastline and has the second largest coastline next to Gujarat among the 13 coastal states and Union territories of the Indian sub-continent. This coastline is located in southeast coast of Indian Peninsula, forming a part of Coromandel Coast adjacent to the Bay of Bengal. The coastal region of Andhra Pradesh spans across nine districts and 670 villages and is home to about 3.43 million population [1]. With such a huge coastal population, adequate coastal monitoring and management is of paramount importance in safeguarding these coastal communities from various coastal hazards such as sea level rise, storms, floods as well as the movement/change in shorelines.

Shoreline is commonly described as "the line where a body of water and shore meet". The changes in shoreline can simply be credited to the mutual adjustment of topographical features along the coast and the dynamic fluid activity, which also organizes transportation of sedimentation. These dynamics along the coastline can be described as the effects of the fluctuations in sea level, changes in circulation patterns and change in patterns of waves and tides. This physical interface between land and sea can be defined using temporal as well as spatial terms. One of the main aspects of planning a coastline and its protection, thus, starts with understanding how the dynamics of the coastline or shoreline work in relation to many physical factors such as processes of sea level changes, change in sedimentation rates and transportation of sediments on the shoreline.

Shoreline change detection and prediction is considered a very important task in monitoring and assessment of coastal regions [2]. Estimating the shoreline change rates through a time series not only provides an insight into the changing coastal patterns over time but also helps to forecast the future scenario of the coastline [3]. Such studies help in identifying areas of erosion for facility planning during times of sea level rise and help safeguard communities from damages caused during sea level rise, storms and floods [4, 5]. GIS and Remote sensing-based spatio-temporal studies of shorelines and subsequent analysis and assessment of erosion and attrition rates have proved to be very useful in monitoring and planning of coastal environments. The integrated use of Remote sensing and GIS technologies have proved to be of paramount importance for accurate prediction, planning, monitoring coastal dynamics and coastal zone management.

Shoreline assessment for an area of different years can be calculated as percentage rate of change or movement of shoreline and as coastal vulnerability index. Shoreline assessment is done using DSAS toolbox, which is purely a statistical approach. It calculates various rates such as End Point Rate (EPR), Linear Regression Rate (LRR) & Weighted Linear Regression (WLR) and Net Shoreline Movement (NSM).

These rates can be used to predict future shoreline movements in terms of erosion and accretion and can be estimated both for long term or short term depending on the rate type used and also with different accuracies. Many researchers have used the DSAS technique in India to quantify shoreline changes. A few of the key works include those of [6], who used the DSAS model to assess the shoreline changes along the Karnataka coast using time-series LANDSAT data. Multiple shoreline locations were utilized to calculate the LRR and WLR statistics. The results were categorized into three classes showing the locations with attrition, low rates of erosion and high rate of erosion. They concluded that 70% of the study coastline was accreting while 30% of the coastline was experiencing erosion. Estimations of rate of shoreline change along the coast of Andhra Pradesh in GIS and DSAS environment were investigated by [7]. Multi-resolution data from Landsat, IRS-P5 and IRS-P6 satellites were used to estimate short- and long-term changes in the coastline. The shoreline was finally classified as stable, erosional and accretional. The erosional and accretional parts were further classified into low, medium and high categories. The authors estimated that around 275 km of Andhra shoreline was facing erosion, 417 km was undergoing accretion while 153 km long shoreline was stable with no changes. The changes in geomorphology and shoreline shift were estimated using DSAS along the Vishakhapatnam coast of Andhra Pradesh [8]. Multitemporal Landsat, LISS-III and LISS-IV were used to assess the LPR and EPR rates in the study area. They estimated that around 74.6 km of the coastline was accreting at an average rate of +1.08 m/yr. Around 34.8 km of the coast line was eroding with an average erosional rate of −1.40 m/yr., while 41.4 km of the coastline was estimated as being stable with no changes observed. Similarly, the shoreline dynamics along the Odisha coast were investigated [9]. The researchers used the LRR and EPR techniques to assess the changes in net shoreline between a decadal time span using LANDSAT 5 and LANDSAT 8 images. The high coastal erosion was attributed to extreme coastal processes such as high frequency of tropical cyclones and sea level rise. The rates of accretion and erosion along the Vishakhapatnam district coast in Andhra Pradesh were calculated using DSAS [10]. They calculated the LRR, EPR and WLR rates to estimate the rate of shoreline change and demarcated the taluks with high rates of erosion and accretion. An average erosion rate was estimated to be 1.16 m/yr while the accretion in the study area was estimated at a rate of 1.62 m/yr. The mean shoreline change rate for the Vishakhapatnam Coast was estimated at 0.27 m/yr.

Subdivisions along the shoreline are entirely based on the discontinuities, whether it is artificial or natural. With the help of these subdivisions, transects can be generated along the shoreline. Some errors (such as pixel, rectification, tidal, seasonal and digitizing) are used together to analyze the shoreline positional uncertainty, which also helps in finding the accuracy of the model. Shoreline extraction can be achieved by both automatic and manual techniques. While manual digitization is more accurate, it requires a lot of human input and effort and can sometimes cause grave problems during processing. Automatic digitization is less accurate and is difficult to operate as it extracts shoreline with serious errors. The image used to extract the shoreline should be geo-rectified if necessary and image should be enhanced to create better results. Proper enhancement of the images can then be followed by estimation of shoreline changes by calculating the various statistical rates of erosion and accretion and make future predictions of the shoreline.

7.2 Study Area

Machilipatnam is a port city in Krishna district of Andhra Pradesh located between longitude 16.17°N and 81.13°E latitude, on the south-eastern coast of India. The average elevation of the city is 14 m, i.e. 45 ft from the sea level. Figure 7.1 shows the study area along the Machilipatnam coast in Andhra Pradesh.

7.2.1 Climate

A large aspect of the annual rainfall received by Machilipatnam is credited to the southwest monsoon. The climate ranges from hot summers to moderate winters. The period from April to June experiences the hottest temperatures with May witnessing the highest temperatures. The aggressive summer spell is followed by monsoon period, which extends between the months of June to September. October to December months mark the post-monsoonal period while January to February months witness the winter season in the area. The monsoon period usually starts in mid-June and lasts up to mid-October with good rains. The mean daily maximum temperatures in the district range from 38 °C in May month to about 20 °C in the months of December–January. Post-mid-February, the temperatures in the district start rising and continue till May [11]. The temperature decreases to about 20 °C on the onset of monsoon and is fairly uniform in the monsoonal period. The relative humidity in the mornings is around 80% throughout most part of the year. The average normal rainfall of the district is around 959 mm (37.8 in) and being a coastal area Machilipatnam is susceptible to cyclones due to frequent high surges in the sea [12].

7.2.2 Geology

Krishna district houses different geological formations varying from the oldest Archaean's to Alluvium being the recent formations. The formations found in the area are classified hydrologically, as follows:

(i) Consolidated or hard
(ii) Semi consolidated
(iii) Unconsolidated or soft.

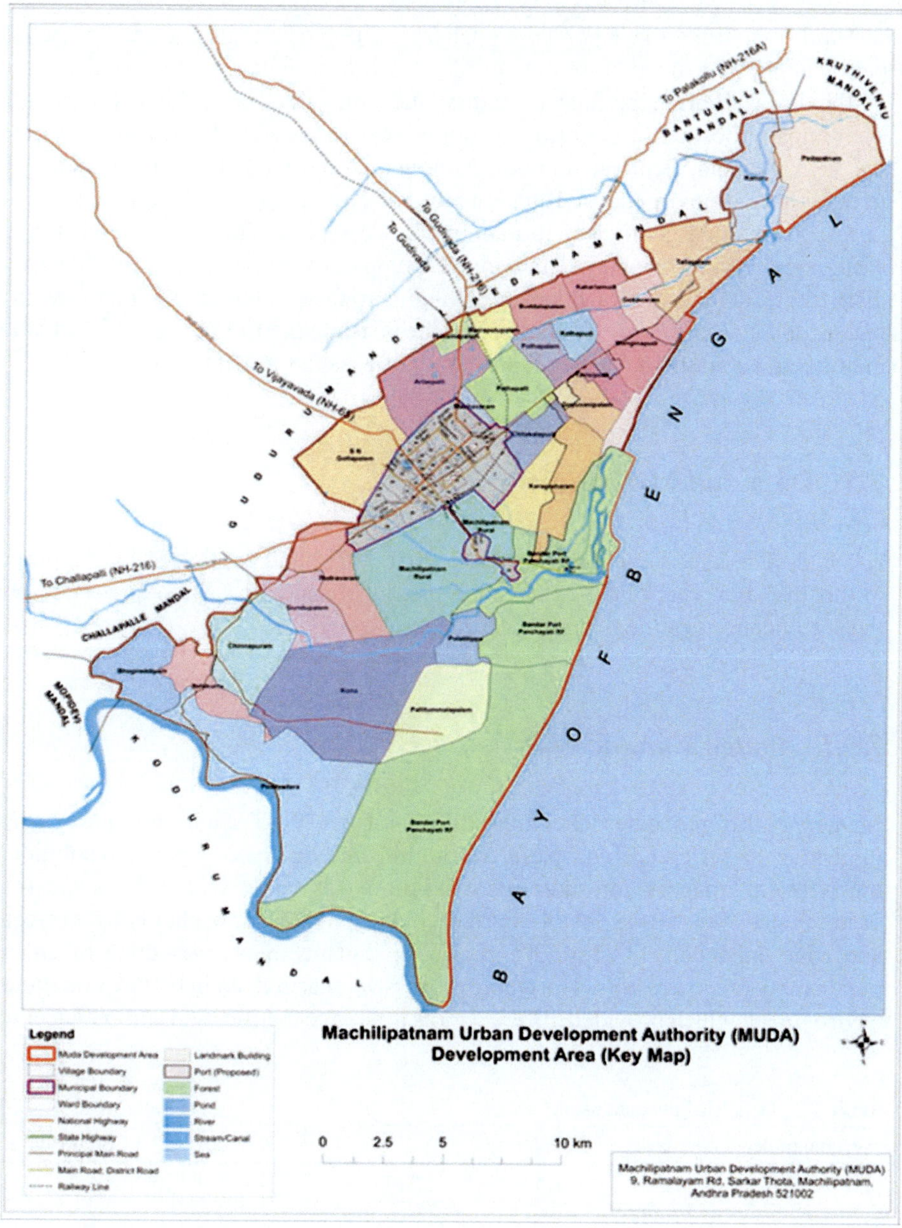

Fig. 7.1 Study area (Image Source: Machilipatnam Urban Development Authority, MUDA)

Hard crystallines like Khondalites, Charnockites and granite gneisses of Archaean period along with the meta-sediments like dolomites, shales, phyllites and quartzites of the Pre-Cambrian periods constitute the consolidated geological formations. They mostly occur in the north and north-west regions of the Krishna district, in the Jaggayyapeta, Penuganchimolu, Nandigama and Chandralapadumandals. The tertiary formations made of Rajahmundry & Gollapalli sandstones represent the semi-consolidated geological formations of the region. These are found in north-eastern regions of Musunuru, Nuzividu, Bapulapadu and Gannavaram mandals of the district. The deltaic alluvial deposits belonging to the Quaternary period make up the unconsolidated or the soft formations of the Krishna district. These soft formations majorly make up the delta area in the south of Krishna district [11].

7.3 Data and Methodology

LANDSAT images of years 2000, 2005, 2010, 2015 and 2019 used for the purpose of this research were obtained from the official website of USGS. Table 7.1 shows the specifications of the downloaded images.

7.3.1 Image Enhancement

Imagery of different dates taken from different LANDSAT satellites required resampling due to difference in spatial resolution among them. Spatial resolution of the different imagery for shoreline extraction was taken within the same frame of reference, i.e. 30 $\times$ 30 m. Further images were atmospherically corrected, and edge enhancement (3 $\times$ 3 kernel edge enhancement) was done in order to obtain more precise results. The whole process was carried out in ERDAS Imagine 14 environment, which provided different tools for image rectification and enhancement.

Table 7.1 Details of the data used

Serial number	Year	Satellite	Bands	Composite
1	2000	LANDSAT 7 ETM +	1, 2, 3	RGB
2	2005	LANDSAT 4–5 MSS TM	1, 2, 3	RGB
3	2010	LANDSAT 4–5 MSS TM	1, 2, 3	RGB
4	2015	LANDSAT 8 OLS	2, 3, 4	RGB
5	2019	LANDSAT 8 OLS	2, 3, 4	RGB

7.3.2 Shoreline Extraction

Shoreline extraction from an RGB composite satellite imagery can be easily carried out from RGB composites that help to identify the boundary between land and water. RGB composite creates a sharp boundary between these two, and from these composites, the shoreline was digitized manually in ArcGIS 10.5 version very precisely. Shoreline can also be extracted automatically using unsupervised classification techniques in ERDAS Imagine but it can be difficult sometimes and can also give some grave errors. Baseline was extracted taking the last shoreline (year 2000) and its orientation was delineated using the DSAS v5 toolbox in ArcGIS 10.5. Transects were created using the cast transect tool in DSAS after providing the information about the baseline and then for shoreline using the metadata. Transects using DSAS v5 were created till the last or youngest shoreline.

7.3.3 Rate Calculation

Rates such as End Point Rate (EPR), Linear regression Rate (LRR), Weighted Linear Regression (WLR) and Net Shoreline Movement (NSM) were calculated after the transects have been created using DSAS v5 toolbox. These rates are associated with the changes in the shorelines starting from the baseline. A DSAS Summary file is generated while calculating the rates, which indicate the erosion and accretion of the shorelines for a pair or some number of transects. Negative values indicate that erosion and positive values show accretion for a number of transects per rate calculated. Also, the summary provides the average distance/rate, minimum and maximum distance/rate depending on the type of rate calculated. Each rate is further used to predict the future movement of the shorelines with respect to the baseline.

7.3.4 Beta Forecasting

Beta forecasting is a tool in DSAS toolbox, which is used to predict future shoreline movement in 10 and 20 years ahead. This tool gives out a point and a polygon layer and an uncertainty band is also received, which is considered as the uncertainty associated with the prediction.

7.4 Results and Discussion

It is clear that the coastline/shoreline is a site of dynamic activities, and the rates of increment or decrement in accretion of sediments depend upon coastal activities such as wave and tidal phenomena, rate of sediment supply, morphological properties of the study area and changes in the sea level. The variability in the shoreline position can be defined spatially as well as temporally.

In our study, we looked at the coastline of the port city, Machillipatnam in Andhra Pradesh. The study area had a total of 220 transects along the coastline for the Shoreline Change Envelope (SCE). The Net Shoreline Movement (NSM) was calculated, which resulted in 183 transects with negative distance and 37 transects with positive distance. The average negative distance is −328.39 m against the 93.13 m of average positive distance. The result for NSM shows 83.13% of negative distance indicating high rates of erosion along the shoreline.

The average End Point Rate came out to be −13.52 with the rate of uncertainty being reduced to ±0.74. The statistical significance of erosion is 82.27%. The parameter, Linear Regression Rate, shows the statistical rate of erosion to be 74.77% and that of accretion to be 3.21%. The average Linear Regression Rate (LPR) is −13.55 ± 3.1. The resulting higher negative values are an affirmation to the above-indicated erosional environment. The WLR or the Weighted Linear Regression lays more emphasis on finding the best fit line using extremely reliable data. In our study, the values of LRR and WLR are similar, both of which indicate slow landward movement of the shoreline (erosional).

In the following figures, we show a graphical interpretation of the project revelations. The values obtained for different rates, namely, (SCE), (LRR), (EPR) and (NSM) are plotted graphically in Figs. 7.2, 7.3, 7.4 and 7.5. The data from the aforementioned statistical parameters are split into different groups so as to understand the rates of accretion and erosion in different groups of transects.

Further to the graphical representation of the movement rates, Beta Shoreline Forecasting tool in Digital Shoreline Analysis System (DSAS) was used to forecast the position of the shoreline in the next 10 and 20 years. Our earlier results from Shoreline Analysis affirm the findings of Beta Shoreline Forecasting Tool. If the erosion and accretion rates along the coastline of Machillipatnam continue at the

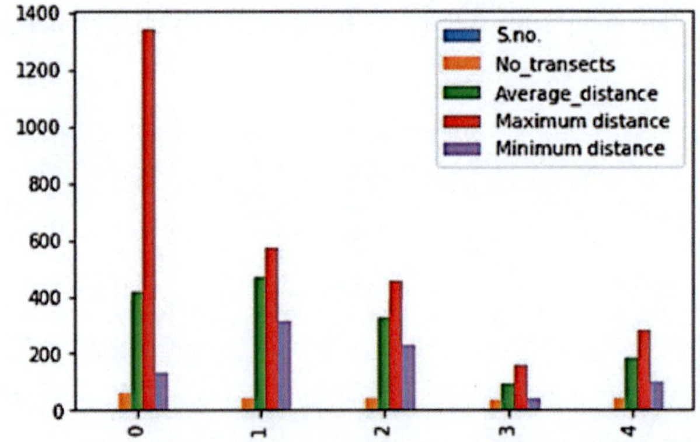

Fig. 7.2 Shoreline change envelope (SCE)

Fig. 7.3 End point rate (EPR)

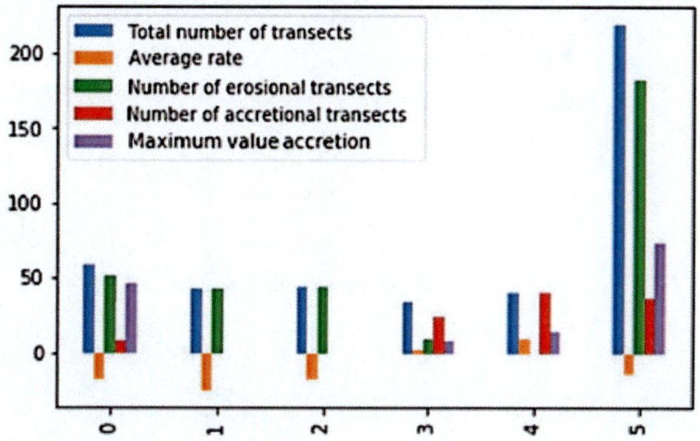

Fig. 7.4 Linear regression rate (LPR)

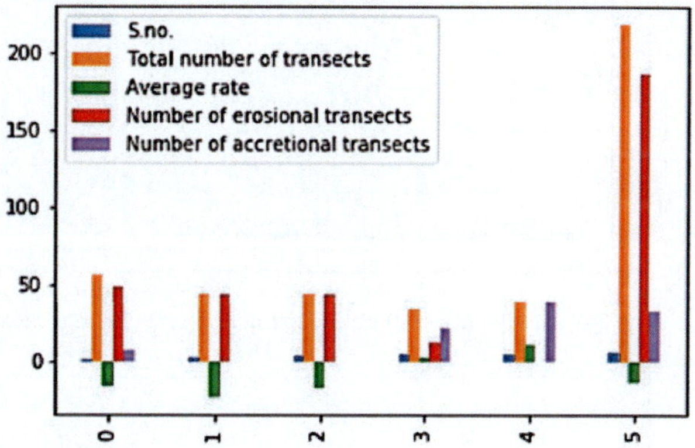

Fig. 7.5 Net shoreline movement (NSM)

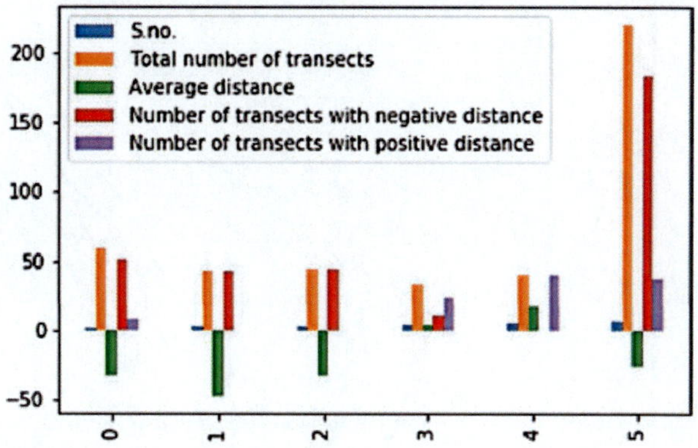

same rate, then there is a high tendency of the landward movement of the shoreline in coming years. Figure 7.6 shows the shorelines for different study years and the forecasted shoreline position in the next 10 and 20 years.

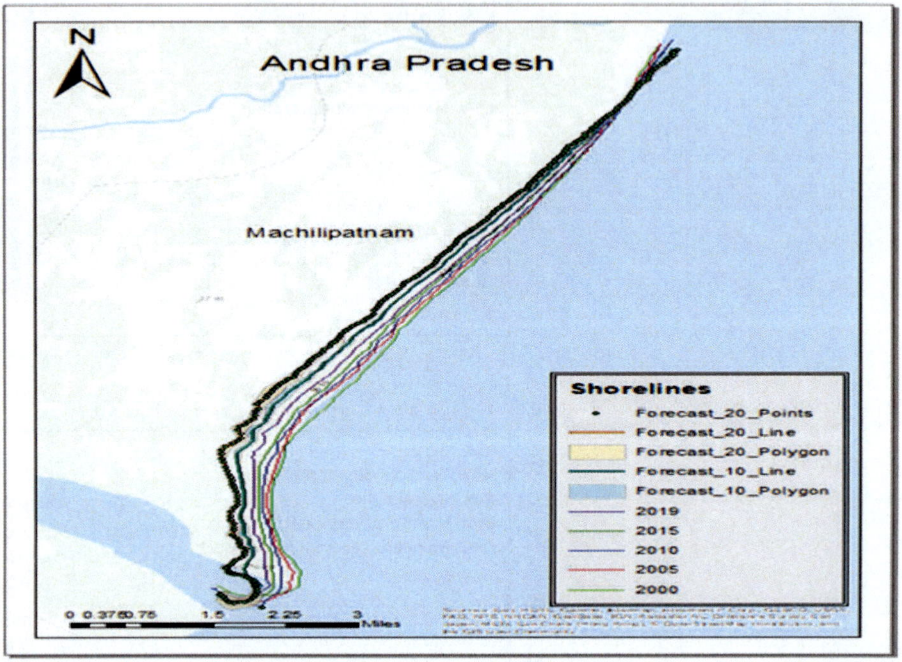

Fig. 7.6 Study period shoreline extraction and future forecast

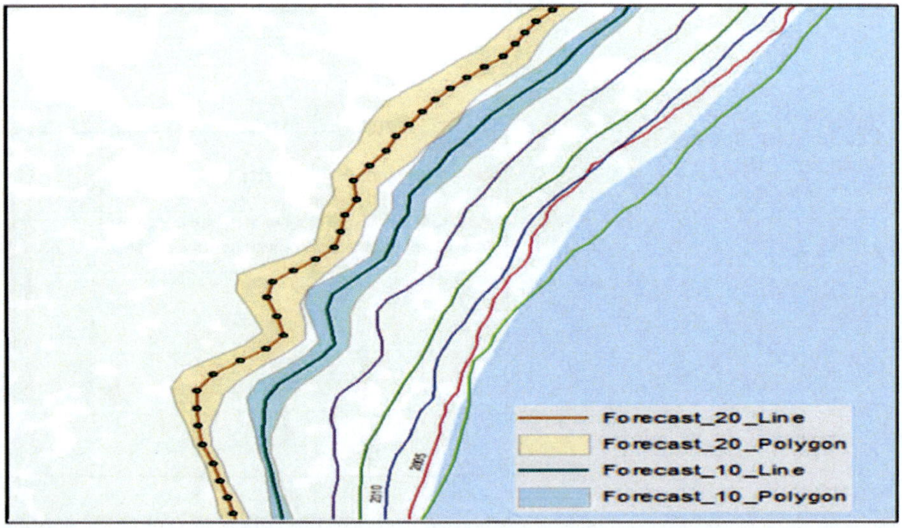

Fig. 7.7 Future Shoreline Movement

In the above figure (Fig. 7.7), the blue line with the blue buffer shows the predicted position of the shoreline after the next 10 years, and the dot-on line with orange buffer predicts the position of the shoreline on the coast in the next 20 years.

7.5 Conclusion

Our results in this project were expected on the assumption that the inferences made about the periodic changes in the placement of the shoreline can be used to predict the placement of shoreline for the upcoming years. Using the DSAS tools and the in-built methods has led us to interpret that the shoreline along the coast of Machillipatnam, in Krishna district of Andhra Pradesh has witnessed a significant percentile of erosional activities. In the present time, there is no landform that has not been subjected to adverse aftereffects of human activities. The landward movement of the shoreline is observed to be continuous and persistent since the early 2000s. These changes can be credited to the globally suspected sea level rise and the local geology. Erosion in one area leads to accretion in another. Some of the ways that can be used to prevent or keep the coastal erosion in check can be construction of seawalls on the coast or some soft engineering ways like coastal plantation, beach nourishment, etc. In the wake of current state of global crisis; natural as well as economic, the alterations caused in the geomorphological systems on the coast will lead to a similar faulting in the anthropogenic structures existing around the coastline as well.

References

1. K.K. Basheer Ahammed, A.C. Pandey, J. Cartogr. Geogr. Inf. **69**, 285 (2019)
2. Machilipatnam Urban Development Authority (MUDA) master plan data. Accessed at https://muda.ap.gov.in/Muda/UserInterface/Home/MasterPlanData.aspx
3. E. Armenio, F.D. Serio, M. Mossa M., A.F. Petrillo, Nat. Hazards Earth Syst. Sci. **19**(9), 1937 (2019)
4. A. Mukhopadhyay, S. Mukherjee, S. Mukherjee, S. Ghosh, S. Hazra, D. Mitra D., Eur. J. Remote. Sens., **45**(1), 201 (2012)
5. E.R. Thieler, E.S. Hammar-Klose, U.S. Geological Survey Open-File Report 2000–178, (2000)
6. S.C. Selvan, R.S. Kankara, B. Rajan B., Indian J. Mar. Sci. **43** (2014)
7. R.S. Kankara, S.C. Selvan, V.J. Markose, B. Rajan, S. Arockiaraj, Procedia Eng. **116**, 855 (2015)
8. R. Kannan, A. Kanungo, M.V.R. Murthy, J. Remote Sens. GIS **5**(1), 1000157 (2016)
9. K.K. Barik, R. Annaduari, P.C. Mohanty, R.S. Mahendra, J.K. Tripathy, D. Mitra, Indian J. Mar. Sci. **48**(12), 1990 (2019)“
10. M.R.I. Baig, I.A. Ahmad, Shahfahad, M. Tayyab, A. Rahman, Ann. GIS **26**(4), 361 (2020)
11. District survey report—Krishna district; Department of mines and geology, Government of Andhra Pradesh (2018). Available at https://www.mines.ap.gov.in/miningportal/downloads/applications/krishna.pdf
12. Ground water brochure, Krishna district, Andhra Pradesh Central Ground Water Board report, (2013). Available at http://cgwb.gov.in/District_Profile/AP/Krishna.pdf, https://doi.org/10.3133/ofr00178

Chapter 8
A Geospatial Approach to Monitoring Land Use and Land Cover Dynamics: A Review

Ashish Aggarwal

Abstract Land Use and Land Cover Change commonly referred to as LULC reflects the patterns of land use in a region and plays an important role in land use monitoring and proper administration of land resources, along with improved understanding of global climate change and sustainable development. Timely and precise land use/land cover classification is a critical component for monitoring the changes in biodiversity. Remote sensing and GIS techniques have provided valuable tools at the hands of scientists and land planners not only to gain a comprehensive knowledge on spatial and temporal patterns of land use patterns but also to understand and analyze the driving factors responsible for these changes and hence provide quantitative and qualitative assessment of LULC dynamics at regional to local levels. The present paper reviews the progress in geospatial technology over the last decades from traditional methods to machine learning-based approaches, for addressing the phenomenon of land use and land cover change and tries to highlight the importance of geospatial technologies in providing data and solutions with increased prediction accuracies.

Keywords GIS · Land cover · Land use · Monitoring · Remote sensing

8.1 Introduction

Land use/land cover mapping and monitoring plays a vital role in sustainable development of our ecosystem. The terms land use and land cover are often used interchangeably but differ in the sense that land cover refers to the physical land type or the surface cover such as vegetation, built-up area, water bodies, barren land, wetlands, forests and fallow land and are important for thematic mapping and change detection analysis of various land cover types [1, 2]. On the other hand, land use defines how

A. Aggarwal (✉)
Petroleum and Earth Sciences Department, University of Petroleum and Energy Studies (UPES), Energy Acres Building, Bidholi, Dehradun 248007, Uttarakhand, India
e-mail: ashi_aggarwal@hotmail.com

A. K. Sahu et al. (eds.), *Energy Storage and Conservation*, Springer Proceedings in Energy, https://doi.org/10.1007/978-981-99-2870-5_8

the land is being used by people for development, conservation or other uses such as recreation, wildlife sanctuaries or agriculture. Land use/land cover (LULC) patterns depend on spatio-temporal usage of these natural resources by humans. Thus, any alteration to land use pattern will affect the land cover type cause and vice-versa [3].

Many developing countries lack a robust land use monitoring system to administer their land resources due to the paucity of infrastructural and technological capabilities as well as unorganized scientific capacities. The growing human population has changed the land cover patterns due to surge in demand of habitable settlements, intensification of agriculture and over utilization of water resources. An accurate and up-to-date information on changes in LULC is necessary to understand and assess their impacts on environment [4]. Such studies also help to analyze the extent of human interventions on natural environment. Satellite-based remote sensing data and geospatial techniques play an important role in dynamic monitoring, qualitative/ quantitative assessment, modeling, forecast and planning of land use at regional/ local levels [2, 5].

The timely availability of accurate land use/land cover maps is critical for addressing multiple issues, which include agriculture development and food administration, environmental impact assessment, ecology, forestry, geology and hydrology [6]. Land use activities have a major impact on soil cover, soil quality, soil fertility and can trigger detrimental effects such as soil erosion, soil salinization, water logging or desertification, which are critical for agriculture management such as in crop yield planning and food security [7].

Land cover maps are important for foresters to assess past and future deforestation rates, develop sustainable harvesting plans, conservation of biodiversity, reduce impacts on flora and fauna and minimize forest degradation. Land use, land-use change and forestry referred to as (LULUCF) focuses on emission and removal of greenhouse gases in the atmosphere produced from anthropogenic interventions on land use such as human settlements and commercial establishments. Land use change is responsible for changes in carbon dioxide levels and contributes to global climate change [8].

The dynamics of land use induced by anthropogenic activities affect the hydrological response in different river catchments [9]. Both land use patterns or their alterations and administration of hydrological resource are mutually interdependent. Spatial and temporal changes in land use/land cover alter waterflow path and its dynamics and have affected the natural water balance [10]. Land use changes modify the patterns of land erosion, sedimentation, pollutant transport, groundwater discharge and surface runoff and have a direct impact on water resource development programs such as water supplies to urban and rural sectors, hydropower generation, irrigation, aquaculture, etc. [11]. Land cover maps are important for river morphometric analysis and are also used in conjunction with various hydrology models to analyze flood risks and coastal erosion and help in decision-making process by various public or private stakeholders [5, 12, 13]. The present work tries to review the techniques employed over the years ranging from traditional methods relying heavily on ancillary information to current machine learning and deep learning approaches to improve the classification approaches and evaluate the classification accuracies using various statistical methods.

8.2 Approaches to Land Use/Land Cover Mapping

Traditionally, land use mapping was done by compiling the revenue records from the Economics and Statistical Department of the country. Survey of India provides topographic maps showing land use categories at a scale of 1:50,000 to 1:25,000. The categories were mapped using ground information obtained from ancillary sources supported with few aerial photographs. The traditional ground-based methods of land use mapping required huge manpower and were time-consuming. In India, though efforts have been put forward by various state and national governments to provide information of land use/land cover through thematic maps, technical figures and statistical reports, but the efforts are unstructured leading to random and incomplete data collection and duplicated results. The increased time gap between monitoring, collection and availability of data often made it difficult to gather up-to-date information on LULC dynamics as well as its underlying processes, and hence could not accurately represent the present condition of the different LULC classes. The usage of different classification methods by different collection agencies further weakened the utility of such data for accurate representation of the categories. The availability of national spatial databases on agricultural and forest ecosystems or surface water bodies were negligible, leading to limited studies being carried out on land use practices [14].

In recent years, satellite-based earth observation techniques have been developed, which have proved very useful in providing accurate and timely information on land resources to prepare highly efficient land use/land cover maps at different resolutions and scales. These techniques provide a time and cost-effective solution to land use mapping especially in inaccessible areas where remote sensing remains the only viable option [15]. Various national and international Earth observation agencies are providing spatio-temporal datasets that have fostered advancement in land use classification methods. Earth observation (EO) datasets are now available in public domain as well in commercial segments. Especially, the availability of these free datasets in both optical and microwave spectrum has fostered quick leaps in advancing techniques required for improving image classification. Remote sensing approach because of providing multi-spatial, multi-spectral and multi-temporal capabilities have given way to the development of more accurate geographic data for understanding, modeling and forecasting changes in land cover [16, 17]. Automatic classification systems have been developed that have not only drastically reduced the time for data collection and resource requirements but have also improved the accuracy and consistency of the data outputs. The remote sensing-based classification systems have increased the scalability of land use mapping with their utilization now spread to a greater number of earth's resource-based applications. The accuracy of remote sensing approach is limited by numerous factors such as image resolution, atmospheric conditions, thematic content, type of data as well as processing and analytical techniques employed and affect the derived LULC information content from remote sensing data [18]. Different types of land cover/land use classified products may be required for different applications under different environment conditions, which makes generalization and image classification process a challenging process [19]. The major purpose of image classification is to categorize and label the pixels in the captured mages with meaningful information classes such as cadastral information, vegetation types and other land cover categories [20].

Several methods have been developed for analyzing remote sensing data for land use/land cover, such as image differencing, temporal image classification, image rationing, vegetation index differencing, post-classification comparison, principal components analysis and change vector analysis [21]. While supervised classification methods use algorithms such as maximum likelihood classifier, nearest neighbor classifier, Mahalanobis distance, parallelepiped classifier, minimum distance to mean, support vector machines and random forest classifiers; unsupervised classification techniques include algorithms of K-means, ISODATA clustering, Fuzzy C-means clustering and neural network techniques [22]. However, newer approaches like Fuzzy C-means clustering, artificial neural network and object-based classification that have been put forward for land use classification procedures require specialized training and expertise.

Landsat images that have a spatial resolution of 30 m are the most widely used earth observation data and have proved to be extremely useful for mapping and predicting land use/land cover dynamics [1]. The Sentinel-2 mission that started in 2014 also provides global data on land resources in the optical region with a higher spatial resolution of 10 m and is also used lately to map land use and land cover with a higher precision.

Schuft et al. [23] investigated different sampling methods and landscape metrics to characterize the stream networks in riparian environment using aerial photography and GIS. The sample area was derived from morphological characteristics of chosen points along the stream network. The extent of different land cover types was interpreted from aerial photographs and was examined using incremental buffers with a maximum distance upto 300 m. The study helped determine landscape metrics such as the composition, extent and connectivity in woody and non-woody vegetation; and provided an efficient framework to identify indicators for riparian ecosystem at different spatial scales.

Nikhil and Azeer [24] investigated the spatio-temporal changes in LULC of Bharathapuzha river basin in South India. Landsat images from 1973 to 2005 were used to identify the different categories of land use and were analyzed in remote sensing and GIS environment to reveal the land use patterns in the study area. A decrease in natural vegetation and wetland agriculture areas was observed with a decrease percentage of 31% and 8.7%, respectively. The urban area cover showed a remarkable increase and increased by 32% during the study period.

Kawy et al. [25] also used Landsat images to investigate the land use patterns in western Egypt. The Landsat images from four different years, namely, 1984, 1999, 2005 and 2009 were processed using supervised classification algorithm to show the historical and present land use scenario in the western Nile region. For pre-processing of Landsat images; geometric correction, gap filling of ETM + data and image enhancement were carried out. To check the accuracy of the classified LULC maps, error matrix was generated. The classification results showed substantial transition of barren lands into agricultural lands. The authors estimated that about 28%, 14% and 9% of barren land transitioned into agricultural land during the three study periods of 1984 to 1999, 1999 to 2005 and 2005 to 2009, respectively. They also suggested that land degradation was happening in the area mainly because of anthropogenic activities such as construction of quarries, free water bodies and other land reclamation projects. On the basis of the identified causes of the changes in LULC, the authors suggested few policy recommendations for better administration of land resources.

Rimal [26] used supervised classification method to classify Landsat images of 1976, 1990, 1999 and 2009 into different land use categories and predict urban growth in Biratnagar sub-metropolitan city, Nepal. They identified four land use classes, namely, urban (built-up), water body, forest cover and cultivated land. Change in urban land use for the year 2019 was modeled using a Markov chain technique and concluded that there was a gradual increase in urban land use, which might threaten the areas falling under the categories of forest and agricultural lands.

Mallupattu and Reddy [27] analyzed the changes in LULC using remote sensing and GIS in Tirupati, India. They utilized the topographic maps obtained from Survey of India and satellite data from LISS-III and PAN data from IRS-ID. Their results demonstrated that amongst the classified LULC categories, there was an increase in urban area, open forests and plantation during the study period, while the area under agricultural land, water bodies and dense forests showed a significant decrease. They concluded that rapid urbanization has led to the decrease in agriculture, water bodies and dense forest categories.

Kindu et al. [28] used object-based classification approach for LULC change detection in Munessa-Shashemene region of the Ethiopian Highlands. They studied the LULC patterns over a period of 39 years from 1973 to 2012 using Landsat images of 1973, 1986, 2000 and Rapid Eye image of 2012. The results from the change analysis demonstrated that there was a gradual reduction in woodland cover during the three different study intervals. The woodland cover decreased from 81.8% during 1973–1986 to 52.3% during 1986–2000 and decreased further to 36.1% in 2000–2012. The forest cover also showed similar decreasing trend with a declining cover of 26.1%, 21.1% and 24.4%, respectively, during the three study periods. However, an increase was witnessed in the cropland category with the percentages increasing from 13.1 to 31.5 and 22.7% respectively, during the three study periods.

Welde et al. [10] investigated the effects of LULC on dynamics of hydrological response namely stream flow and sediment yield in the Tekeze dam catchment of northern Ethiopia. They used the SWAT model along with GIS to study the land use in the region. The catchment was divided into 47 subcatchments, and multiple hydrologic response units (HRU) were defined for simulation and sensitivity analysis. The validation of SWAT model was done and later the LULC patters were analyzed for hydrologic response for three scenarios, which were climate of 2000s and LULC of 2008, climate of 2000s and LULC of 1986 and climate of 1980s and LULC of 1986. They concluded that the shift of grass and shrubland to agricultural land had a positive impact on the watershed response. The increase in agricultural land in the Tekeze Dam watershed led to an increase in both yearly and seasonal stream flow and also increased the sediment yield in the watershed. The mean annual stream flow increased from 129.20 to 137.74 m^3/s, which was around 6.02%, and the sediment yield increased by 17.39% (12.54–15.18 t/ha/yr.), and this increase was attributed to the changed LULC patterns. The LULC dynamics were seen to impact the hydrologic response majorly in the months of August to October with lesser impact in the other months of the year.

Heidarlou et al. [29] performed LULC change detection in Zagros forest area using a series of Landsat images between 1992 and 2016 using supervised maximum likelihood classification algorithm and classified the satellite images into different categories of forest, vegetation, rangeland and built-up area. Atmospheric and geometric errors were removed in preprocessing and postprocessing was done to enhance the accuracy of the classified product. The suitability maps were produced using CA-Markov chain model. The dependent and independent variables were calibrated through Land Change Modeler (LCM) using multi-layer perceptron, and future predictions of land use were made up to the year 2024. The maps of land cover between the period 2002 and 2012 were used for calibration and then the LCM predicted LULC map was validated with the actual land cover map of 2016.

Jamali [30] evaluated eight different machine learning algorithms for classifying land use in north of Iran. The classifications were performed individually using eight different algorithms, namely, Random Forest, Decision trees, DTNB, J48, Multilayer Perceptron, Lazy IBK, Non-Nested Generalized Exemplars (NNge), as well as Simple Logistic. The results from each classification were evaluated in terms of RMSE, overall accuracy and mean absolute error. He concluded that though all algorithms produced an accuracy of more than 99%, but NNge algorithm produced the highest accuracy while the algorithms J48 and DTNB gave the worst results for classifying land use in the study area.

Recently, many researchers have used machine learning algorithms to classify land use (30, 31, 32, 34). While the supervised machine learning classifiers include algorithms such as support vector machine, random forest, radial basis function, decision tree, naive Bayes, maximum likelihood classifier, spectral angle mapper, etc. [31–33], the unsupervised classification algorithms include ISODATA, K-means, Fuzzy C-means, etc. [34, 35].

Nguyen [16] used parametric and non-parametric classifiers to classify land use from multi-temporal images of Sentinel 2 in Dak Nong province of Vietnam. They used 446 images between 2017 and 2018, to perform logistic regression and support vector machine, k-nearest neighbor and random forest to classify the land use in the area. The accuracy of the classification methods was gauged from the confusion matrix, and they estimated an overall accuracy between 63.9% and 80.3%, and the Kappa coefficient was calculated in the range of 0.611–0.813.

Abdi [32] compared the land use classification accuracy using non-parametric methods, namely, support vector machines, random forests, extreme gradient boosting (XGBoost) and deep learning in Sweden region with mixed land use settings. They used multi-temporal Sentinel 2 images from different seasons. The training and testing subsets were chosen using stratified random sampling. McNemar's chi-square test was utilized for comparing the classification of each class, and two proportion Z test was used to evaluate the classification accuracy of pixels. Their result showed that the highest classification accuracy was obtained in the range of 0.758 ± 0.017 using support vector machine, which was followed by XGBoost, random forests and lastly deep learning algorithm. The results from McNemar's test showed that almost 62% of predictions at class levels showed significance at 5% significance level.

Nehzak [36] assessed the support vector machine (SVM) and artificial neural network (ANN) algorithms for classifying land use categories. They used three different band combinations, namely, (1) reflective bands, (2) reflective and thermal bands and (3) reflective, thermal and slope bands from Landsat 8 images, and analyzed them for their accuracy assessment for land use classification. The two machine learning algorithms chosen for the research work were applied on Landsat 8 OLI/TIRS images to extract land use maps of the study area. They concluded from their work that the band combination of reflective, thermal and slope bands produced the maximum classification accuracy. While the residential areas showed the highest classification error using ANN, the lowest error was observed in water bodies with SVM.

8.3 Conclusion

The changing land use/land cover dynamics whether due to human or natural factors have a direct impact on our ecosystem and may lead to adverse climatic effects such as floods or droughts. Studies on historical and present LULC patterns as well as prediction of future land use scenarios are, therefore, critical in evaluating the impact of the changing LULC on the natural biodiversity and ecosystem. There are various remote sensing and GIS-based techniques available for monitoring and assessment of LULC ranging from traditional digital image classification to recent machine learning and deep learning algorithms to object-based image analysis. Irrespective of the technique utilized, the method employed depends on the scale, resolution, accuracy levels and the type of scientific application to which the method is applied to. Whatever the application, the studies on land use/land cover changes give a lot of insightful information to the public and private stakeholders, urban planners, hydrologists, geologists and/or forest officials to make informed decisions on safeguarding the LULC environment.

References

1. J.S. Rawat, M. Kumar, Egypt. J. Remote. Sens. Space Sci. **18**(1), 77 (2015)
2. S. Chowdhury, D.R. Peddle, M.A. Wulder, S. Heckbert, T.C. Shipman, D.K. Chaoa, Int. J. Appl. Earth Obs. Geoinf. **94**, 10224 (2021)
3. A.K. Hua, J Environ Public Health, art 7515130 (2017)
4. C. Giri, Z. Zhu, B. Reed, Remote Sens. Environ. **94**, 123 (2005)
5. P.L.M. Tolentino, A. Poortinga, H. Kanamaru, S. Keesstra, J. Maroulis, C.P.C. David, C.J. Ritsema, PLoS ONE **11**(10), e0163941 (2016)
6. Q. Weng, Int J Remote Sens **22**(10), 1999 (2001)
7. D. Saah, G. Johnson, B. Ashmall, G. Tondapu, K. Tenneson, M. Patterson et al., Environ. Model. Softw **118**, 166 (2019)
8. I. Tzamtzis, P. Ganatsa, Int. J. Glob **22**, 1,111 (2020)
9. J.M. Bosch, J.D. Hewlett, J. Hydrol. **55**(1–4), 3 (1982)
10. K. Welde, B. Gebremariam, Int. Soil Water Conserv. Res. **5**(1), 1 (2017)
11. T.C.C. Huang, K.F.A. Lo, Environments **2**(1), 32 (2015)
12. P. Gong, J. Wang, L. Yu, Y. Zhao, L. Liang, Z. Niu, X. Huang, H. Fu, S. Liu, C. Li, X. Li, W. Fu, C. Liu, Y. Xu, X. Wang, Q. Cheng, L. Hu, W. Yao, H. Zhang, P. Zhu, Z. Zhao, H. Zhang, Y. Zheng, L. Ji, Y. Zhang, H. Chen, A. Yan, J. Guo, L. Yu, L. Wang, X. Liu, T. Shi, M. Zhu, Y. Chen, G. Yang, P. Tang, B. Xu, C. Giri, N. Clinton, Z. Zhu, J. Chen, J. Chen, Int. J. Remote Sens. **34**(7), 2607 (2013)
13. K. Hibbard, A. Janetos, D.P.V. Vuuren, J. Pongratz, S.K. Rose, R. Betts, M. Herold, J.J. Fedemma, Int. J. Climatol. **30**(13), 2118 (2010)
14. NRSA Project Report, 2007. Document no. NRSA/RSGIS-AA/NRC/NLULC-AWIFS/PROJREP/R01/JUN07
15. J.F. Olorunfemi, Environ. Int. **9**(1), 27 (1983)
16. H.T.T. Nguyen, T.M. Doan, E. Tomppo, R.E. McRoberts, Remote Sens. **12**(9), 1367 (2020)
17. S.I. Toure, D.A. Stow, H.C. Shih, J. Weeks, D.L.L. Carr, Remote Sens. Environ. **210**(8), 259 (2018)
18. R.R. Regmi, Phd thesis, (2015). https://shodhganga.inflibnet.ac.in/handle/10603/51039
19. R. Nedd, K. Light, M. Owens, N. James, E. Johnson, A. Anandhi, Land **10**, 994 (2021)
20. B. Usman, Elixir Comp. Sci. Engg. **63**, 18671 (2013)
21. M. Mohajane, A. Essahlaoui, F. Oudija, M.E. Hafyani, A.E. Hmaidi, A.E. Ouali, G. Randazzo, A.C. Teodoro, Environments **5**(2), 131 (2018)
22. S.M. MohanRajan, A. Loganathan, P. Manoharan, Environ. Sci. Pollut. **27**, 29900 (2020)
23. J.S. Schuft, T.J. Moser, P.J. Wigington Jr., D.L. Stevens Jr., L.S. McAllister, S.S. Chapman, T.L. Ernst, Photogramm. Eng. Remote Sens. **65**(10), 1157 (1999)
24. P.P. Nikhil, P.A. Azeer, J. Geogr. Inf. Syst. **2**(04), 185 (2010)
25. O.R. Kawy, J.K. Rod, H.A. Ismail, A.S. Suliman, Appl. Geogr **31**(2), 483 (2011)
26. B. Rimal, Eng. Sci. Technol. an Int. J. ENG SCI TECHNOL, **2**, No. 1 (2012)
27. P.K. Mallupattu, J.R.S. Reddy, Sci. World J. **2013**, art 268623 (2013)
28. M. Kindu, D. Schneider, T. Teketay, T. Knoke, Remote Sens. **5**(5), 2411 (2013)
29. H.B. Heidarlou, A.B. Shafiei, M. Erfanian, A. Tayyebi, A. Alijanpour, Land Use Policy **81**, 76 (2019)
30. A. Jamali, S.N. Appl, Sci. **1**, 1448 (2019)
31. H.C. Shih, D.A. Stow, Y.H. Tsai, Int. J. Remote Sens. **40**, 1248 (2019)
32. A.M. Abdi, GISci. Remote Sens. **57**(1), 1 (2020)
33. S. Talukdar, P. Singha, S. Mahato, Shahfahad, S. Pal, Y.A. Liou, A. Rahman, Remote Sens., **12**(7), 1135 (2020)
34. A.W. Abbas, N. Minallh, N. Ahmad, S.A.R. Abid, M. A. A., Sindh Univ. Res. J. (Sci. Ser.), **48**(2), 315 (2016)
35. E. Paradis, Int. J. Appl. Earth Obs. Geoinf. **107**, 102675 (2022)

36. H.K Nehzak, M. Aghaei, R. Mostafazadeh, H.Rabiei, In book: Computers in Earth and Environmental Sciences, Chapter 5 - Assessment of machine learning algorithms in land use classification, p. 97–104 (2022)

Chapter 9
A Comprehensive Review of Potential Sites for CO_2 Sequestration in India

N. P. Nayak and Bhavesh Venkat

Abstract Decarbonization is the need of the hour and with agreements to reduce the carbon footprint to ~33% of the year 2005 levels by 2030 and completely become a net-zero emitting country by the year 2070, India has committed itself to mammoth tasks, which can be achieved only by taking huge steps. While switching to green energy is a gradual transition, the reduction of current CO_2 levels in the atmosphere is also being considered, and this process of carbon dioxide sequestration has already been successfully performed in the Western Hemisphere. Four storage sites have been classified, and they are Depleted Oil & Gas fields, Coal Seams, Saline Aquifers and Basalts. While the first two storage sites fall under the CCUS (Carbon Capture Utilization and Storage) wherein the injected CO_2 is used for enhanced recovery of the fuels, the latter two fall under the CCS (Carbon Capture and Storage) category. Although several studies have already been performed on assessing the theoretical storage capacity of the potential storage sites in India using different methodologies, India still doesn't possess a working CCS storage facility yet. With just about 10% of the total administrative districts contributing nearly around 60% of the total CO_2 emissions, some have also performed CCS source-sink studies that would facilitate in the transportation of the greenhouse gas for long-term storage. This article aims to provide a comprehensive review on the latest developments that have taken place in this essential decarbonization technology in the subcontinent.

Keywords Carbon capture · Sequestration · Saline aquifer · Basalt

9.1 Introduction

With rising temperatures due to global warming, the need for switching to greener energy and limiting the emissions of greenhouse gases are slowly being realized by the population. CO_2 being one of the most important GHG (Green House Gas)

N. P. Nayak (✉) · B. Venkat
Energy Cluster, UPES, Dehradun 248007, India
e-mail: npnayak@ddn.upes.ac.in

A. K. Sahu et al. (eds.), *Energy Storage and Conservation*, Springer Proceedings in Energy, https://doi.org/10.1007/978-981-99-2870-5_10

responsible for the rising temperatures has for long time been considered for sequestration. Countries are called upon to actively monitor their CO_2 emissions and frame goals to reduce the emissions. The Paris Climate Agreement (PCA) 2015 which saw an active participation of several major countries had set up a goal of limiting the temperature rise below 1.5ÌC [1]. Keeping this goal in consideration, the Intragovernmental Panel on Climate Change [2] framed a carbon budget of 2900 Gt so as to track the performance of the PCA participating countries and their progress in achieving the goals. However, almost 75% of the total budget had already been consumed, which has caused a rise of about 1.1ÌC and a further addition of 500 Gt of CO_2 will cause us to cross the agreed limit of 1.5ÌC [3] estimated that to minimize the temperature rise by the agreed 1.5ÌC as per the PCA 2015, around 640–950 Gt of CO_2 has to be removed from the atmosphere considering a business as usual scenario. Since the beginning of the industrial revolution, ~15–40% additional anthropogenic carbon emissions have been reported, and it is also found that 51 Gt of CO_2 is released in to the atmosphere annually [4]. India being the third largest emitter of CO_2 the country has acted responsibly by signing the PCA and agreeing to reduce the emission intensity to 33–35% from the levels of 2005 by 2030 [5]. Moreover, India has also committed itself to become a net-zero carbon emitter by 2070. However, these are some mammoth tasks that the country has committed to, as it is found that an investment of about 3.4 times the size of India's current GDP is required to meet the goals of PCA.

This is where the decarbonization technologies of CCUS and CCS become important, as these are two important means of achieving the ambitious goals that the major countries have committed themselves to. As per an IEA2019 report, CCS can help in the reduction of cumulative emissions of about 13% by 2060 considering a Clean Technology scenario. Currently, there are 65 commercial CCUS facilities worldwide, of which 26 are operating, 21 in the early development phase, and 2 are suspended facilities (Global CCS Institute, GCCSI 2020). The cumulative storage potential of these facilities is merely 40 MT of CO_2 which when compared to the annual emission (2019) of 52 BT of CO_2 is a very small number (16). When considering a zero further emission scenario, the current storage facilities would require 130 years to sequester the required amount of CO_2 to meet the PCA agreed temperature limit of 1.5ÌC and as per IPCC, in order to achieve this target by the end of the century, we must sequester 10BT of CO_2 by 2050(16) and up to 1200 Gt of CO_2 by 2100 to meet the climate model pathways proposed by the IPCC [5]. The storage capacity of various potential landforms of India is discussed in the following sections based on assessments made by previous studies.

9.2 Geological Sequestration

9.2.1 *Sequestration in Depleted Oil and Gas Fields*

Considered the most doable form of CO_2 sequestration due to the availability of required data already, storing CO_2 in depleted oil and gas fields especially in tight reservoirs already has the required trapping structures and conditions so as to prevent the escape of the gas. Provided these reservoirs have the necessary conditions to keep the gas in supercritical state (i.e., >800 m depth) (16). This form of sequestration falls under the CCUS (Carbon Capture Utilization and Storage) category, as the injection of CO_2 into a depleted field is actually an effective enhanced recovery method used to remove the oil and gas in place that weren't recoverable through primary methods. CO_2-EOR (Enhanced Oil Recovery) has been in practice for a long time now and as of 2016 there are 136 reported CO_2-EOR projects across the globe [6].

India consists of 26 sedimentary basin, which has been categorized into three categories based on the evaluated hydrocarbon productivity. Of these 26 basins, the seven basins belonging to Category-I housing proven hydrocarbon reserves have relatively better data availability when compared to the less explored Category II and III basins. This limited data availability caused the studies on CO_2-EOR and CO_2 Sequestration in India to be focused largely on the Category I basins. Previous studies performed in estimating the holding capacity of Oil and Gas fields of India as a whole through EOR reported the capacities to be 2 Gt of CO_2 [7], 1.1 Gt of CO_2 [8] and 7 Gt of CO_2 [9]. However, the latest study by [5] concluded a theoretical capacity of 3.4 Gt of CO_2 by using the data concerning the detailed volumes of hydrocarbon in place and the ultimate recoverable reserves taken from the latest report by the DGH. This theoretical estimate is similar to the estimate by Ref. [10].

9.2.2 *Sequestration in Coal Seams*

CO_2 sequestration in coal seams too falls under the CCUS category as here the injection of CO_2 into the coal seams aids in the enhanced recovery of the unconventional resource of Coal Bed Methane (CBM). While CBM itself isn't being exploited commercially in India yet, the country has a CBM potential of 1000 billion m^3, which translates a 2.5 Gt of CO_2 theoretical storage capacity. Previous storage estimates for ECBMR (Enhanced Coal Bed Methane Recovery) reported theoretical storage capacities of 2 Gt of CO_2 [7], 0.345 Gt of CO_2 [8] and 5 Gt of CO_2 [9]. Considering the adsorption ratios of CH_4: CO_2 to be 1:3 (for bituminous and sub-bituminous ranks of coal prevalent in India), the similar results obtained from the industry favored Kim's and Langmuir's equations gave an estimate of 3.5–3.7 Gt of CO_2 as per the latest study [5]. However, the ECBMR method to be implemented in India will take decades as even the primary recovery of CBM in the country hasn't started yet (Wuppertal Institute 2012).

9.2.3 Sequestration in Saline Aquifers

Saline aquifers is a very good option for geological sequestration and previous studies estimated a storage capacity of 100–10,000 gt of CO_2. However, if the necessary storage conditions of a trapping structure are considered this number comes down to 200–320 gt of CO_2 (16). Due to the limited availability of data, the estimates for category ii and iii basins were calculated based on the patchy amount of data that was available. The US DOE (United States Department of Energy) methodology was used, and the total storage volume of the formations and storage efficiency factors were considered [5]. While the previous estimates for storage of co_2 in saline aquifers of India were 102, 360 and 142 gt of CO_2 [7], the latest study by [5] suggests a cumulative storage capacity of 291 gt of CO_2 in the saline aquifers of India.

9.2.4 Sequestration in Basalts

A CO_2 sequestration method wherein the gas is injected for long-term storage, CCS in basalts involves mineral carbonation as the gas gets converted into carbonate minerals that cannot escape and is permanently immobilized. Suitable structures like sills and dikes, which are impermeable in nature, would allow CO_2 sequestration in basaltic formations. However, the slow mineral conversion rates and due to large uncertainties, this form of storage was excluded by some studies [11, 12]. As per the case studies from other countries [13], it is estimated a storage potential of 0.12–0.6 Mt of CO_2/km of basalt is possible, which is ought to differ considering the local geology. The Deccan and Rajmahal Traps of India are the two main basaltic formations of the country and the estimated storage potential of the Basalts as suggested by Singh et al. [9] was 200 Gt of CO_2, whereas the latest estimate by [5] was reportedly between 97 and 316 Gt of CO_2 considering the net thickness and porosity.

9.3 Source-Sink Mapping

The sources of CO_2 in the country is largely concentrated to only ~10% of the total number of administrative districts (i.e., 64 of 641 districts) as they contribute to about 60% of the total CO_2 emissions ((16). The sinks that have been classified and categorized are then matched with the sources, so as to tap the co_2 gas into the subsurface for sequestration. The major sources of co_2 are thermal power plants, cement and iron and steel industries and refineries. These sources belonging to the power and industry sector contribute about 60% of the total emissions. On a global average, a coal-based thermal power plant emits 3.96 mt of co_2/year, a cement industry emits 0.79 mt of CO_2/year, an integrated iron and steel mill emits 3.5 mt of CO_2/year and a refinery emits 1.25 mt of co_2/year [1]. Considering this, a detailed CO_2 source-sink studies were carried out for the western and eastern parts of India and pipelines were proposed for co_2 sequestration. The author [14] concluded that for the western coast of India, the major CO_2 sources are coal-based thermal plants and cement industries, which contributed 187 and 68 mt of CO_2/year respectively. Four CCS hubs were proposed, of which three are in the state of Maharashtra and one in Gujarat. As far as the sinks are concerned, the oil fields are feasible for CO_2-EOR while the gas fields are still under primary production and hence will be ready for sequestration only when they are depleted. These sinks can help reduce the emissions by 168 mtpa of CO_2 from the states of Maharashtra and Gujarat and the source-sink distance too is lesser or equal to 150 km, which is far below the estimated feasible pipeline distance of 500 km [10] as per the economic analyses by Dooley et al. [7].

A similar CO_2 source-sink study [15] was conducted for the Eastern part of India, which hosts some of the most important coal fields wherein the CO_2 emissions from coal-based thermal plants of the region were matched to the nearby coal seams as potential sinks for sequestration. A threshold value of >5Mt of CO_2/year was assumed and seven sources (Thermal Power Plants) were identified. Three coalfields, namely, Talcher, IB and Birbhum Coal Fields and Krishna-Godavari Basin identified as the possible sinks for sequestration. Pipeline distances between source and sink were calculated along the railway tracks, and it was proposed to sequester the CO_2 in the coalfields initially, and once when the coal seams get saturated, the rest of the emissions can be sequestered in the relatively farther Krishna-Godavari Basin (Fig. 9.1).

9.4 Conclusion

To achieve the targets that the country has committed to, a good amount of research and development carried out in the decarbonization technologies. Both carbon capture and storage are eminent decarbonization technologies that can reduce the cumulative carbon emissions by 13% by the year 2060 [16]. The lack of required data is hindering the research that needs to be carried out in accurately estimating the exact storage capacity of basins and potential landforms of India. A pilot project should be commissioned to understand the feasibility of the real time, as this is an inevitable method that helps us in keeping track of carbon-neutral future.

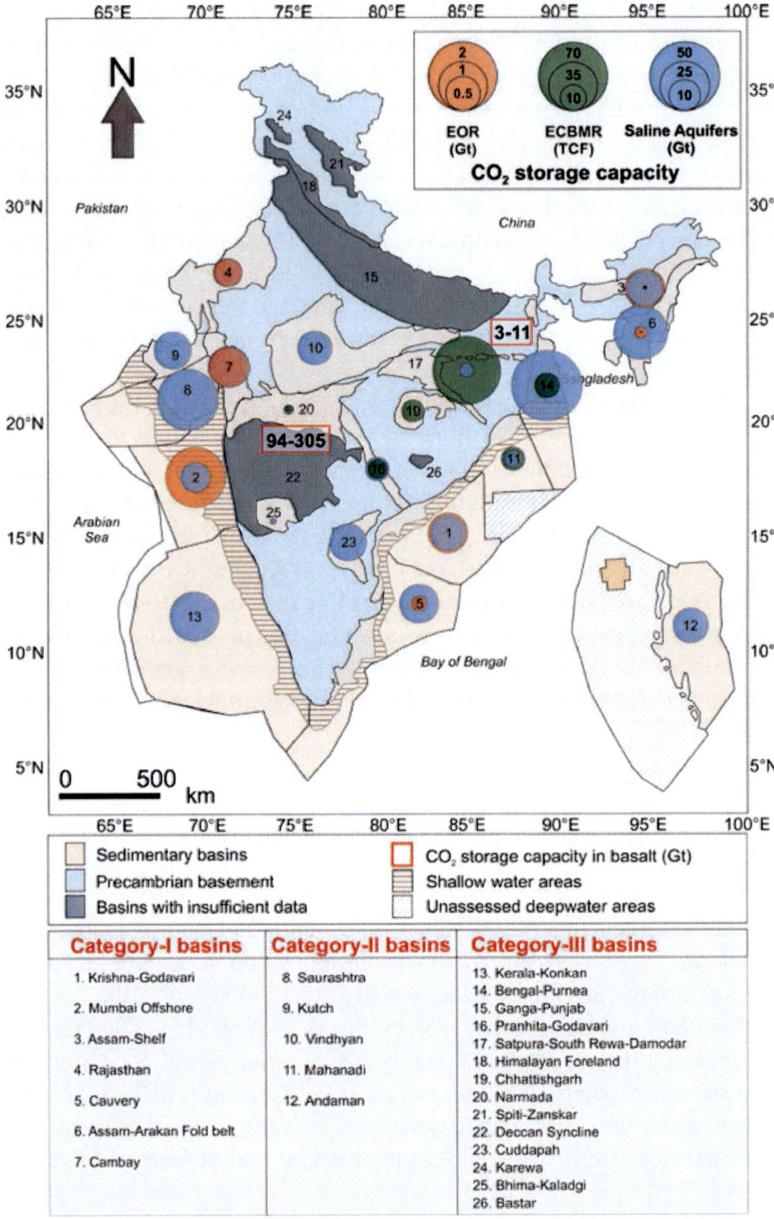

Fig. 9.1 Potential CO_2 Storage Sites of India. (*Source* Vishal et al. [5])

References

1. J. Rogelj, D. Shindell, K. Jiang, S. Fifita, P. Forster, V. Ginzburg, C. Handa, H. Kheshgi, S. Kobayashi, E. Kriegler, others (2018) Mitigation pathways compatible with 1.5 C in the context of sustainable development, in *Global warming of 1.5 C* (pp. 93–174). Intergovernmental Panel on Climate Change
2. M. Collins, R. Knutti, J. Arblaster, J.-L. Dufresne, T. Fichefet, P. Friedlingstein, X. Gao, W.J. Gutowski, T. Johns, G. Krinner, & others (2013) Long-term climate change: projections, commitments and irreversibility, in *Climate change 2013-The physical science basis: Contribution of working group I to the fifth assessment report of the intergovernmental panel on climate change* (pp. 1029–1136), Cambridge University Press
3. P. Arias, N. Bellouin, E. Coppola, R. Jones, G. Krinner, J. Marotzke, V. Naik, M. Palmer, G.-K. Plattner, J. Rogelj, others, *Climate Change 2021: The Physical Science Basis. Contribution of Working Group14 I to the Sixth Assessment Report of the Intergovernmental Panel on Climate Change; Technical Summary* (2021)
4. R. Shaw, S. Mukherjee, The development of carbon capture and storage (CCS) in India: a critical review. Carbon Capture Sci. Technol. 100036, (2022)
5. V. Vishal, Y. Verma, D. Chandra, D. Ashok, A systematic capacity assessment and classification of geologic CO2 storage systems in India. Int. J. Greenhouse Gas Control **111**, 103458 (2021)
6. A. Ettehadtavakkol, Storage of CO2 in depleted/producing oil reservoirs, in *Geologic Carbon Sequestration* (Springer, 2016), pp. 185–209
7. J.J. Dooley, S.H. Kim, J.A. Edmonds, S.J. Friedman, M.A. Wise, A first-order global geological CO2-storage potential supply curve and its application in a global integrated assessment model, in *Greenhouse Gas Control Technologies*, vol. 7 (Elsevier, 2005), pp. 573–581
8. S. Holloway, A. Garg, M. Kapshe, As assessment of the potential for CO_2 storage in the Indian subcontinent. Greenh Issues **89**, (2008)
9. A.K. Singh, V.A. Mendhe, A. Garg, CO2 storage potential of geologic formations in India, in *8th Greenhouse Gas Technology Conference, Trondheim, Norway* (2006)
10. A. Esken, S. Höller, D. Vallentin, P. Viebahn, *CCS global: prospects of carbon capture and storage technologies (CCS) in emerging economies; final report. Part III: Country study China* (2012)
11. D. Large, L. Patey, (2010) *Caught in the middle: China and India in Sudan's transition* (Issue 2010: 36). DIIS working paper
12. N. Nakicenovic, *World energy outlook 2007: China and India insights.* IEA/OECD (2007)
13. B.P. McGrail, H.T. Schaef, A.M. Ho, Y.-J. Chien, J.J. Dooley, C.L. Davidson, Potential for carbon dioxide sequestration in flood basalts. J. Geophys. Res. Solid Earth **111**(B12) (2006)
14. H.K. Bokka, K. Zhang, H.C. Lau, Carbon capture and storage opportunities in the west coast of India. Energy Rep. **8**, 3930–3947 (2022)
15. P. Jain, K. Pathak, S. Tripathy, Possible source-sink matching for CO2 sequestration in Eastern India. Energy Procedia **37**, 3233–3241 (2013)
16. S. Rani, E. Padmanabhan, B.K. Prusty, Review of gas adsorption in shales for enhanced methane recovery and CO2 storage. J. Petrol. Sci. Eng. **175**, 634–643 (2019)

Chapter 10
Reducing Carbon Emissions Through CO_2 Capture, Transport and Storage: A Review

Rohit Sharma, Barasha Deka, Sameer Muhammed, and Sai Dinesh Maganti

Abstract The aim of this paper is to comprehensively review different techniques used to capture, transport and store CO_2. The increase in CO_2 content due to man-made as well as natural causes are extremely detrimental, and there is an urgent need to reduce carbon emissions into the atmosphere, and to achieve this, enormous attention is now being given to CO_2 capture, transport and storage which is going to significantly contribute to change in the overall energy mix of world. This review describes the different methods of the same and their future prospects. Upon further analysis and discussions, most appropriate technique for each stage is described by taking the risk assessment into account. The article concludes with providing suggestions regarding effective techniques for all the categories of CCUS. This article is of interest for industries, research institutes and academicians involved in carbon capture and sequestration (CCUS) including petroleum industry.

Keywords CO_2 capture · CO_2 sequestration · CO_2 transport · CCUS · Hydrocarbon reservoirs

---

R. Sharma (✉)
Department of Petroleum Engineering and Earth Sciences, School of Advanced Engineering, UPES Dehradun, Dehradun, India
e-mail: rohitismpetro@gmail.com; rohitsharma@ddn.upes.ac.in

B. Deka
Department of Petroleum Engineering, IIT (ISM), Dhanbad, India

WASM: Minerals, Energy and Chemical Engineering, Curtin University, Bentley, Australia

B. Deka · S. Muhammed · S. D. Maganti
Department of Petroleum Engineering, Presidency University, Bengaluru, Karnataka, India

A. K. Sahu et al. (eds.), *Energy Storage and Conservation*, Springer Proceedings in Energy, https://doi.org/10.1007/978-981-99-2870-5_11

10.1 Introduction

Human exercises have prompted an enormous expansion in CO_2 outflows, which is an essential ozone-depleting substance that is adding to environmental change with higher than 1 °C in an Earth-wide temperature boost than that of the pre-modern level. To reduce the CO_2 emissions, new methods to capture, transport and sequestration of the captured CO_2 were evolved after the Paris Agreement [1], which is an international treaty on climate change, it was adopted on December 2015, the goal is to limit global warming to well below 1.5 °C when compared to pre-industrial levels (Sims, R., IPCC report Transportation, 2018). There are numerous strategies that have been developed to tackle the climate change, some of them are nuclear and wind energy to reduce the combustion of fossil fuels linked with the emission sources. Climate change in today's world is a crucial step, if not properly followed to mitigate CO_2 emissions, could result in shifting weather patterns (many parts of the world are currently facing the weather crisis), rising sea levels that will eventually leads to floods, and humans and animals could be faced many more challenges for the survival because of climate change such as heat waves, melting glaciers and warming oceans [2]. So, to mitigate the climate change action, rapid development and implementation of carbon dioxide capture, utilization and storage (CCUS) technology promises to significantly reduce the amount of greenhouse gases entering into the atmosphere. CCUS encompasses mainly three steps: first is the capture of CO_2 from various sources such as capturing CO_2 in thermal power plants, coal-fired plants and various industrial sources. The capturing can be done before combustion of fossil fuels or after the combustion by placing adsorbents near the chimney that absorb CO_2, capturing techniques also include direct air capture (DAC), which is still in experimental stages concerning the cost in implementing such technologies worldwide [3]. However, in this paper, an attempt has been made to discuss the various carbon capture methods. The next step involves the transportation of the captured CO_2 via pipelines in liquid form and injecting into the geological formations such as depleted oil and gas fields, deep coal beds or its utilization into enhanced oil recovery etc. [4]. The main motivation of this review article is to present the complete guide of CCUS projects and all the aspects on how the CO_2 is captured to successfully sequestered in the formations: it involves the different capturing techniques and transportation of CO_2 and also discusses briefly sequestrating methods for successful storing CO_2 over a long period of time, and the experiments carried out by many authors are discussed and their methodology in sequestering CO_2 is elaborated briefly in this paper.

10.2 Methodologies

10.2.1 Capture

Carbon capturing is one of the major techniques that is been followed by both petroleum as well as non-petroleum industries to reduce the emission of CO_2, which is a major greenhouse gas that has been mainly emitted to atmosphere or environment by man-made activities. Overall interest in CCS is mainly from three variables.

First is developing acknowledgment that enormous decreases in worldwide CO_2 emanations are expected to keep away from serious environmental change influences, electric power plants hold a large portion of the outflows. Second is the acknowledgment that huge emanation decreases can't be accomplished effectively or rapidly essentially by utilizing less energy or by supplanting petroleum derivatives with elective energy sources that transmit next to zero CO_2. Actually, this present reality depends on petroleum products for more than 85% of its energy use (as does the U.S.). Changing that image emphatically will require some investment. Finally, energy economic models show that adding CCS to the other reduction measures significantly lowers the cost of mitigating climate change. Such studies also indicate that by 2030 and beyond CCS is a major component of a cost-effective portfolio of emission reduction strategies [5]. As a result, carbon capture is a recently evolved technology that can provide a present solution to mitigate CO_2 by 50% by the year 2050 [4]. Along with this, it can achieve a 14% reduction of greenhouse gas emissions by the year 2050, by capturing them using simple strategy to collect the carbon from the power plants. The three major technologies used for carbon capture are listed below:

2.1.1. Pre-combustion: a capturing method where CO_2 is removed from fossil fuels before the completion of combustion process. For example, gasification of coal by reacting it with oxygen with or without steam to form a mixture of carbon monoxide and hydrogen commonly referred as syn gas [6]. Followed by a water shift reaction to convert it into carbon dioxide and hydrogen as a by-product, where the concentration of carbon dioxide lies in the range of 15–50% [7]. Later this carbon dioxide is separated for storage or utilization purposes through various absorption process [8]. Even though it's been an efficient process of capturing but when compared to traditional pulverized coal power plants, this process becomes more expensive.

2.1.2 Oxy-Fuel Combustion: a capturing method where combustion of exhaust gas in an oxygen-enriched atmosphere, which mainly consists of carbon dioxide and water vapor. Followed by providing a condensation temperature greater than the ambient conditions carbon dioxide can be easily separated as the water vapor gets condensed and it's the most efficient and promising method of carbon capture. By providing an air separation unit, carbon-based fuel is combusted in the re-coursed flue gas and unadulterated oxygen in a boiler. Then the flue gas is shipped off molecule expulsion unit followed by the condenser and cooler for water evacuation. Finally, CO_2 is compressed and dehydrated for storage and later for transport purposes.

2.1.3 Post-combustion: It's the process of collecting CO_2, which is emitted by the combustion of fossil fuels. This method can be implemented on large-scale emission plants like cement industries, coal and gas-powered plants. In this process, with the help of liquid solvents, the exhaust gases out from the combustion of fossil fuels are captured in a container and from where carbon is separated and absorbed as they flow through the chimney. By this, nearly 80–90% of carbon which is emitted into the atmosphere can be prevented. Monomethylamine absorption is considered to be the most commercial and common method of post-combustion approach [9].

$$CO_2 + N_2 \rightarrow \text{MEA (absorbs } CO_2 \text{ and releases } N_2) \rightarrow CO_2 \rightarrow \text{storage (absorbed } CO_2) \quad (10.1)$$

Other capture technologies include:

2.1.4 Direct Air Capture (DAC): a method in which the CO_2 is directly gathered from the air, the gathered CO_2 can be put away forever in profound topographical arrangements, in this way assists in accomplishing with negative outflows. This is been carried out by two approaches: (1) **Liquid DAC**, where CO_2 is removed as air passes through hydrogen peroxide solution and (2) **Solid DAC**, which makes use of solid sorbent filters that chemically binds with CO_2 when they are heated and placed, they discharge concentrated CO_2, which is then caught and later sequestrated. Significant benefit of DAC is that it restricts the land and water impression and wipes out the requirement for significant distance transport. But as the CO_2 in atmosphere is much diluted, this leads to DAC's energy and cost relatively higher. However, there are many operations and demonstrations being conducted to make the project viable in terms of policy-making and other considerations. As of the data 2020, at least 26–30 commercial scale capture projects are operating around the world in which 13 are in advanced level, and they are demonstrated in real life like coal-fired power generation plants, natural gas processing, and these gave significant results in carbon capture [8].

10.2.2 Transport

Transport is a stage that interlinks between capture and storage sites by considering the physical, environmental and social factors. Transportation of carbon can be done through pipelines in the form of solid, liquid and gaseous form. If it is transported to closer to atmospheric pressure then gaseous form with large facilities is preferred, while for longer distances CO_2 is compressed to liquid and dry carbon does corrode carbon–manganese steel but solid conversion is a costly process of transportation. Long-distance pipelines are instrumented at intervals so that the flow can be monitored as pigging and corrosion is a major concern. Computers control a significant part of the activity, and manual mediation is vital just in strange upsets or crisis conditions. In some cases, marine transportation is done where captured carbon is stored on land, and it is transported via ships in liquified form to the desired destinations but this is a costly method therefore it's not practiced in common. After the capturing of carbon is done, it needs to be transported and sequestrated based upon the geological formations and site location. Also, this captures carbon dioxide that can be utilized by using it for enhanced oil recovery (EOR) operations. To transport the captured CO_2, it is initially compressed from atmospheric pressure and temperature to 11 MPa to 38 °C, which requires 111 kWh/tCO_2 of electrical work, and later it is transported and injected into the storage system at a pressure range between 10 and 153 MPa, which will lock the CO_2 permanently within the pore spaces of geological formations [10].

For transmission over large distances, dense form like liquid or super critical regime using refrigeration energy for the liquefaction of CO_2 at −25 °C and 153 MPa is preferred as less thermodynamic power requirement by using water as a coolant to nearly 45.83% reduction than conventional process, And also for safer transportation of CO_2 over long distances through pipelines, boosting pump stations are required considering the heat exchange between the CO_2 in pipe and the pipeline surroundings to restore the inlet pressure of 153 bar for the given inlet pressure. An increase in ambient temperature yields a higher velocity for transportation as the CO_2 density reduces, thus, during this phenomenon, pressure drop needs to be monitored as it can lead to the development of choke conditions, which yields no increment in flow rate as the fluid velocity reaches sonic values due to large pressure drops in the pipelines, which adds up the costs due to the additional recompression stations needed to be installed. Therefore, the pipelines are usually buried underground at a depth of 1.2–1.5 m and welded together to provide more environment stability and stable temperature. The cost of pipeline transport at consistent limit increments straightly with distance, yet nonlinearly with limit. While huge scope transport of CO_2 is supposed to be fundamentally overwhelmed by pipeline transport, shipping becomes cost-serious and for lower limits of carbon. Additionally, marine vehicle utilizing ships are supposed to be more practical than having across the nation pipelines, and they very well might be utilized for bringing in CO_2 for geographical sequestration [10]. During transportation for enhanced oil recovery, 9 MPa of minimum pressure is preferred, and booster stations are installed to restore pressure up to 153 MPa.

10.2.3 Storage

Carbon capture and sequestration is an essential method to prevent the CO_2 from reaching the atmosphere. Generally, it is stated that CO_2 is sequestrated at depths in the range of 800–1000 m [10]. The main sequestration methods that are practiced in the field of petroleum as well as other industries are discussed below:

2.3.1 Biological Sequestration: a storage process where biological organisms capable of synthesizing its own food such as autotrophs creates stored energy in plants like cellulose and starch to form glucose in their tissues generated from light energy. This creates organically synthesized carbon, which is sequestered from the environment. And this sequestration occurs from the environment through Photosynthesis, the same process has been followed in natural, terrestrial and aquatic sequestration process. Future research works on biological sequestration will create an effective way to create low carbon in the future [11].

2.3.2 Terrestrial Sequestration: a storage process where the carbon dioxide is captured from the atmosphere and gets stored in the soils or in wetlands and vegetation. The largest terrestrial sequestration occurs through the forest which accounts 4.1 billion hectares area, thus, providing a significant large reserve for carbon storage. Future works on terrestrial sequestration is the long term and effective storage option, which can be practiced through proper forest management activities [12].

2.3.3 Underground Sequestration: Underground sequestration or geological sequestration is the process where CO_2 caught from the flue gases is infused into the sedimentary rocks. This training appears to be more practical than other sequestration procedures and has a long history of commercialization like in enhanced oil recovery. This method has a huge potential capacity that can be easily accessed later, which can be utilized for EOR or CO_2 injection into the natural gas fields and oil fields. Underground sequestration can occur in depleted hydrocarbon reservoirs, thus making use of the spent and unutilized resources of petroleum industry, which possess already built infrastructure. Also, the presence of magnesium or calcium containing rocks in reservoir formations could neutralize the injected CO_2 to prevent massive changes in formation PH. Extensive research activities are underway including identifying suitable locations for sequestration, characterizing the location in terms of maximum storage capacity and the expected stability of the injected CO_2 [13], thus, making it one of the most promising sequestration techniques.

2.3.4 Saline Aquifer Sequestration: Among all the underground storage sites, it is best suited for salt sink for storage of CO_2 as it contains porous and permeable reservoir rocks, which holds saline fluid and can provide numerous storage capacity for carbon. It provides long-term storage potential for CO_2. Many industrial projects have been carried out to know the viability, potential storage sites and mechanism behind the sequestration in saline aquifer [14].

2.3.5 Capillary Trapping: alluded to as remaining stage catching, this cycle traps CO_2 principally after infusion pauses and water starts to guzzle into the CO_2 tuft. The following edge of the CO_2 is immobilized, easing backup-plunge movement. Here, sequestration is more compelling in plunging springs that don't have underlying conclusion. As of late examinations demonstrated the way that total CO_2 in a tuft can be immobilized along these lines [15].

2.3.6 Agricultural soil for CO_2 sequestration: an experimental method of storage where the researchers and scientists estimated or experimented that this method can store additional billion amount of carbon each year. Through a proper farming pattern by planting crops not needed to re-harvest for the next subsequent year, and which can grow automatically as parent crops, with deep roots capable of storing more amount of CO_2, will definitely help in mitigating climate change. With deeper crop roots, larger sequestration will occur. This method is a negative emission technology that creates a massive reduction in greenhouse gas emissions. Research has been carried out for future long-term carbon storage and climate mitigation with farming pattern suitable for smaller cultivation areas [16].

2.3.7 Oceanic Sequestration: provides the largest possible sink for carbon storage where CO_2 is injected from offshore platforms into water depth below 1 km, Thereby the injected carbon dissolves with water due to density differences followed by organic process reaction from bacteria. Through this, nearly 90% of carbon can be sequestrated, which will help reduce the global mean temperature and climate mitigation. Extensive research has been carried out to minimize the ocean acidification and to reduce their significant effect in marine aquaculture [11]. There are several sequestration projects that are currently in planning stage and some have been operational, the project details are discussed herewith: The project named Captain in UK, which can store about 3.8 MT/Year through pipelines to offshore deep saline formation and this project is still in planning stage (James et. al 2013). There are some other projects still in planning stage a project named Don Valley in the UK, which estimates to store 4.9 Mt/year the fate of CO_2 can be sequestered through pipelines in offshore deep saline formations. These two projects can be known as significant in terms of their storage potential. The project named Pembina in Canada, which is operational since 2005 can store 50 t/day in the cardium formation operated by company Penn West.

2.3.8 In situ mineralization: an arising way to deal with eliminate carbon from air and store it as minerals fundamentally as calcite and magnesite. Under in situ mineralization process, CO_2-rich liquid is infused on mafic and ultramafic rock arrangements, which are molten rocks comprising of Iron and Magnesium silicates with adequate SiO_2 content. It is a dominant trapping method for sequestration in deep sedimentary formations, and this trapping will be worthy even decades after injection stops. Recently, field experiments have been carried out by injecting water-saturated supercritical CO_2 in deep formations. Ongoing Carb-Fix project has the goal of storing CO_2 and H_2S produced from power plants.

2.3.9 Microbiologically Enhanced sequestration: a trapping mechanism where injected carbon incorporates into the solid carbonate mineral. This microbe precipitates calcium carbonate under high pressure subsurface conditions, thereby creating a biomineralization barrier constituting favored flow paths in the vicinity of CO_2 injected which reduces the upward migration of CO_2 along with its mineralizing. This sequestration can reduce the carbon leakages as it facilitates trapping in non-labile forms along with the net emission of carbon from fossil fuel combustions. Recently, many laboratory and field studies are carried out in investigating general process that involves injection of CO_2 into saline solution developments as it lessens the pH of the salt water by 1.5–4 pH units through the disintegration of carbon relying upon the chemistry, lithology and temperature of saline solution [17].

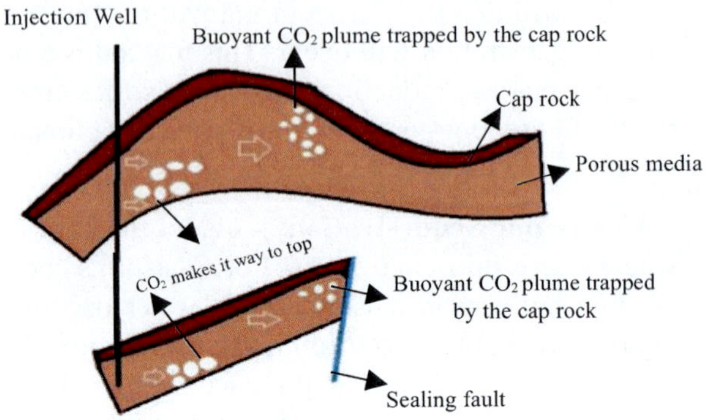

Fig. 10.1 The injected CO_2 as a result of structural trapping which forms the anticlinal structure

2.3.10 Solubility Trapping: where the disintegration of CO_2 and other pipe gas pollutants into the pore water prompts the trapping, which relies upon factors like pressure, temperature, and saltiness of the brine solution. As CO_2 relocates through the development, a few breaks up in the development water and a piece of the non-wetting CO_2 are isolated from the continuum, it is caught in view of slim strain and improves storage security, solubility trapping is of extraordinary interest in aquifer storage. Future research are carried out to understand the change in the interfacial tension and its influence in trapped gas saturation [18] (Fig. 10.1).

2.3.11 Tar-Sand Sequestration: a sequestration method in which at 200 bar pressure and 400 °C temperature CO_2 is compressed and injected to the deep (600–1000 m) sea oil–bitumen sand bed, and they exist as supercritical fluid and rises till it encounters a seal as they are buoyant in saline formation water. Here, bitumen can be extracted from the seams as it becomes liquid as it's soluble in CO_2. Currently, works are going for development of more cost-effective technology for this sequestration approach [19].

2.3.12 CO_2 fixation by Photo-Bio Reactor System: a system that uses the photosynthetic change of CO_2 into starch is finished in a reactor in the presence of microorganisms or miniature green growth under a controlled climate to change over light, intensity and carbon dioxide to helpful items, like sugars, hydrogen and oxygen. This process is environmentally friendly. Production of liquid hydrocarbon fuels from CO_2 and water by using a concentrated solar energy source is an established technology, which is one the major future research going on [19].

2.3.13 Mineral Trapping: a mechanism that occurs when dissolved CO_2 reacts directly or indirectly with minerals in the geologic formation, promoting precipitation of carbonate minerals. It can immobilize CO_2 for longer periods. Researches have been carried out regarding the dissolution of silicate minerals as the process is thought to be comparatively slow [15].

10.3 Conclusion

Different capturing techniques of CO_2 were discussed in this article, among which the post-combustion approach is considered for carbon capture process more feasible because of its high absorption rate and low solvent cost [20]. There are many sequestration techniques and methods to store CO_2 out of which ocean sequestration qualifies as the largest possible sink for CO_2 storage as three-fourth of earth is covered by water, which provides more adaptive habitat for phytoplankton, or algae, to grow and suck carbon dioxide out of the air and it is dissolving it in water through photosynthesis [21]. It can be injected via pipelines and from moving ships into the geological formations and other places. The best-salted sink for storage of CO_2 among all geological sequestration qualifies to be saline aquifer formations, which accounts to store about 103 GT of CO_2 in Alberta deep saline basin [22]. Based on our analysis, pipeline transportation in liquid form is found to be the most economical and energy efficient way to transport large volume of CO_2 to longer distances as they can provide sufficient compression with less interstage coolers. There can be several hazards like leakage in pipelines, intermediate losses or exposure to the jet of gas at low temperatures, which affects the life of humans, animals as well as plant and vegetation in the area of gas released and speeded through the clouds. So the risk needed to assess for each stage of capture, transport and storage. Many companies are working towards their net-zero goal, and these preliminary projects show some significant results to achieve this goal. Thus, this work provides a comprehensive guide for explaining the technological advancements towards a low carbon economy, where several technologies for carbon capture, transport and sequestration were discussed in detail and were compared.

References

1. Y. Gao, X. Gao, X. Zhang, The 2 C global temperature target and the evolution of the long-term goal of addressing climate change—from the United Nations framework convention on climate change to the Paris agreement. Engineering **3**(2), 272–278 (2017)
2. B.R. Singh, O. Singh, Study of impacts of global warming on climate change: rise in sea level and disaster frequency. Glob. Warm. Futur. Perspect. (2012)
3. E.S. Rubin, H. Mantripragada, A. Marks, P. Versteeg, J. Kitchin, The outlook for improved carbon capture technology. Prog. Energy Combust. Sci. **38**(5), 630–671 (2012)
4. A.I. Osman, M. Hefny, M.I.A. Abdel Maksoud, A.M. Elgarahy, D.W. Rooney, Recent advances in carbon capture storage and utilisation technologies: a review. Environ. Chem. Lett. **19**(2), 797–849 (2021)
5. D. Jansen, M. Gazzani, G. Manzolini, E. van Dijk, M. Carbo, Pre-combustion CO2 capture. Int. J. Greenhouse Gas Control **40**, 167–187 (2015)
6. P. Madejski, K. Chmiel, N. Subramanian, T. Kuś, Methods and techniques for CO2 capture: review of potential solutions and applications in modern energy technologies. Energies **15**(3), 887 (2022)
7. R. Kumar, R. Jilte, K.C. Nikam, M.H. Ahmadi, Status of carbon capture and storage in India's coal fired power plants: a critical review. Environ. Technol. Innov. **13**, 94–103 (2019)

8. M. Fasihi, O. Efimova, C. Breyer, Techno-economic assessment of CO2 direct air capture plants. J. Clean. Prod. **224**, 957–980 (2019)
9. M.D. Aminu, S.A. Nabavi, C.A. Rochelle, V. Manovic, A review of developments in carbon dioxide storage. Appl. Energy **208**, 1389–1419 (2017)
10. F. Kazemifar, A review of technologies for carbon capture, sequestration, and utilization: cost, capacity, and technology readiness. Greenh. Gases: Sci. Technol. **12**(1), 200–230 (2022)
11. D.J. Farrelly, C.D. Everard, C.C. Fagan, K.P. McDonnell, Carbon sequestration and the role of biological carbon mitigation: a review. Renew. Sustain. Energy Rev. **21**, 712–727 (2013)
12. J.T. Litynski, S.M. Klara, H.G. McIlvried, R.D. Srivastava, An overview of terrestrial sequestration of carbon dioxide: The United States Department of Energy's fossil energy R&D program. Clim. Change **74**(1), 81–95 (2006)
13. A. Yamasaki, An overview of CO2 mitigation options for global warming—emphasizing CO2 sequestration options. J. Chem. Eng. Jpn. **36**(4), 361–375 (2003)
14. S.A. Jikich, W.N. Sams, G. Bromhal, G. Pope, N. Gupta, D.H. Smith, Carbon dioxide injectivity in brine reservoirs using horizontal wells, in *National Energy Technology Laboratory, United States Department of Energy (eds) Proceedings for the 2nd national conference on carbon sequestration* (pp. 5–8) (2003)
15. S.M. Benson, D.R. Cole, CO2 sequestration in deep sedimentary formations. Elements **4**(5), 325–331 (2008)
16. B. Henderson, J. Lankoski, E. Flynn, A.S. Sykes, F.P. Payen, M. MacLeod, Soil carbon sequestration by agriculture. (2022)
17. A.C. Mitchell, K. Dideriksen, L.H. Spangler, A.B. Cunningham, R. Gerlach, Microbially enhanced carbon capture and storage by mineral-trapping and solubility-trapping. Environ. Sci. Technol. **44**(13), 5270–5276 (2010)
18. T. Suekane, T. Nobuso, S. Hirai, M. Kiyota, Geological storage of carbon dioxide by residual gas and solubility trapping. Int. J. Greenhouse Gas Control **2**(1), 58–64 (2008)
19. B.L. Salvi, S. Jindal, Recent developments and challenges ahead in carbon capture and sequestration technologies. SN Appl. Sci. **1**(8), 1–20 (2019)
20. D. Aaron, C. Tsouris, Separation of CO2 from flue gas: a review. Sep. Sci. Technol. **40**(1–3), 321–348 (2005)
21. H. Herzog, What future for carbon capture and sequestration? Environmental Science and Technology-Columbus **35**(7), 148A (2001)
22. H. Liu, B.G. Tellez, T. Atallah, M. Barghouty, The role of CO2 capture and storage in Saudi Arabia's energy future. Int. J. Greenhouse Gas Control **11**, 163–171 (2012)

Chapter 11
Analysis of First-Order Differential Equations in Temperature and Heat Transmission Problem

Bhanu Priya, R. K. Poonia, and Abhilasha Saini

Abstract Differential equations represent some of the mathematics that plays an important role in understanding the natural sciences in physics, engineering, chemistry, and many other disciplines. To solve a specific problem, it was necessary to develop mathematical models. The main purpose of this paper was to show how first-order ordinary differential equation techniques can be used to solve temperature problems and heat transmission problems like heat conduction in solids, heat convection in fluids, and radiation of heat in space. In this paper, we have used the Newton's laws of cooling to find the first-order DEs in the temperature problems and those are studied by the separation of variables. We have also used the first-order ordinary differential equations (ODEs) to determine the heat transferred by convection in a fluid problem. To solve these problems, we will first talk about the solutions of ordinary and inhomogeneous DEs before applying the solutions of first-order ODEs to heat transfer, especially thermal convection in fluids.

Keywords Differential equation · Newton's law · Temperature · Convection · Heat transfer

11.1 Introduction

Differential equations refer to equations in which the derivatives of an unknown feature play a critical role in finding an equation that is satisfied by an unknown feature. The differential equation that includes the simplest first-order derivatives is referred as a first-order differential equation, and it has a good deal of software in mathematics, engineering, physics, and plenty of different subjects.

In mathematics, the differential equation is invented by German mathematician Gottfried Leibniz and English physicist Isaac Newton. Later on, Sasser [1] also discussed the History of ordinary differential equations in 1992.

B. Priya · R. K. Poonia (✉) · A. Saini
University Institute of Science, Chandigarh University, Gharuan, Punjab 140413, India
e-mail: rkpooniae7260@cumail.com

A. K. Sahu et al. (eds.), *Energy Storage and Conservation*, Springer Proceedings in Energy, https://doi.org/10.1007/978-981-99-2870-5_12

Research on common methods of integrating DEs began in 1671 when Isaac Newton divided the first-order DEs, known as flow equations, into three classes

$$\frac{du}{dv} = f(v) \tag{11.1}$$

$$\frac{du}{dv} = f(v, u) \tag{11.2}$$

$$v\frac{dz}{dv} + u\frac{dz}{dv} = z \tag{11.3}$$

Equations (11.1) and (11.2) are the usual derivatives of one or more dependent variables for independent variables, known as ODEs. Partial derivatives of dependent variables are also known as partial DEs, which was represented by Eq. (11.3), which was also discussed by M.D Raisinghania in 2013 [2]. In this paper, we only focus on simple first-order DEs.

Differential equations were first studied in 1675 by Gottfried Wilhelm von Leibniz, who wrote the equation:

$$\int x dx = \frac{1}{2}x^2 \tag{11.4}$$

11.1.1 History

A historical review of the law of cooling that was given by Newton up to the early twentieth century was used as an educational tool. It starts with a description and interpretation of Newton's work in 1701 and a look at research done in the eighteenth century that either agreed or disagreed with Newton's law. This is the beginning of this history. Research done after 1850 looked into the laws of thermal radiation and natural and man-made convection. Dulong and Petit's work in 1817 was the most important. Besson [3] also presents an historical overview of the research on the cooling law in 2012.

11.1.2 Objective of the Study

The main goal of this paper is to talk about issues that related to first-order DE in both temperature and heat transfer issues, especially heat convection in fluid. In both methods, we need to apply Newton's law of cooling. This paper is divided into five sections. In the first section, we describe the introduction part, and Sect. 11.2 describes the preliminaries' part of this paper. Sections 11.3 and 11.4 focus on temperature issues and first-order DEs in the thermal convection of fluids. These two sections also discuss Newton's law, heat transfer, and some numerical problems related to them. Section 11.5 is dedicated to observation and a list of references.

11.2 Preliminaries

Differential Equation (DEs):—For one or more independent variables, a DE is an equation that contains the derivatives and differentials of one or more dependent variables.

Ordinary Differential Equations (ODEs):—A DE containing an unknown function and its derivatives called ODEs [2]. Also, DEs that contain only the first derivative are called first-order DEs.

We also discuss the separation of variables and solutions of ODEs, which are used to find the first-order DEs in the temperature problem and heat convection in fluids:

Separation of variables:—The variables are said to be separable if the equation can put all of the x and dx functions on one side and all of the y and dy functions on the other.

The equation in the form $\frac{dy}{dx} = f_1(x)f_2(y)$, is a given equation of the function x and y.

Then the general solution is $\int \frac{1}{f_2(y)}dy = \int f_1(x)dx + c$. Where c is a constant of integration.

Solution of first-order ODEs:—Linear DE solutions can be divided into two types. One is in the form of $y' + p(t)y = 0$, and equally $y' = -p(t)y = f(t)$ is known as a linear differential equation of the same order. These solutions were also defined in the ordinary differential book by William's et al. [4], and it gives the solution $y = \frac{c}{h(x)}$ (c is the integration constant) and h (x) is known as the integral coefficient obtained from the equation

$$h(x) = e^{(\int -k(x)dx)}$$

The integrating factor in the equation is defined by the formula $h(x) = e^{(\int a(x)dx)}$. The first order is non-homogeneous linear DEs of the type $y' + p(t)y = f(t)$. The derivatives of the product $y(x)h(x)$ are obtained by multiplying the integrating factor h(x) by the left side equation. Jayaraja and Karthikeyan [5] discussed the first-order DEs and their solution. The general solution to DEs is as follows: $y = \frac{\int h(x).f(x)+c}{h(x)}$, where c is an arbitrary constant.

11.3 Differential Equation in Temperature Problem

Understanding differential equations has many amazing interrelationships in the study of differential equations, as it provides a basic tool for analysing very important phenomena, starting with basic physical principles and solve the applications of DEs to the temperature problem. In 2018, Zakari and Hassan [6] discussed the application of first-order differential equation in temperature problems and stated that Newton's law is very important to solve these problems.

Law of Cooling: The rate of cooling of a hot body (or loss of heat) by radiation is directly proportional to the difference in temperatures between the body and its surroundings, depending on the type and surface of the thermal substance. The Law of cooling is very useful in the research of hot water because it can show how fast hot water cools in a pipe. These things were discussed by Cheng and Fujii [7] in 1998.

For this sort of problem, we assume that the inquiry includes the temperature (T_t) of a particular object that is in a constant temperature medium (T_S) and when time (t) changes such that T (T denotes the rate of change in relation to t.). k is a positive proportionality constant that depends on the object's surface and properties.

The cooling law of Newton is stated by

$$\frac{\mathrm{dT}}{\mathrm{dt}}\alpha(T_t - T_s)$$

$$\frac{dT}{dt} = -k(T_t - T_s) \tag{11.5}$$

Newton's law requires the negative symbol to make $\frac{dT}{dt}$ (time rate of body temperature change) negative in the cooling process when the body temperature is higher than the ambient temperature.

$$\frac{dT}{dt} < 0$$

And when you put it in the heating process positive, the body's temperature will be lower than the ambient temperature.

An object with an initial temperature of T_o °C is cooled in air maintained at a constant temperature of T_s ° C, and C is assumed to be an arbitrary constant. As a result, Newton's law of cooling is written as:

$$T(t) = T_s + (T_o - T_s)e^{-kt}$$

or

$$T(t) = T_s + Ce^{-kt}$$

If $k < 0$, $\lim_{t\to\infty} e^{-kt} = 0$ and $T = T_s$,

We can also state that the body's temperature approaches the ambient temperature over time. The curve drawn between body temperature and time is called the "cooling curvature" [8]. The rate of temperature reduction is indicated by the slope of the line with the curve at each point.

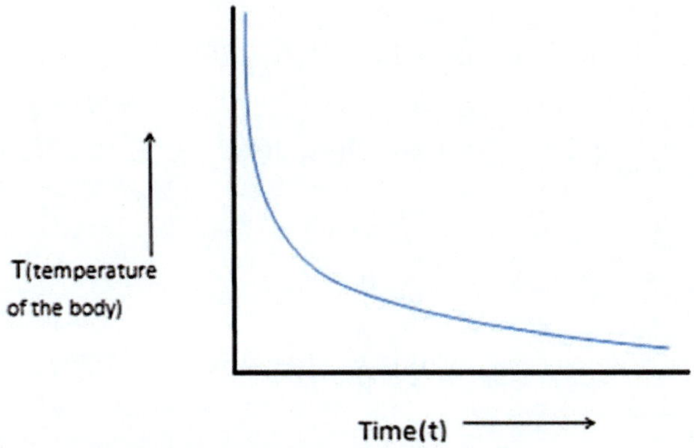

This law can be used to determine how quickly a substance will cool in a particular atmosphere at a particular temperature. It also helps us to understand how the speed with which an object cools is influenced not only by the temperature difference between the substance and the surrounding environment but also by the substance's constant.

3.1 Numerical Problem: The coffee in the brewing pot is 180 °F, while the ambient temperature is 76°. After 5 min, the coffee's temperature has reached 168 °F.

(a) Construct an exponential equation to describe this circumstance.
(b) How long does it take for coffee to get up to 155 °F, the serving temperature?

Result:

Given:

(a) Oil temperature after 6 min, $T(t) = 168\,°\mathrm{C}$,

$$T_s = 76\,°\mathrm{C},$$

$$T_o = 180\,°\mathrm{C},$$

$$T = 5\,\mathrm{min}$$

When the given values are substituted into Newton's cooling equation, the following results are obtained:

$$T(t) = T_s + (T_o - T_s)e^{-kt}$$

$$168 = 76 + (180 - 76)e^{-kt}$$

$$168 - 76 = 104e^{-kt}$$

$$\ln\left(\frac{92}{104}\right) = \ln e^{-kt}$$

$$\ln\left(\frac{92}{104}\right) = -5k$$

$$-0.1226023220 = -5k$$

$$k \approx 0.0245204644$$

$$T(t) = 76 + (104)e^{-0.0245204644t}$$

(b) How long to reach 155 °F?

$$T(t) = 76 + (104)e^{-0.0245204644t}$$

$$155 = 76 + (104)e^{-0.0245204644t}$$

$$\frac{79}{104} = e^{-0.0245204644t}$$

$$\ln\frac{79}{104} = \ln e^{-0.0245204644t}$$

$$\ln\frac{79}{104} = -0.0245204644\text{t}$$

$$t \approx 11.21279933\text{ min.}$$

We can also rewrite our equation as shown below:

$$T(t) = 76 + (104)e^{-0.0245204644t}$$

$$T(t) = 76 + 104(e^{-0.0245204644t})$$

$$T(t) = 76 + 104(0.97577772)^t$$

$$T(t) = 76 + (104)(1 - 0.02422228)^t$$

The above equation is similar to transforming $A = P(1 + r)^t$. And the negative value of r indicates that the function is decreasing. It also shows that every minute, the temperature difference decreases by 2.422228%. The variable x tends to infinity, the function approaches 76.

11.4 First-Order Des for the Heat Convection in Liquid

Heat transfer: The study of heat transfer caused by temperature differences is known as heat transfer. In 2020, both Aliu et al. [9] and Rehan [10] discussed heat transfer. It is the energy that moves molecules. Molecules with high thermal energy move fast, but in the case of low thermal energy, they move slowly. Molecules accumulate, move faster, spread, and expand objects. This is known as thermal expansion.

Heat will constantly move when there are two objects with different temperatures, and it also moves from the warm to the cold object. This heat transfer continues until the object reaches the same temperature. Heat can be transferred (or moved) in three ways.

- Heat transfer by conduction in a solid.
- Heat transfer by convection of liquid.
- Heat transfer by radiation in space.

As we can see in Fig. 11.1 that the solid is in contact with the liquid at various temperatures.

Energy exchange between objects is caused by one or a combination of these modes. Conductivity is the transfer of heat through a solid or liquid. Lienhard and et al. [11] discussed the concepts of convection and conductivity. In solids, heat flows through conductivity. Fourier's law can solve this problem. Heat will always flow from hot to cold in solids. Convection is a heat transfer method that involves the movement of fluids. In liquids, heat flows by convection. We can calculate this with Newton's law of cooling. Radiation does not need a heat transfer medium. This mode uses electromagnetic radiation emitted from heat exchange objects [5].

Convection is the way heat is transported through a liquid. Any substance with freely moving molecules that may easily travel from one location to another is referred to as a liquid. Fluids include liquids and gases. Convection can be divided into two types: (1) natural convection (2) forced convection. The flow of blood in human blood

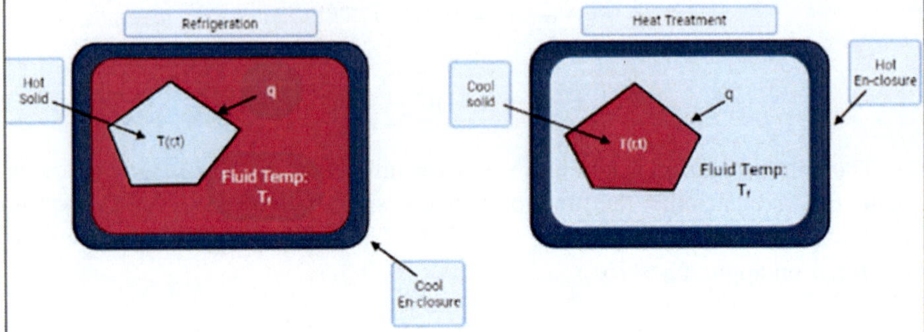

Fig. 11.1 Heat transferring solid submerged in fluid

vessels or the flow of air and water on human skin are examples of heat transfer by convection. It is also important to study the movement of liquids near the surface.

11.4.1 Mathematical Modelling of Small Solids in Cooling and Heating

In the process of convection, it is easy to introduce a heat transfer coefficient h for heat transfer from an interface exposed to a relatively low-velocity fluid stream as defined by Eq. 11.5, which is Newton's law. The surface temperature is T_a, while the characteristic fluid temperature is T_b. The cooling law gives the rate of convection heat transfer Q to the surface as

$$q \propto (T_a - T_b) = h(T_a - T_b) \tag{11.6}$$

Equation (11.6) shows that heat flows from point T_a to point T_b with $T_a > T_b$, which expresses the heat flux between two points, A and B (Figs. 11.2 and 11.3). Thus,

$$q = h\left[T_s(t) - T_f\right] = h[T(t) - T_f]$$

In the above equation, the transformation between solid and bulk fluid is satisfied by the coefficient h. The technique can also be used to change the temperature of a solid $T(t)$ over time t, according to the first rule of thermodynamics.

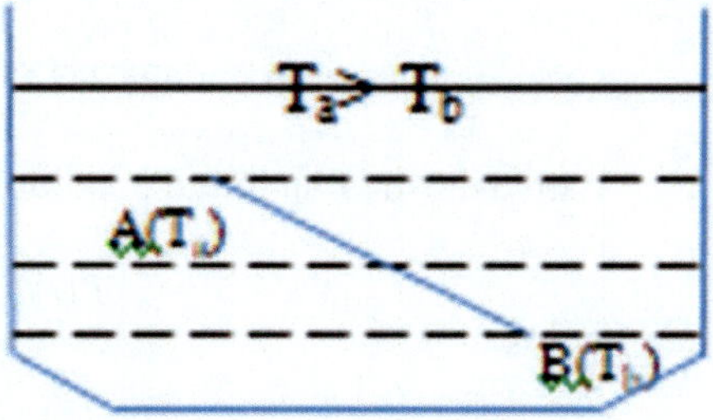

Fig. 11.2 Heat flux between two points

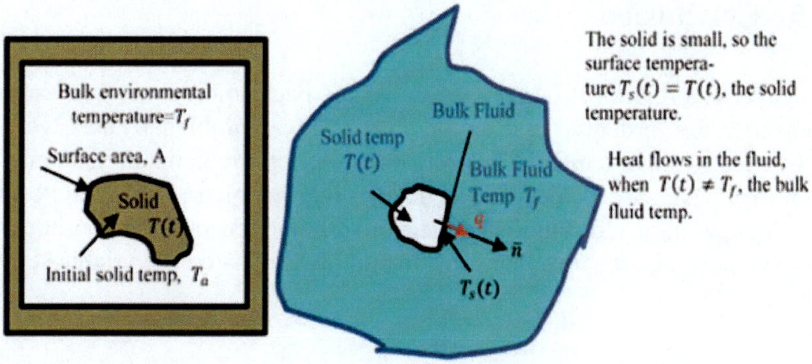

Fig. 11.3 Mathematics model for heat transfer

$$-pcV\Delta T(t) = Q = qA\Delta t = hA[T(t) - T_f]\Delta t \tag{11.7}$$

Equation (11.7) gives

$$\frac{\Delta \mathrm{T(t)}}{\Delta \mathrm{t}} = -\frac{h}{pcV}A[T(t) - T_f] \tag{11.8}$$

In Eq. (11.8), *h*, *p*, *c*, and *V* are constant so we equate the ratio of these constants with α

$$\propto = \frac{h}{pcV} \tag{11.9}$$

Equation (11.9) is further expressed as

$$\frac{\Delta \mathrm{T(t)}}{\Delta \mathrm{t}} = - \propto A[T(t) - T_f] \tag{11.10}$$

As a result, T(t) is a continuous temperature change in the solid w.r.t., time t, i.e., ΔT → 0, and the generic symbol A replaces the constant surface area. Equation (11.10) can also be expressed as first-order DEs [12].

$$\frac{dT(t)}{dt} = - \propto A[T(t) - T_f]$$

With the use of an initial condition

$$T(t)|_{t=0} = T(0) = T_0$$

11.5 Conclusion

This paper is a review of DEs in which we had been studied many research articles based on the first-order ODEs in temperature and heart problems. These two problems can be used to solve temperature or heat convection problems in math and physics. The primary DEs are also described in this review, which is divided into two parts: variable separation and solutions of first-order homogeneous and non-homogeneous DEs. The best way to solve Newton's law of cooling is to use variable separation.

References

1. J.E. Sasser, History of ordinary differential equations: The first hundred years. Proc. Midwest Math. Hist. Soc. **1** (1992)
2. M.D. Raisinghania, *Ordinary and Partial Differential Equations*, (S. Chand Publishing, 2013)
3. U. Besson, The history of the cooling law: When the search for simplicity can be an obstacle. Sci. Educ. **21**(8), 1085–1110 (2012)
4. W.A. Adkins, G. Mark, Davidson. First order differential equations, in *Ordinary Differential Equations* (Springer, New York, NY, 2012), pp. 1–100
5. N. Karthikeyan, A. Jayaraja, Application of first order differential equations to heat transfer analysis in solid. Int. J. Eng. Innov. Technol. (IJEIT) **5**(8), 5–8
6. A. Hassan, Y. Zakari, Application of 1st Order Differential Equation in Temperature Problem. Annuls. Comput. Sci. Ser., 16th Tome 1st Fasc. (2018)
7. K.C. Cheng, T. Fujii, Heat in history Isaac Newton and heat transfer. Heat Transf. Eng. **19**(4), 9–21 (1998)
8. Y. Praroopa, A. Shou Reddy, P. BabuRao, Review concept of application of Differential equations. Int. J. Latest Eng. Res. Appl. (IJLERA) (2016)
9. S. Aliu, et al., Mechanisms of heat transfer and boundary layers. *Applications of Heat, Mass and Fluid Boundary Layers* (Woodhead Publishing, 2020), pp. 23–53
10. Z. Rehan, Application of first-order differential equation to heat convection in fluid. J. Appl. Math. Phys. **8**, 1456–1462 (2020)
11. I.V. Lienhard, H. John, *A Heat Transfer Textbook* (Phlogiston press, 2005)
12. T.-R. Hsu, *Application of first order ordinary differential equation in mechanical engineering analysis* (San Jose State University, USA, 2005)

Chapter 12
Environmental Impact of Drilling Fluid Waste and Its Mitigation Techniques

Mayank Agnihotri, Uday Bhan, V. R. Nagalakshmi, Nitu Kumari Thakur, Somenath Ganguly, Anamika Kushwaha, and Lalit Goswami

Abstract To overcome the demand of energy, there is a rise in the hydrocarbon exploration activities. Different types of drilling fluid are used, in which oil-based drilling fluids are much more effective in terms of efficiency. Used drilling fluid discharges are hazardous to human health and environment due to the presence of potentially harmful substances. Hydrocarbon industries produced a huge amount of used contaminated drilling fluid with different compositions and characteristics depending on formation of drilling operation. Therefore, a remediation technique is required to remove the pollutants from the drilling fluid before disposal. This review paper provides a description of the fundamental elements behind drilling fluids, as well as their roles, sources, characterisation and environmental potential constituents of drilling fluid waste material. Additionally, it draws attention to some environmentally significant elements, such as the various minerals that are present in drilling fluid waste by-product and the remediation techniques, which can be utilised for further processing before disposal to save the environment and human health.

12.1 Introduction

The rising population is being attributed to the rising need for energy sources. Oil shines out from the other major energy sources because of how widely it is used. The expansion of proven global reserves encourages the development in this source of energy. As oil is still a significant source of energy for the globe. Report by

M. Agnihotri · U. Bhan (✉) · N. K. Thakur · S. Ganguly
Department of Petroleum Engineering and Earth Science, University of Petroleum and Energy Studies, Dehradun, India
e-mail: ubhan@ddn.upes.ac.in

V. R. Nagalakshmi
Godavari Institute of Engineering and Technology, Rajahmundry, India

A. Kushwaha · L. Goswami
Department of Chemical Engineering, Chungbuk National University, Cheongju 28644, South Korea

A. K. Sahu et al. (eds.), *Energy Storage and Conservation*, Springer Proceedings in Energy, https://doi.org/10.1007/978-981-99-2870-5_13

International Energy Outlook (IEA), World Energy Outlook [1] said that oil demand will increase at a slower rate until 2040 and meet 106 mb/d, leading to environmental impact by the exploration and production industries. Due to their superior performance, there are many different uses for drilling fluids that are based on oil (OBDF), and these fluids have been used in drilling activities, and as easy hydrocarbon discoveries are declining and complicated wells have increasingly become more prevalent, where OBDF is essential to use [2]. Abundant greasy drilling cuttings will be generated throughout the oil and gas drilling operation. If it is not correctly treated, it will result in significant issues with protection of environment. They mostly manifest themselves in three areas [3–5]: (1) contaminating groundwater, (2) diminishing permeability of the soil and wettability,and (3) several petroleum chemicals are reactive and dangerous.

Research has revealed that high concentrations of dissolved salts, heavy metals, and hydrocarbon remnants in drilling fluids are dangerous to the environment and soil quality as well as for the health of plants. These findings come from the research that has been done on the effects of dumping used drilling fluid on aquatic and terrestrial ecosystems [6]. May be because of one or a combination of mentioned factors like: micronutrient addition, rise in pH values, and improvement of soil properties—other researchers observed positive or no influence from drilling fluid supplied at reduced rates in coarse-textured soils in dry climates 7, 8]. Whereas a few research have examined the effects of hazardous substances released from oil well drill cuttings on terrestrial ecosystems [9]. Cuttings and resultant fluids from oil exploration are disposed away, which pollutes both the marine and terrestrial environments. These waste products may include trace levels of contaminants that become toxic to the exposed species if they bio-accumulate and bio-concentrate. According to Siddique et al. [10], when chemicals are spilled on the surface, the liquid component of the substance eventually makes its way through the soil, killing all living things inside and contaminating the groundwater. In order to effectively address environmental contamination, this study emphasises the origins and characterisation of drilling wastes, the ecologically critical compounds contained in drilling wastes, as well as the treatment technology currently in use.

Waste generation, its potential environmental impact, waste migration mechanisms and pathways, effective waste management strategies, disposal methods, contaminated site remediation strategies, and regulation are all issues of focus. Drilling and production create a lot of waste, which is why the industry has come up with many technical and scientific ways to reduce or get rid of the damage to the environment.

12.2 Overview of Drilling Fluid

The earliest drilling fluids were created in the 1890s, and were water-based drilling fluids (WBDF), which are made of water and clay and are used to stabilise well walls. But drilling challenges, particularly in shale formation [11] and under extreme pressure and temperature conditions, reduced the efficiency of water-based fluids. As consequence, OBDFs started to be employed in the 1960s, and they demonstrated improved drilling efficiency. When compared to WBDFs, the OBDF's drilling performance was excellent, nevertheless, due of their high toxicity, they have an adverse effect on the environment. SBDFs were created in 1990 as an alternative to OBDF since they are less poisonous and more biodegradable, however, the price of formulating them is incredibly expensive [12]. In the 1990s, pneumatic fluids such as air, foam and gas were also developed. In certain mature fields, the application of this fluid to low-pressure reservoirs, particularly in rocks that are hard and dry, has shown to be successful. Pneumatic drilling fluids can be generically categorised as air, mist, foam, and gas and are having limitations. To make the drilling process sustainable and profitable, drilling fluids should have a variety of physical qualities, such as rheology. In addition, after the drilling process is finished, the drill cuttings are either trapped, suspended, or absorbed in the drilling fluids, all of which occur without causing any change in the physical properties of the drilling fluids [13]. Natural radioactive substances and dissolved gases may be present in these fluids, as well as many other types of dissolved minerals, oily products, salts and metal ions. Before being disposed of in a landfill, these fluids need to go in treatment process for an appropriate level in order to comply with environmental standards. Authentic and reliable physiochemical characterisations of drilling wastes are required to identify the presence of problematic materials to find the most suitable treatment method [14].

12.3 Sources and Environmentally Relevant Elements in Drilling Waste

While an oil well is being drilled, the drilling fluids that are used are continually circulated inside the well. As the drill makes its way further into the reservoir, contaminants in the oil are drawn up by the drilling fluid and brought to the surface along with them. The residue that is left behind contains a number of potentially harmful substances, particularly contamination from reservoir oil [15].

Along with other types of process industries, the oil and gas sectors have a negative impact on the surrounding environment. How hazardous the ecologically important components in the produced drilling wastes depend on each compound, its level of concentration throughout exposure, the biological condition at the discharge point, as well as the length of exposure. Figure 12.1 illustrates the common kind of drilling wastes and some of its probable components.

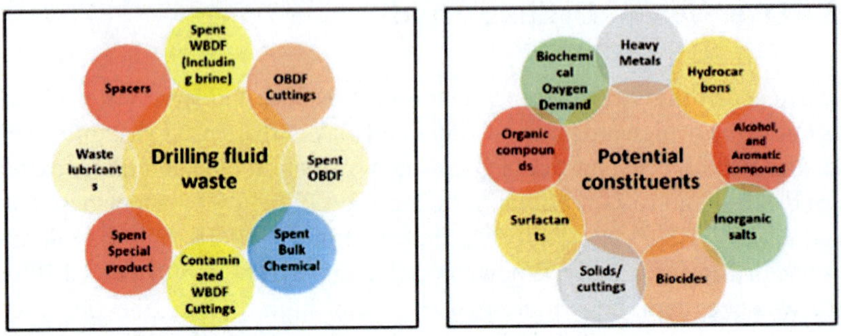

Fig. 12.1 Drilling fluid waste and their potential constituents, which can harm the environment [16]

The waste disposal raises serious environmental concerns since it is believed that a few of the metals are present in amounts that are much greater than the naturally found levels of the sediments [16]. According to the literature, ecologically relevant compounds include phenanthrene, arsenic, zinc, naphthalene, nickel, diuron, copper, chromium, anthracene, fluoranthene, and pyrene [17].

Drill cuttings and wasted mud are the two primary forms of waste generated as a by-product of the drilling activities that take place throughout the exploration and production in the industry. The types of drilling wastes are mostly determined by the cuttings depending on the well's depth, the lithological structure of the borehole, and the fluid used during drilling [14]. The amount of drill cuttings that accumulate is estimated to be between 130 and 560 m^3 per well. Waste lubricants, contaminated WBDF, mineral oil, formation solids, wetting agents, OBDF cuttings, bulk chemicals like fluid loss additives viscosifiers, barite, cement, thinners, bentonite, and used different products like scavengers, and, defoamers are some potential sources and constituents of waste [16].

The oil sector produces a large amount of residual drilling fluid. According to Daae et al. [18], the Norwegian oil and gas industry was responsible for generating 260,000 tons of drilling by-products, the majority of which consisted of cuttings polluted with OBDF and was carried to land-based facilities for treatment in 2016. According to statistics provided by Daneshfar and Ardjmand [19], drilling debris produced by the Iranian Offshore Oil Company (IOOC) is around 3,000 cm^3 annually, some of which are OBDF, which are more detrimental to the environment than other forms of fluid. According to Pereira et al. [20], 118,000 tons of harmful solid waste were produced in 2019, out of which 5,000 tons were wasted drilling fluids and contaminated cuttings, which are categorised as greasy waste.

Majorly, petroleum industries leave these leftovers in a well itself or transport them to landfills since they are both inexpensive end destinations. To establish the environmental hazards, it is necessary to take into account risk evaluations for concerns such as human health, groundwater contamination, and knowledge of the chemicals utilised and the reactivity of the waste produced [21]. In order to draw a conclusion, it is important to note that around 53% of the chemicals used in drilling operations are released into the environment as waste.

When drilling waste is disposed of, the precise composition of the waste provides information about the drilled subsurface strata and the amount of chemical compounds utilised to create drilling fluid that has remained chemically bonded to the cuttings. Information of great significance on the long-term impacts of drill cutting deposition on geochemical as well as hydro-geological conditions are provided by temporal patterns in drilling activity. Different groups' results supported the existence of different metals in drill cuttings and related possible environmental impact. Drill cuttings mostly include the following metals: As, Al, Ba, Cd, Cr, Co, Cu, Fe, Mn, Ni, Pb, V and Zn [22].

Some studies have focused on the toxicity of waste generated by drilling activities. Gbadebo, Taiwo and Eghele [23] studied environmental impacts and toxicities of OBDF and WBDF waste at their disposal location. The research studied wastes from Ondo State/Igbokoda Nigeria's onshore wells. The authors evaluated total petroleum hydrocarbon (TPH) and polycyclic aromatic hydrocarbons (PAH) from wastes using GC–MS. TPH values were 7156.57 mg/kg^{-1} for WBDF, which are higher than 6449.55 mg/kg^{-1} of OBDF waste. Most samples exceeded PAH limits. PAH are mutagenic and carcinogenic, causing health issues. The scientists also drew attention to the significant amounts of heavy metals that were found in the waste that was produced. These metals have the potential to induce bioaccumulation in aquatic organisms.

Onwuka et al. [24] confirmed that drilling mud had a pH range of 9.35–12.9 in the Niger Delta's X-gas field. This was greater and above the permissible limits set by both the Department of Petroleum Resources (DPR) and the Federal Ministry of Environment (FME). Researchers also determined that the high pH was brought on by the caustic soda and cement that was present in the drilling mud. Cu concentrations of 15.6 mg/l from samples obtained at 3000 and 4000 ft and 7.22 mg/l for Pb evaluated at 1000 ft are above DPR/FME allowed limits and have been linked to the reliability and composition of drilling fluids in the subsurface. The Pb values were measured at high lead levels may have resulted via weathering of pyrite (iron sulphide), the process of metal or iron oxides being dissolved, or other factors. Cd poses a risk to the public's health since it surpasses the DRP/FME limit.

The issues with workers' exposure to high quantities of hazardous components from landfills were discussed by Daae et al. [18]. They conducted a number of experiments and found that there were biological and chemical airborne compounds. The majority of them seem to represent a health risk to employees who are exposed to a polluted workplace.

While assessing the effects of OBDF samples utilised during drilling operations on human exposure to hazardous substances and heavy metals, Okoro et al. [25] reported that the drilling crew's cancer risk was between 1.1×10^{-3} and 7.7×10^{-3}, which they claimed was over the risk acceptance range taken into account by the safety and environmental government entities. Additionally, they said that the heavy metals Lead, Cadmium (Cd), Mercury, Copper, Arsenic (Ar), Chromium (Cr), and

Nickel (Ni) had carcinogenic risks that revealed Ni to be the most probable primary contributor to cancer risk, followed by Ar, Cr, and Cd.

During the course of research investigation on the chemical properties of drilling waste at five different drill locations in Poland, Mikos-Szymańska et al. [26] discovered that wasted drilling fluids include significant levels of Ni and Hg, which are 270 and 8.77 mg/kg, respectively. This is much higher than the maximum permitted values recommended by the European Commission (EC) regulations for the safety of soils, which are 75 mg/kg for Ni and 1.5 mg/kg for Hg. ICP-OES was used to analyze the drilling waste's mineral content, while CV-AAS was used to analyze the waste's mercury (Hg) content.

Waste that contains high levels of heavy metals, hydrocarbons, as well as other products of drilling wastes are expected to have a negative influence on cancer risk, despite the scarcity of epidemiological data, which is probably owing to poor research power. The fact that there is already a connection between heavy metals and an increased risk of cancer, as shown by the results of previous research, lends credence to the hypothesis that pollution caused by hydrocarbons and heavy metals also plays a significant part in the development of the disease. Additionally, it is evident from the reviewed literature that despite the existence of non-threshold genotoxic carcinogens in drilling waste, hydrocarbons nonetheless constitute a threat to the environment in which they are dumped and perhaps to human health. Wastes that cannot be disposed of naturally need to be handled. It is applied to lessen the quantity and toxicity of wastes to be disposed of. The wastes can be treated in a variety of ways and disposed of in different ways.

12.4 Remediation Methods of Drilling Fluid Waste

12.4.1 Bioremediation Technique

Used drilling fluids are one sort of waste that can be treated via bioremediation. This treatment technique uses a process called organic matter biodegradation to turn waste's organic content into humus and some other constituents including carbon dioxide and water. The process of the aerobic breakdown of n-alkane is shown in Fig. 12.2. It was feasible to see that the usual degradation processes include dehydrogenation, hydroxylation, reduction, and oxidation. Energy stimulation is needed for these processes, which is often given by enzymes like monooxygenase and NADH (nicotinamide-adenine-dinucleotide). With or without oxygen, or via aerobic or anaerobic mechanisms, the biodegradation process may take place [27, 28]. About 25 hydrocarbon-degrading microorganisms were described by [28], and the majority of them demonstrated an efficiency of hydrocarbon biodegradation exceeding 80%. The capacity to biodegrade huge amounts of material and low cost are the key benefits of employing bioremediation to handle oily wastes.

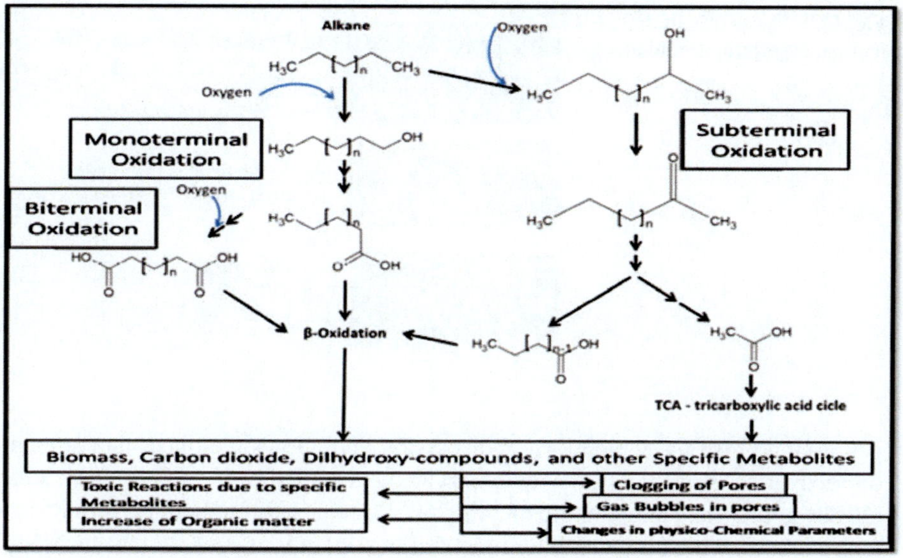

Fig. 12.2 Aerobic degradation of n-alkane [29]

However, it is difficult to increase biodegradation rates in-situ since several variables need to be under control including the levels of oxygen, nutrients, temperature, and salinity in the water [29].

12.4.2 Thermal Treatment

The bulk of the drill cuttings' liquid phase hydrocarbons are separated and condensed using the thermal desorption technique when the heating process is carried out in an oxygen-free environment. The first step involves the evaporation of water, which results in the production of steam and a subsequent decrease in the temperature at which the oil boils. Because of this, the procedure is carried out at a temperature that is lower than the maximum temperature at which the oil is capable of evaporating. Thermal desorption technique consists of different unit types as mentioned in Fig. 12.3. All of the aforementioned devices are capable of lowering the amount of oil in the finished by-product to around 1% and extracting the oil by condensation.

But at the other side, it may also be divided into low- and high-temperature systems based on the working temperature. Light oil wastes can be treated with a low temperature system, whereas wastes with heavy oil content are best treated with a high temperature system.

Because of the massive heating surface necessary for indirect thermal separation, the system is not only bulky but also consumes a significant amount of heat. While these issues are well-solved by the TCC, it additionally breaks solid particles with

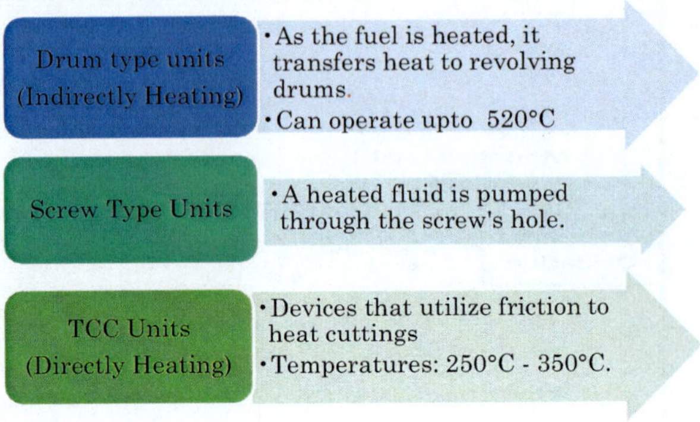

Fig. 12.3 Types of unit in thermal desorption technique

percussion crushing, which lowers dispersion and allows the maximum amount of oil recovery. As of the UKCS region, TCC has handled over 15,000 metric tons of OBDF cuttings at a temperature of 245–260 °C and at a throughout performance of 6 tons per hour, which is enough for most drilling operations. In Kazakhstan, MI-TCC SWACO's system processes around 50,000 tons of oil-based mud annually [30].

This treatment is helpful because it results in a significant reduction in the volume of the residue and also allows the stored energy to be used for heating, the formation of steam, and the generation of electric energy.

12.4.3 Solvent Extraction/Supercritical Fluid Treatment

Typically, supercritical fluids like CO_2 and propane are used to recover oil-based drill cuttings. In a supercritical state, a fluid is neither a gas nor a liquid. As a gas, it has a high permeability, but as a liquid, it has a high density and high solubility. The most common extraction method for cuttings is supercritical CO_2 technology. Waste OBDF is mixed with supercritical fluid in the fundamental process of supercritical Carbon dioxide extraction technology before the oil is recovered. In order to reuse the extracted oil, both solvent and decompression have to be completed. However, in order to successfully complete the operation, it is necessary to maintain a high pressure of 14.5 MPa and a temperature of 40 °C. Ma et al. [31] assessed the effectiveness of SC-CO_2 in the process of oil extraction from OBDF waste with 25% oil content. The extraction efficiency rises considerably from 10 to 60 min but not between 60 and 240 min. They also studied pressure's effect on extraction efficiency. Increasing the pressure all the way up to 220 bar improves the extraction efficiency. The extraction efficiency drops above that pressure. The increasing temperature had the same effect. Extraction efficiency improved until 308 K. The scientists determined that at 308 K, 200 bar, and duration of 60 min ensured the highest effective oil extraction of 98.6% from OBDF Waste.

12.4.4 Incineration

Incineration turns OBDF cuttings' organic components into inert residues at 1200–1500 °C. Incineration technique can dispose of practically any trash, avoid waste contamination, and dispose of many cuttings. Incineration has drawbacks: (1) It cannot handle metals in OBDF cuttings; (2) Metals in steam require dust-removal equipment; (3) It uses excessive energy; (4) Large volumes of CO_2 and NO_2 are produced.

12.4.5 Solidification and Stabilisation

The stabilisation and solidification processes, respectively, both include the transformation of waste leftovers into less movable, soluble, and hazardous forms as well as the incorporation of the waste contained inside a single mass (a monolith) that has excellent structural integrity. Together, these procedures lessen the possibility that waste contaminants may damage people. To jump-start the curing process, chemically stabilise the waste, and solidify it, binders like cement, lime, clay, and other such compounds are added to waste. Normally, this process takes around 28 days [32].

12.4.6 Cutting Dryer

The coal industry has transformed, and so has the cuttings drier. It works by using a high-speed spinning centrifugal device to squeeze the oil out of drill cuttings that contain oil. There are both horizontal and vertical versions of drill cutting dryers, and these categories are determined by the orientation of the centrifugal's axial axis. For instance, the Hutchison Hayes Dusters is a horizontal drier, but MI-Vertis SWACO's GTM cuttings dryer is vertically installed [33]. The production of new OBDF can be accomplished with the use of drill cuttings dryer oil. The Gulf of Mexico uses drill cutting dryers extensively. Several studies report therapeutic effects. According to one portion of study, the retention on cuttings (ROC) was reduced by drill cutting dryers from 11.8 to 2.1% [34]. In another research, [33] used waste from 23 wells to demonstrate that treating drill cuttings with a dryer lowered the ROC on average from 11.47 to 3.99%, although the ROC for a single well ranged from 2 to 6%.

12.5 Conclusion

Classification and characterisation of drilling waste is a straightforward first step in order to meet the current strict guidelines regarding the disposal, reuse, or reprocessing of this waste or the recovery of valuable components, for example specific metals. Because the method must be chosen in accordance with the demands of field applications, it cannot be used to process all varieties of drill cuttings. Large-area processing techniques, for instance, cannot be used when dealing with drill cuttings made of oil that are generated by offshore drilling activity due to the size restriction of the deck. Therefore, TCC technology is always used for offshore oil-based drill cuttings. Additionally, combining various treatment techniques can help meet strict emission policy requirements while also increasing the process efficiency of cuttings made of oil. Different waste treatment plans, including bioremediation, thermal treatment, supercritical fluid treatment, incineration, solidification and stabilisation, cutting dryers, and encapsulation of pollutants, can be implemented based on this information.

12.6 Future Scope

There is a one-of-a-kind assortment of drilling fluid wastes at each and every one of the oil and gas industry's operation sites. These wastes are distinct as a result of the variations in drilling activity, including the use of materials in drilling fluid with varying concentrations and the variations in geological formations, so none of the specified drilling fluid waste characterisation instruments is available to determine the composition and nature of the wastes. When deciding how to dispose of drilling fluid cuttings, guidelines from the government, operator regulatory standards, price, safety, transportation, and other factors need to be taken into consideration when comparing the primary methods of disposal.

References

1. IEA, World Energy Outlook. *No Title* (2020)
2. H. He et al., A combined technology of normal temperature cleaning and microbial treatment for oil-based drilling cuttings. Nat. Gas. Ind. **5**, 122–127 (2016)
3. Z. Chen et al., Continuous supercritical water oxidation treatment of oil-based drill cuttings using municipal sewage sludge as diluent. J. Hazard. Mater. **384**, 121225 (2020)
4. Z. Chen et al., A laboratory evaluation of superheated steam extraction process for decontamination of oil-based drill cuttings. J. Environ. Chem. Eng. **6**(5), 6691–6699 (2018). https://doi.org/10.1016/j.jece.2018.10.040
5. I.P. Júnior et al., A novel method to determine total petroleum hydrocarbon (TPH) and water contents in reservoir drill cuttings. J. Petrol. Sci. Eng. **195**, 107760 (2020)

6. A.K. Wojtanowicz, Environmental control of drilling fluids and produced water, in *Environmental Technology in the Oil Industry*. (Springer, 2016), pp. 101–165. https://doi.org/10.1007/978-3-319-24334-4_4
7. L. Yao, M.A. Naeth, Soil and plant response to unused potassium silicate drilling fluid application. Ecol. Eng. **73**, 461–468 (2014)
8. L. Yao, M. Anne Naeth, Soil and plant response to used potassium silicate drilling fluid application. Ecotoxicol. Environ. Saf. **120**, 326–333 (2015). https://doi.org/10.1016/j.ecoenv.2015.06.021
9. M.O.L. Magalhães et al., The effects of oil well drill cuttings on soil and rice plant development (oryza sativa) under two redox conditions. Bull. Environ. Contam. Toxicol. **92**(3), 311–316 (2014). https://doi.org/10.1007/s00128-014-1196-7
10. S. Siddique, et al. Oil Based Drilling Fluid Waste: An Overview on Environmentally Persistent Pollutants, in *IOP Conference Series: Materials Science and Engineering*. (IOP Publishing, 2017), p. 12008. https://doi.org/10.1088/1757-899X/195/1/012008
11. U. Bhan et al., Mineralogy, organic richness and macerated microbial studies of the Rohtasgarh shales in the Vindhyan basin, India: implications for gas generation potential. J. Geol. Soc. India **98**(4), 567–575 (2022)
12. C.C. Corrêa et al., Evaluation of the potential use of bioglycerin as a base for formulation of aqueous drilling fluids for oil and gas wells. Quim. Nova **40**(4), 378–387 (2017). https://doi.org/10.21577/0100-4042.20170019
13. D. Zhou et al., Applied properties of oil-based drilling fluids with montmorillonites modified by cationic and anionic surfactants. Appl. Clay Sci. **121**, 1–8 (2016)
14. K. Piszcz, J. Luczak, J. Hupka, Mobility of shale drill cuttings constituents. PhysChem. Probl. Miner Process **50**(2), 795–810 (2014). https://doi.org/10.5277/ppmp140230
15. A.M.H. Elnenay et al., Treatment of drilling fluids wastewater by electrocoagulation. Egypt. J. Pet. **26**(1), 203–208 (2017)
16. S.I. Onwukwe, M.S. Nwakaudu, Drilling wastes generation and management approach. Int. J. Environ. Sci. Dev. **3**(3), 252–257 (2012). https://doi.org/10.7763/ijesd.2012.v3.226
17. M. Khodja, et al., Drilling fluid technology: performances and environmental considerations. *Products and services; from R&D to final solutions* (2010), pp. 227–256
18. H.L. Daae et al., Occupational exposure during treatment of offshore drilling waste and characterization of microbiological diversity. Sci. Total Environ. **681**, 533–540 (2019)
19. M.A. Daneshfar, M. Ardjmand, Selecting a suitable model for collecting, transferring, and recycling drilling wastes produced in the operational areas of the Iranian offshore oil company (IOOC) using analytical hierarchy process (AHP). J. Environ. Manage. **259**, 109791 (2020)
20. L.B. Pereira et al., Environmental impacts related to drilling fluid waste and treatment methods: a critical review. Fuel **310**, 122301 (2022)
21. J. Ma et al., Bioremediation enhances the pollutant removal efficiency of soil vapor extraction (SVE) in treating petroleum drilling waste. Water Air Soil Pollut. **227**(12), 1–10 (2016)
22. D. Pozebon et al., Heavy metals contribution of non-aqueous fluids used in offshore oil drilling. Fuel **84**(1), 53–61 (2005)
23. A.M. Gbadebo, A.M. Taiwo, U. Eghele, Environmental impacts of drilling mud and cutting wastes from the Igbokoda onshore oil wells, Southwestern Nigeria. Indian J. Sci. Technol. **3**(5), 504–510 (2010). https://doi.org/10.17485/ijst/2010/v3i5/29744
24. O.S. Onwuka et al., An assessment of the effectiveness of drilling waste treatment process in X-gas field, Niger Delta, Nigeria. Geol. Ecol. Landscs **2**(4), 288–302 (2018). https://doi.org/10.1080/24749508.2018.1473751
25. E.E. Okoro et al., Risk assessment of human exposure to radionuclides and heavy metals in oil-based mud samples used for drilling operation. Int. J. Environ. Health Res. **32**(5), 972–983 (2022)
26. M. Mikos-Szymańska et al., Characterization of drilling waste from shale gas exploration in Central and Eastern Poland. Environ. Sci. Pollut. Res. **25**(36), 35990–36001 (2018). https://doi.org/10.1007/s11356-018-2365-8

27. M. Dubey, U. Bhan, R. Kumar, Catalytic remediation of chlorinated organic compounds (COCs) in wastewater, in *Emerging Trends to Approaching Zero Waste* (Elsevier, 2022), pp. 133–151
28. A. Imam et al., Analytical approaches used in monitoring the bioremediation of hydrocarbons in petroleum-contaminated soil and sludge. TrAC Trends Anal. Chem. **118**, 50–64 (2019)
29. M. Hassanshahian, S. Cappello, Crude oil biodegradation in the marine environments. Biodegrad-Eng. Technol. **5**, 102–135 (2013)
30. A.J. Murray, et al., Friction-based thermal desorption technology: Kashagan development project meets environmental compliance in drill-cuttings treatment and disposal, in *Proceedings—SPE Annual Technical Conference and Exhibition.* OnePetro (2008) pp. 3432–3437. https://doi.org/10.2118/116169-ms.
31. B. Ma et al., Experimental study on harmless disposal of waste oil based mud using supercritical carbon dioxide extraction. Fuel **252**, 722–729 (2019)
32. H. Boutammine, Z. Salem, M. Khodja, Petroleum drill cuttings treatment using stabilization/solidification and biological process combination. Soil Sediment Contam. **29**(4), 369–383 (2020). https://doi.org/10.1080/15320383.2020.1722982
33. R.W. Cannon, D. Martin, Reduction of synthetic based fluid discharges offshore by the use of vertical basket centrifuges, in *SPE—EPA Exploration and Production Environmental Conference Proceedings.* OnePetro (2001). https://doi.org/10.2118/66535-ms
34. H.R. Melton, et al., Environmental aspects of the use and disposal of non aqueous drilling fluids associated with offshore oil & gas operations. in *SPE International Conference on Health, Safety, and Environment in Oil and Gas Exploration and Production.* OnePetro (2004)

Chapter 13
Comparative Study of Different Flooding Mechanisms in 5-Spot and 7-Spot Pattern Using Simulation

Nithin Joseph Thomas and Shailendra Naik

Abstract The paper describes the different flooding mechanisms mainly enhanced oil recovery processes involved in recovery of hydrocarbons from the reservoir. Oil recovery is usually done by primary, secondary and tertiary recovery mechanisms. Natural drive of the reservoir aids in recovery by primary mechanism, gas or water is injected to increase reservoir pressure or displace oil is by secondary mechanism and use of thermal and chemical enhanced oil recovery processes is by tertiary mechanism. The study comprises of recovery of oil by different flooding mechanisms in 5-spot and 7-spot pattern using CMG simulation software. The study analyzes cumulative oil and water production using different flooding mechanisms and suggests the most feasible flooding mechanisms and pattern of flooding for efficient recovery.

Keywords Enhanced oil recovery · Volumetric sweep efficiency · Alkaline-surfactant-polymer flooding

13.1 Introduction

The main objective of oil production is the recovery of oil from the reservoir underground. In order to meet the demands of the energy requirements, various techniques are used other than the conventional methods. Oil recovery is primarily separated into primary, secondary and tertiary recovery mechanisms. Primary recovery is the method in which natural drive of the reservoir is used to recover oil. In primary recovery, reservoir pressure is higher than bottom-hole pressure and declines as the production takes place. Methods like artificial lift are used to increase the pressure differential and are considered as primary recovery. Secondary recovery is the method in which external fluid like gas or water is injected. The main objective of secondary

N. J. Thomas (✉) · S. Naik
School of Petroleum Engineering, MIT World Peace University, Pune 411038, India
e-mail: nithinpeediakal@gmail.com

S. Naik
e-mail: shailendra.naik@mitwpu.edu.in

A. K. Sahu et al. (eds.), *Energy Storage and Conservation*, Springer Proceedings in Energy, https://doi.org/10.1007/978-981-99-2870-5_14

recovery mechanism is to maintain reservoir pressure and displace hydrocarbons to the wellbore. Gas or water is injected to sweep oil from the reservoir that was not recovered by primary mechanism.

Tertiary mechanism or enhanced oil recovery (EOR) is used to recover hydrocarbons that were not recovered by primary and secondary mechanisms. It involves recovery by thermal and chemical methods. Thermal methods like steam flooding and fire flooding are used to decrease the viscosity of oil and decrease mobility ratio and result in better recovery of hydrocarbons. Chemical enhanced oil recovery methods like polymer flooding and alkaline-surfactant-polymer flooding are used to increase viscosity of water and alter interfacial tension between oil and rock surface. The oil production rate for different mechanisms varies with regard to the type of reservoir and its characteristics. This paper studies the cumulative oil production, cumulative water production of different flooding mechanisms in 5-spot and 7-spot pattern using reservoir simulation.

13.2 Objective

(1) To study and analyze different flooding mechanisms of enhanced oil recovery.
(2) To create a reservoir model for normal 5-spot and normal 7-spot pattern and compare its production using Water flooding, Steam flooding, Polymer flooding, Alkaline-Surfactant-Polymer flooding.
(3) To determine the best possible flooding mechanism for normal 5-spot and normal 7-spot model respectively using CMG.

13.2.1 Types of Enhanced Oil Recovery

Enhanced oil recovery is a tertiary recovery method where the main objective is to reduce the residual oil saturation after the secondary recovery methods. Enhanced oil recovery includes mechanisms, which involve injection of a fluid or other agents that can alter the fluid or reservoir parameters and aid in recovery of hydrocarbons. The principals involved in enhanced oil recovery has the concept of increasing mobility by decreasing oil viscosity or increasing water viscosity. Other parameters like wettability can be altered when Interfacial tension decreases [3].

- Steam Flooding—It is a thermal enhanced oil recovery process where reservoir is heated to reduce oil viscosity. By heating the crude oil in the formation, the steam injection reduces viscosity and partially vaporizes some of the oil, increasing mobility. The improved reservoir seepage conditions, lower surface tension and increased oil permeability are all the benefits of reduced viscosity. Oil vaporization makes it possible for oil to flow more easily through the reservoir and after it has condensed to produce superior oil [5].

- Polymer Flooding—It is a type of chemical enhanced oil recovery process where polymer solution is added to water through injection wells to increase the viscosity of the water to improve volumetric sweep efficiency for better recovery. The purpose of polymer flooding is to improve volumetric sweep efficiency by increasing viscosity of water, decrease in water and mobility ratio and decrease in volume of capillary trapped oil. Several reservoir parameters like mobility ratio, effective porosity, permeability, mobile oil saturation, volumetric sweep efficiency should be analyzed before any polymer flooding operation. Several reservoir pre-screening should be done before polymer flooding operation that includes:

  (a) Lithology—Sandstones are preferred.
  (b) Wettability—Water wet
  (c) Current oil saturation—Above residual oil saturation
  (d) Porosity type—Matrix preferred
  (e) Gas cap
  (f) Aquifer—Edge aquifer tolerated
  (g) Salinity
  (h) Dykstra-Parsons—$0.1 < DP < 0.8$
  (i) Clays—Low
  (j) Water cut
  (k) Flooding pattern and spacing—Confined small spacing [4].

- Alkaline-Surfactant-Polymer (ASP) Flooding—It is another type of chemical enhanced oil recovery process where the combination of alkaline, surfactant and polymer is used to recover oil efficiently. It is one of the most reliable flooding mechanisms since it has the synergy of three components together to work efficiently. Polymers are used to increase the viscosity of injected water, which improves macroscopic volumetric efficiency by stabilizing the displacing front. Surfactants are used to decrease the interfacial tension between injected water and crude oil, which results in the improvement of microscopic displacement efficiency. Alkaline chemicals generate soap like reaction when contacted with crude oil which results in reduction of adsorption to grain surfaces [1]
- Water Flooding—It is a secondary recovery mechanism in which water is injected to the reservoir using injection wells to displace residual oil which was not recovered by natural drive of the reservoir. Water seeps through the pore spaces where oil is trapped and is displaced to the well bore. The main disadvantage of water flooding is that since water is not viscous compared to oil it can easily seep through permeable spaces and not efficiently recover oil.

13.2.2 Parameters Affecting Recovery

- Mobility ratio: It is defined as the ratio between mobility of displacing fluid and displaced fluid where mobility can be expressed as:

$$M = \lambda D/\lambda d \tag{13.1}$$

where M is the mobility ratio, λD is the mobility of displacing fluid and λd is the displaced fluid.

- Volumetric Sweep Efficiency: It is defined as the fraction of the floodable pore volume swept or contacted by the injected water. Volumetric sweep efficiency is a combination of areal sweep efficiency and vertical sweep efficiency [2].
- Relative Permeability: It is the ratio of effective permeability of a particular fluid at a particular saturation to absolute permeability of that fluid at total saturation. If a single fluid is present then relative permeability is 1.

13.2.3 Types of Injection Pattern

It is the arrangement of the injection and production wells in reservoir. The size of the field, location of existing wells, reservoir shape and size are considered before deciding an injection pattern,

Basic injection types are as follows:

- Direct line drive
- Staggered line drive
- Two-spot
- Three-spot
- Four-spot
- Five-spot (Normal and inverted)
- Seven-spot (Normal and inverted)
- Nine-spot (Normal and inverted)
- Normal 5-Spot: It is a type of flooding pattern where there are four injection wells and one producer well.
- 7-Spot Pattern: It is a type of flooding pattern where there are six injection wells and one production well.

13.3 Methodology

A reservoir model was made using CMG STARS for 5-spot pattern and 7-spot pattern. The steps that were involved during the generation of the reservoir model with the application of different flooding mechanisms are as follows:

- **Builder**—Model was generated setting simulator type as STARS and starting date as 2005/01/01.
- **Reservoir**—Under this section, Normal 5-spot pattern type was selected with pattern area as 10 acres, 30 m thickness, Porosity as 0.3, and Permeability as 400,400 and 40 for I, J and K, respectively.

- **Component**—In this section, PVT data was generated using Reservoir temperature as 37.778 degree C, Maximum pressure as 12000 kPa, Bubble point pressure as 8576 kPa, Oil density at stock tank API as 21 and Gas density at (Air = 1) as 0.65.
- **Rock-Fluid**—In this section, rock fluid properties were added using Relative Permeability Correlations.
- **Initial Conditions**—In this section, Reference Pressure was assigned as 8576 kPa, Reference depth as 504 m, Water–Oil Contact Depth as 526 m, Gas-Oil Contact Depth as 504 m.
- **Numerical**—In this section, Timestep Control value was assigned as 0.001 day. Geomechanics was ignored since it was a single porosity model.
- **Wells and Recurrent**—In this section, upper layers and bottom layers Perforation of Injectors and Producer were deleted. Constraints were added to injectors and producers.
- For **Steam Flooding**, Temperature and Steam quality was assigned as 325 degree C and 0.8, respectively. Events were copied to all the injection wells and dates were added till 2010/01/01.
- For **Water Flooding**, Temperature was assigned as 30 degree C and steam quality was unchecked. Events were copied for all injectors.
- For **Polymer Flooding and Alkaline-Surfactant-Polymer Flooding**, using Process Wizard polymer mole fraction and alkaline properties were added for polymer flooding and ASP flooding separately. Events were copied for all injectors.
- For **7-Spot pattern**, all the steps were similar to 5-Spot pattern. Under **Reservoir** section, Normal 7-Spot pattern was selected.
- Under **Wells and Recurrent** section, all events for steam flooding, water flooding, polymer flooding and ASP flooding were applied for all six injectors.

13.4 Results and Discussion

13.4.1 5-Spot Pattern

Oil Production

Polymer flooding>Steam flooding>Water flooding>Alkaline-surfactant-polymer flooding

Water Flooding: In the reservoir model, water flooding has a cumulative oil recovery of 4605.01 m^3 till 1826 days.

Figure 13.1 represents the oil production or recovery through different flooding mechanisms where water flooding is significantly less as compared to other enhanced oil recovery techniques. Less production is due to the unfavorable mobility ratio ($M > 1$) that significantly reduces the Volumetric sweep efficiency.

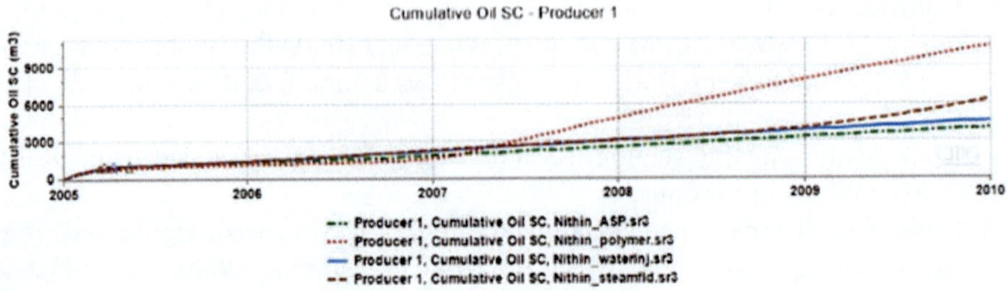

Fig. 13.1 Comparison of Cumulative Oil Produced during Water flooding, Steam flooding, Polymer flooding and ASP flooding

Steam Flooding: In the reservoir model, steam flooding method has a cumulative oil recovery of 6405.82 m^3 till 1826 days.

Figure 13.1 represents the oil production or recovery through different flooding mechanisms. Steam injection has the second-highest recovery when compared to other flooding techniques used. Steam is injected through injection wells and it reduces the viscosity of oil and make it more mobile.

Polymer Flooding: In the reservoir model, polymer flooding has a cumulative oil production of 10,726.1 m^3 till 1826 days.

Figure 13.1 represents the oil produced or recovered through different flooding mechanisms. In this reservoir model, polymer flooding has the highest rate of recovery since the polymer injected increases water viscosity and results in increase of volumetric sweep efficiency as mobility ratio decreases ($M < = 1$).

Alkaline-Surfactant-Polymer Flooding: In the reservoir model, ASP flooding has a cumulative production of 4094.59 m^3 till 1826 days.

Figure 13.1 represents the oil production or recovery during different flooding mechanisms. In the reservoir model generated, ASP flooding has the least recovery compared to other flooding mechanisms. The oil recovery in ASP flooding works with the addition of combination of alkali which reacts with oil to create soap, surfactant that reduces the interfacial tension to alter the wettability and polymers to increase the viscosity of water to increase mobility (Table 13.1).

Polymer flooding has been proven to have a higher cumulative oil recovery of 10,726.1 m^3 than water flooding, steam flooding and ASP flooding.

Table 13.1 Cumulative oil recovery for different flooding techniques

Flooding technique	Cumulative oil recovery (m^3)
Water injection	4605.01
Steam flooding	6405.82
Polymer flooding	10,726.1
ASP flooding	4094.59

Table 13.2 Cumulative water production for different flooding techniques

Flooding mechanism	Cumulative water produced (m^3)
Water flooding	173,192
Steam flooding	172,552
Polymer flooding	66,628.3
ASP flooding	170,730

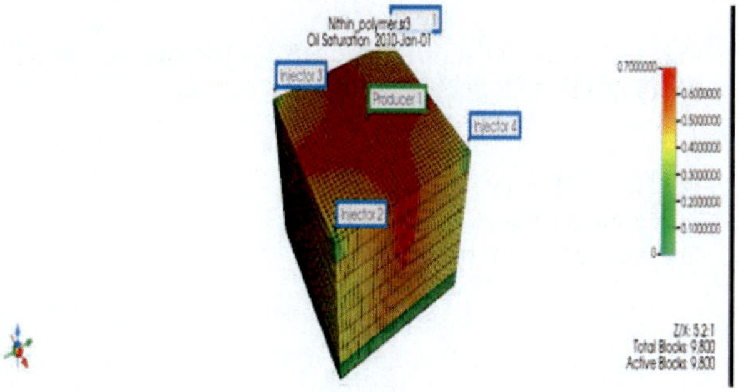

Fig. 13.2 3D view of Oil Saturation during Polymer Injection

Water Production

Table 13.2 shows high production of water in case of water flooding, steam flooding and ASP flooding. Water production is a major problem in the oil and gas industry as water production decreases reservoir pressure that decreases drawdown.

Polymer flooding has less production of water around 66,628.3 m^3 due to favorable mobility ratio, high resistance factor, high residual resistance factor, polymer retention decreases relative permeability of water, and also polymer decreases the mobility of water due to its highly viscous nature, all this contributes to less production of water in case of polymer flooding (Fig. 13.2).

13.4.2 7-Spot Pattern

Oil Production

In 7-spot pattern, different flooding mechanisms were performed including water flooding, steam flooding, polymer flooding and ASP flooding. Cumulative oil and water production rates are compared and analyzed that polymer flooding has a higher cumulative oil recovery compared to other flooding mechanisms in 7-spot pattern.

Figure 13.3 represents the comparison of cumulative oil produced during different flooding mechanisms in 7-spot pattern, and it is analyzed that polymer flooding is the most efficient flooding mechanism with 20,403.7 m^3 in this 7-spot reservoir model (Table 13.3 and Fig. 13.4).

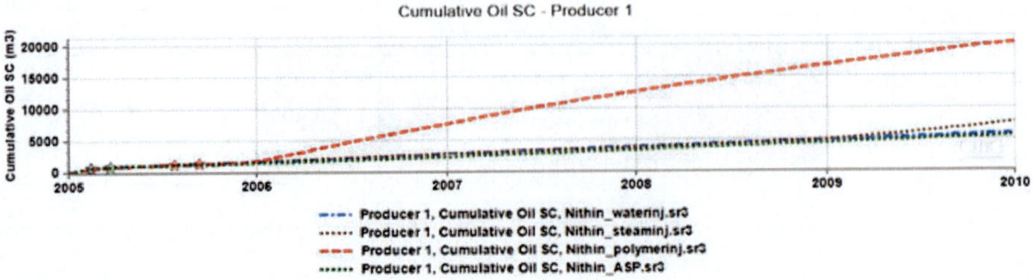

Fig. 13.3 Comparison of Cumulative Oil Produced during Water flooding, Steam flooding, Polymer flooding and ASP flooding for 7-Spot pattern

Fig. 13.4 3D view of oil saturation during polymer flooding for 7-spot pattern

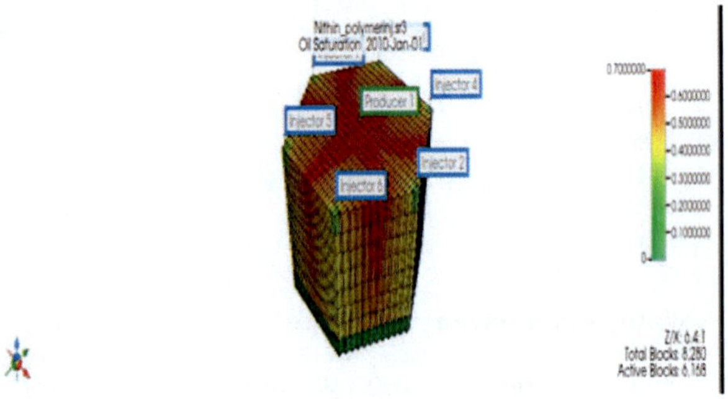

Table 13.3 Cumulative oil production for different flooding mechanisms in 7-spot pattern

Flooding mechanism	Cumulative oil produced (m^3)
Water flooding	5892.49
Steam flooding	7675.93
Polymer flooding	20,403.7
ASP flooding	5564.74

13.4.3 Comparison Between 5-Spot and 7-Spot Pattern

In both 5-spot and 7-spot pattern, polymer flooding is found to be the effective flooding operation to recover maximum cumulative oil production. In Fig. 13.5, we compare and analyze the cumulative production for both patterns for polymer flooding, we can find that 7-spot pattern gives a higher cumulative oil production of 20403.7 m^3 than 5-spot pattern with 10726.1 m^3.

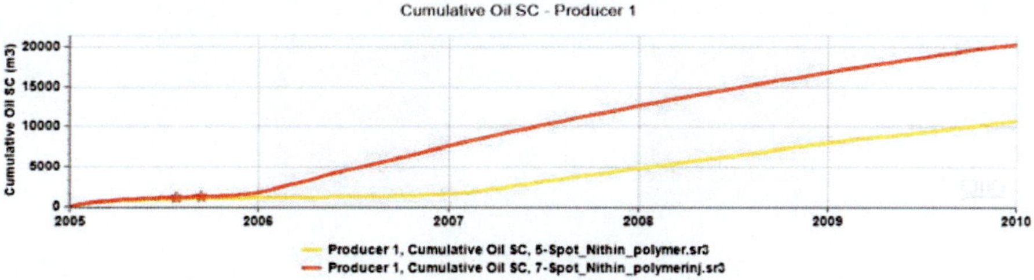

Fig. 13.5 Comparison of Cumulative Oil produced during Polymer flooding for 5-Spot and 7-Spot pattern

13.5 Conclusions

This study investigated through simulation compared different flooding mechanisms in a model in 5-spot and 7-spot pattern. It was analyzed that thermal and chemical enhanced oil recovery methods showed higher recovery than secondary recovery mechanisms like water flooding due to its property of altering viscosity and affecting mobility ratio. The important conclusions from the analysis are:

1. Polymer flooding mechanism showed higher recovery in terms of cumulative oil production and lower water production than all other flooding mechanisms in 5-spot and 7-spot pattern model due to its ability to increase the viscosity of water and increase the volumetric sweep efficiency.
2. We can analyze that the results of cumulative oil production in 5-spot and 7-spot pattern for polymer injection, we could observe that 7-spot pattern shows higher cumulative oil recovery than 5-spot pattern due to its better volumetric sweep efficiency because of its six injection wells.

Acknowledgements I would like to extend a token of gratitude to my research guide and Head of School who helped me throughout the research.

References

1. Y. Alzahid, P. Mostaghimi, M.E. Warkiani, R.T. Armstrong, Alkaline surfactant polymer flooding: what happens at the pore scale? Soc Petrol Eng (2017)
2. W.M. Cobb, F.J. Marek, *Determination of Volumetric Sweep Efficiency in Mature Waterfloods Using Production Data.* (San Antonio, Texas, 1997)
3. P. Druetta, P. Raffa, F. Picchioni, *Chemical Enhanced Oil Recovery and The Role of Chemical Product Design.* (Elsevier, 2019)
4. A. Thomas, *Essentials of Polymer Flooding Technique.* (John Wiley & Sons Ltd., 2019)
5. G. Zerkalov, Steam injection for enhanced oil recovery, in *Coursework for PH240.* (Stanford University,2015)

Chapter 14
Investigation on Risk Assessment by Monte Carlo Analysis (MCS)

M. J. A. Prince, Venkata Ramana Avula, R. Ramesh, and Siddharth Sampathkumar

Abstract Most speculations within the energy industries include imperative hazards, with a wide run of potential results for a specific venture. In any case, numerous financial assessments are based on the "most likely" or "most repeated" variable outcomes that may well anticipated without adequate data given to other conceivable results, and it is well known that introductory outcomes of all these factors have instability. Monte Carlo Simulation (MCS) is an important risk assessment tool that has been applied in different areas, including energy sectors. This paper projects on risk assessment of an energy field with its monetary data through MCS. The overall methodology to conduct MCS with its applications was discussed. The results of the field were portrayed through risk assessment in three categories P10, P50 and P90.

Keywords Monte Carlo simulation · Risk assessment · Probabilistic approach · Stochastic model

M. J. A. Prince (✉)
Department of Petroleum Engineering, JCT College of Engineering and Technology, Coimbatore, Tamil Nadu 641105, India
e-mail: prince466@gmail.com

V. R. Avula
Department of Petroleum Engineering, Godavari Institute of Engineering and Technology (A), Rajahmundry, Andhra Pradesh, India

R. Ramesh
Department of Chemical Engineering, School of Mechanical, Chemical and Materials Engineering, Adama Science and Technology University, Adama, Ethiopia

S. Sampathkumar
Department of Mechanical Engineering, PSN College of Engineering and Technology, Tirunelveli, Tamil Nadu, India

A. K. Sahu et al. (eds.), *Energy Storage and Conservation*, Springer Proceedings in Energy, https://doi.org/10.1007/978-981-99-2870-5_16

14.1 Introduction

An MC model is optimized when the outcomes are based on random variables. This is an application on risk forecasting by ascertaining values on random variables to get different outcomes with best estimates. Based on repeated outcomes, MC depicts the values on the uncertainties with highest probabilities [1]. The premise of an MCS includes assigning multiple values to an uncertain variable to realize numerous outcomes about and after averaging them to get an assess. The frequencies of the various outcomes generated by this simulation form a normal distribution Bell curve. The most likely returns are in the middle of the curve. That is, the actual return is equally likely to be above or below that value [2].

Despite their widespread use, decision trees have a reputation for only being able to address certain types of issues, but there are many situations in which Monte Carlo simulation is applicable. Therefore, rather than just mentioning a few programmes, we give each their own area. To summarise, Monte Carlo simulation is used to provide answers to issues like [3]:

How likely is it that you will lose money?

What is the likelihood of going over budget?

How likely is it that we will finish the well prior to the expected arrival of the icebergs?

The MCS is a design of a stochastic model, which simulates in all essential physical and mathematical aspects. Basically, the method is one of numerical integrations [4]. Some experts apply the name Monte Carlo only to cases that are best illustrated by the use of probabilistic techniques to solve nonstatistical mathematical problems, other experts reserve the Monte Carlo designation only for calculations implementing sophisticated variance-reducing techniques [5]. Statistical sampling procedures and ordered numerical observations of a stochastic nature included in the MCS.

14.2 Methodology

14.2.1 Dice Theory (Contextual Approach)

The possible outcomes of a rolling dice are 6 if the dice has numbers from 1 to 6 [6]. The probability of each number is 1/6 as shown in the figure.

The conditional normal distribution can be calculated as:

The probability can be measured as $P(X = n)$ where $n \in [1, 6]$

The probability of outcome with "1" on dice $= \frac{1}{6}$

The probability of outcome with "2" on dice $= \frac{1}{6}$

The probability of outcome with "3" on dice $= \frac{1}{6}$

The probability of outcome with "4" on dice = $\frac{1}{6}$
The probability of outcome with "5" on dice = $\frac{1}{6}$
The probability of outcome with "6" on dice = $\frac{1}{6}$

The cumulative distribution can be calculated as:

The probability can be measured as P(X < = n) where n € [1, 6]
The cumulative probability of outcome with "1" on dice = P(X <= 1) = $\frac{1}{6}$
The cumulative probability of outcome with "2" on dice = > P(X <= 2) = P(X = 1) + P(X = 2) = $\frac{1}{6} + \frac{1}{6} = \frac{2}{6}$
The cumulative probability of outcome with "3" on dice = $\frac{1}{6} + \frac{1}{6} + \frac{1}{6} = \frac{3}{6}$
The cumulative probability of outcome with "4" on dice = $\frac{1}{6} + \frac{1}{6} + \frac{1}{6} + \frac{1}{6} = \frac{4}{6}$
The cumulative probability of outcome with "5" on dice = $\frac{1}{6} + \frac{1}{6} + \frac{1}{6} + \frac{1}{6} + \frac{1}{6} = \frac{5}{6}$
The cumulative probability of outcome with "6" on dice = $\frac{1}{6} + \frac{1}{6} + \frac{1}{6} + \frac{1}{6} + \frac{1}{6} + \frac{1}{6} = \frac{6}{6} = 1$.

Estimation of Normal Distribution Function

For every iteration, there are control, dependent and independent variables exist. Independent variables are like random variables where the pattern is unrecognizable. Dependent variables are dependent on them.

According to Table 14.1. A is controllable, B is independent and C is dependent variable.

Normal distribution can be estimated as:

Excel function = NORM.DIST(C (dependent variable), mean, standard deviation, FALSE)

Probability density curve

Mass function (Fig. 14.1).

P(x <= 4) = 28%.

Table 14.1 Collection of sample variables

A	B	C
2	9	11
2	7	9
2	5	7
2	10	12
2	8	10
2	7	9
2	7	9
2	6	8

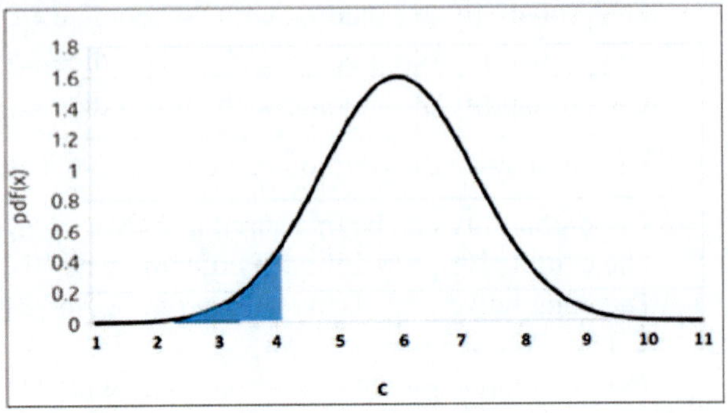

Fig. 14.1 Normal distribution curve with the values less than 4

14.3 Results and Observations

As per Table 14.2, the observed profits were dependent variables from last 5 years. The prediction of profit for next year is dependent on independent variable. MCS technique can be applied directly to dependent variable without considering independent variables.

Normal and cumulative distributions were estimated by standard deviation and mean for the "C" variable as shown in Fig. 14.2. The plot was extended from $ 200,000.00 to $ 1,200,000.00 on X-axis scale. There is a peak variation in the region between $ 600,000.00 and $ 800,000.00, where the average value falls in between.

As per Fig. 14.2(a), the X-axis was considered to be observed or dependent values and Y-axis as the probability function of x. The number of readings observed to be in between 400,000.00 and $ 1,000,000.00, which comes under the area of "Bell" curve. The most likely outcomes for the year 2023 could be in between $ 600,000.00

Table 14.2 Energy data in the form of profits at each date

Time years	A (control variables)	B (independent variables)	C (dependent variables)
2011	10,500	$ 94.88	$ 996,240.00
2012	10,500	$ 94.05	$ 987,525.00
2013	10,500	$ 97.98	$ 1,028,790.00
2014	10,500	$ 93.17	$ 978,285.00
2015	10,500	$ 48.66	$ 510,930.00
2016	10,500	$ 43.29	$ 454,545.00
2017	10,500	$ 50.80	$ 533,400.00
2018	10,500	$ 65.23	$ 684,915.00
2019	10,500	$ 56.99	$ 598,395.00
2020	10,500	$ 39.68	$ 416,640.00
2021	10,500	$ 68.17	$ 715,785.00
2022	10,500	$ 101.62	$ 1,067,010.00

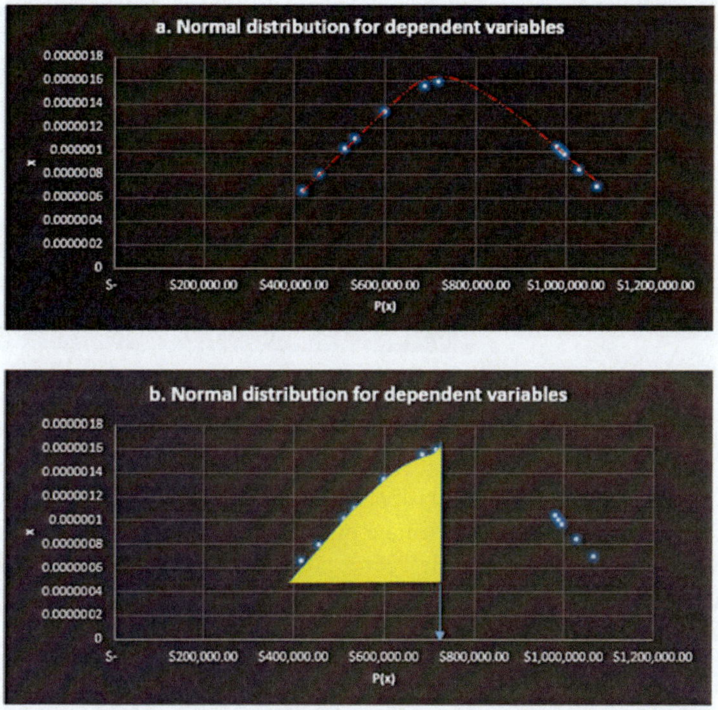

Fig. 14.2 Normal distribution of profits as dependent random variables **a** and with specific values **b**

and \$ 800,000.00, that is the region under "Bell" curve near mean. Figure 14.2(b) represents the area of outcomes covered under "Bell" curve until mean. It shows that the area is increasing while moving towards the mean on probability axis and then decreasing.

If the lower value of profit is subset of higher, then cumulative distribution is observed as shown in Fig. 14.2(b).

Figure 14.2 represents the values of profits, which are independent to each outcome. These values are not linked with one another. The peak observed to fall after \$ 700,000.00 shows that the maximum chance of attaining profit for next iteration could be near to that to estimate the profits, which falls below \$ 700,000.00 can be measured:

P (x < = \$ 700,000.00) = the area covered below \$ 700,000.00 as shown in Fig. 14.3(b).

For cumulative estimation:

Figure 14.3 represents the cumulative distribution for the case above if the values are dependent on each other. The three categories are subdivided into P10, P50 & P90.

P10 and P90 are considered to have less chances of probability. The most likely outcome of probability is considered to be P50. According to Fig. 14.3, P10 and P90

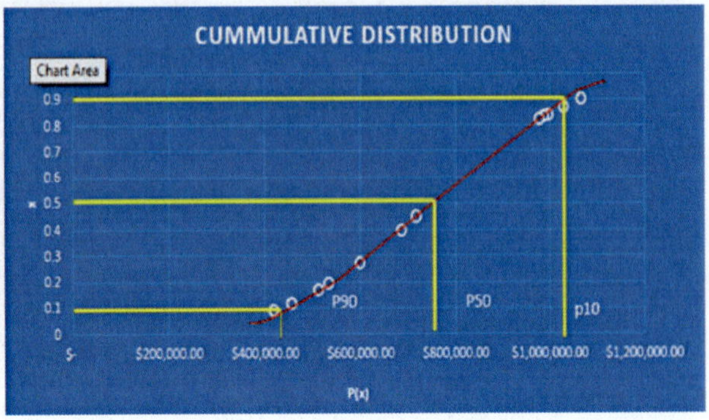

Fig. 14.3 Cumulative distribution of profits as random variables

were observed at $ 500,000.00 and $ 1,100,000.00, respectively. P50 is observed at $ 700,000.00.

The normal and cumulative distributions are matching at $ 700,000.00, which can be considered as the most approximate profit that can be expected in 2023.

The gaps between the area of Bell curve are due to lack of enough data. The number of variables taken here are limited to tens. The same approach can be simulated for thousands and millions for variables for accurate prediction.

14.4 Conclusion

MCS is one of the most reliable tools applied for financial risk assessments for major energy domains, where the scale of risk will be high. The most likely results will be near to the middle of the bell curve generated by normal distribution. The profits of any energy sector are sensible to the price of that product in the market. Fluctuations in the prices could affect the revenue of overall production. Fiscal predictions will become more significant.

For the current investigation, price fluctuations per year were considered as variables to project profit estimations using MCS. Results were observed with the limited data. Another sensitive factor for accurate assessment of variables through MCS requires more data. These assessments can be applied for any energy sector with price and demand fluctuations.

References

1. P. Zakian, N. Khaji, A stochastic spectral finite element method for wave propagation analyses with medium uncertainties. Appl. Math. Model. **63**, 84–108 (2018)
2. P. Zakian, N. Khaji, A stochastic spectral finite element method for solution of faulting-induced wave propagation in materially random continua without explicitly modeled discontinuities. Comput. Mech. **64**(4), 1017–1048 (2019)

3. H.W. Huang, L. Xiao, D.M. Zhang, J. Zhang, Influence of spatial variability of soil Young's modulus on tunnel convergence in soft soils. Eng. Geol. **228**, 357–370 (2017)
4. M.J. Cassidy, M. Uzielli, Y. Tian, Probabilistic combined loading failure envelopes of a strip footing on spatially variable soil. Comput. Geotech. **49**(49), 191–205 (2013)
5. H.Z. Cheng, J. Chen, Z.F. Hu, J.H. Huang, Face stability analysis for a shield tunnel considering spatial variability of shear strength in sand. Rock Soil Mech. **39**(8), 3047–3054 (2018)
6. H. Cheng, J. Chen, R. Chen, G. Chen, Y. Zhong, Risk assessment of slope failure considering the variability in soil properties. Comput. Geotech. **103**, 61–72 (2018)

Chapter 15
Pine Needles as Filter Loss Agent for Water-Based Mud

Nihal Ali, Shri Ram Manda, and Rose Havilah Pulla

Abstract Drilling operations and successful completion of an oil well depend on the properties of a drilling fluid. The traditional process of preparing drilling mud involves a variety of chemical additives that are non-biodegradable and involves cost whereas biodegradable waste material, and agricultural residue are better suited as an alternative to chemical additives. There are different ways to evaluate the performance of drilling mud. One method of evaluating a drilling mud performance is to prevent mud filter loss. The fluid loss control additive can be evaluated by the filter cake characterization. The Indian pine needles (Pinus) were chosen as the agro waste-based material as fluid loss control additive for the experimental study as it is abundant and renewable. To ensure that the results were acceptable, different weights and sizes of pine needles were simulated, and the sieved product is considered. The mud sample is formulated and the filtration characteristics of the mud samples were estimated using a Low Pressure Low Temperature (LPLT) filter press in accordance with the API-recommended practice for field testing for Water-based muds. The results revealed that the pine needle solution reduced filter loss and thin cake is formed.

Keywords Drilling mud · Pine needle · Filtration loss · Cake thickness

15.1 Introduction

Drilling mud production is widely recognized as the most important of all fluid mixtures. Drilling operations heavily rely on the selection of the drilling fluid. In drilling operations, the drilling fluid plays several essential functions [1]. The properties of the mud if estimated incorrectly will definitely affect the safety and economics of the process. For the safe, efficient and cost-effective drilling of oil wells, the drilling fluid is the single most important system [2]. Some companies simply ignore

N. Ali · S. R. Manda · R. H. Pulla (✉)
Chemical Engineering Department, University of Petroleum and Energy Studies, Dehradun, Uttarakhand, India
e-mail: rosehavilah@ddn.upes.ac.in

A. K. Sahu et al. (eds.), *Energy Storage and Conservation*, Springer Proceedings in Energy, https://doi.org/10.1007/978-981-99-2870-5_17

the relationship between mud quality and final return on investment in order to find the cheapest (rather than the most appropriate) mud system, even if the overall cost of the well is much higher as a result. A viscous fluid mixture used in drilling operations is called as drilling mud and is used to carry rock cuttings to the surface while also lubricating and cooling the drill bit. Drilling muds are categorized into three types: oil-based muds (OBM), water-based muds (WBM) and formate brines [3]. Over oil-based drilling muds, water-based muds are preferred due to environmental regulations and their usage in areas where oil-based drilling fluids were previously used due to their difficulties [4]. The main reason for an effective drilling operation is the role of the drilling mud, and the prepared mud should have less filtrate volume, less permeability, and a thin filter cake. The liquid that flows through the medium, leaving the cake behind, is known as the mud filtrate. To isolate formation from drilling fluids, a certain amount of filter cake buildup is required. The filtration that occurs in the initial stages is referred to as spurt loss, and once the mud cake has formed, the filtrate loss is continuous and detrimental to drilling operations [5]. When drilling a well, thin, low permeability cakes are preferred, whereas thick filter cakes cause problems such as differential pressure sticking, lost circulation, logging issues, and high torque and drag issues. Drilling fluid additives are used during formulation to achieve a variety of goals, including viscosity control, rheology control additives, and pH control additives. Filter loss control additives are a type of mud additive that is designed to reduce the amount of filtrate that passes through the filter medium. Filtration additives include bentonite, polymers, starch and its derivatives, cellulosic materials such as polynomic cellulose, and thinners [6]. Environmentalists, including various government and non-government organizations, are always concerned about the conventional drilling mud system because the drilling industry is forced to use immense amounts of chemicals as additives. These chemicals are extremely harmful to both humans and the environment [7]. The current significance of any drilling program is the progress of environmentally friendly drilling fluids, so that the ecosystem is conserved to the greatest extent possible, thereby inhibiting the damage of marine bodies such as fish, coastal areas, and oceans, among others, as well as the terrestrial environment such as plant pollution. In recent years, agro residue has taken on a broader dimension, attracting considerable attention from researchers [8]. The main benefits of using agricultural waste are its less cost, biorenewability, and accessibility. Various authors conducted studies on the usage of various resources as drilling fluid loss control agents at various times. These include starch, tree barks, peanut shells, ground peach seeds, grass, husk, and sugar cane bagasse [9]. Azizi et al. investigated the filter loss using agarwood waste and discovered that it has the prospective to be used as a filter loss additive [10]. Cassava starch was used as a filter loss additive by Egan and Abah et al. [11]. Okn et al. investigated the use of rice husk in reducing filter loss to levels equivalent to polyanionic cellulose (PAC) and carboy methyl cellulose (CMC) [12]. Anawe Paul et al. investigated rice husk and sawdust as filter loss additives and achieved a desirable mud cake thickness [13]. Hossain and Wajheeuddin investigated the use of small grass particles as a filter loss additive and discovered that they are suitable for good filter loss control [14]. Onuh concentrated the study on formulating water-based drilling mud with corncobs and coconut shells as additives to improve mud performance, demonstrating that corncob is a better fluid loss control agent than coconut shell [15]. Agwu et al. studied the filtration characteristics of rice husk and sawdust and discovered that the selected cellulosic materials are favorable filtration loss control additives for water-based drilling muds [16]. The present study aims to consider the consequence of locally available, biodegradable, and environmentally friendly agro-based materials. Uttarakhand is a state in northern India that lies at the foothills of the Himalayan mountain ranges. It has a total land area of 53,483 km^2, of which 86% is mountainous and 65% is forested [17]. In India, the largest pine forests are found in the state of Uttarakhand. The Uttarakhand Renewable Energy Development Agency (UREDA) reports that the state's Chir Pine Forest covers 399,329 ha (16.36 percent of total forest area), with an annual gross availability of pine needles ranging from 1.37 to 2.37 million tons. As a result, pine needles may be a better option for use as an agro-based material for filter loss control [18].

15.2 Materials and Methods

This research and study entail the following steps: (1) collection of pine needles, (2) physical treatment (drying, processing to powder), (3) preliminary and final analysis of the pine needles, (4) mud formation, (5) evaluation of filtrate volume and filter cake features of the mud prepared with pine needle powder as filter loss control agents. Oven, weighing balance, measuring cylinder, beakers, Hamilton beach mixer, spatula, mixer, grinding machine, and low pressure low temperature (LPLT) filter press dead weight assembly were all used. Bentonite, water, and pine needle powder are the raw materials used. The dry pine needles were collected from the Vikasnagar village panchayat in Dehradun as shown in Fig. 15.1.

The pine needle samples were manually cleaned to remove small twigs, dried leaves, and other unwanted materials. According to ASTM standards, the moisture

Fig. 15.1 Pine needle samples

Fig. 15.2 **a** Before sieving **b** After pulverization

content, volatile matter, ash, and fixed carbon were calculated. The additives were then dried up in a drying oven for 2 h to remove moisture before being cleaned and cut into small pieces. A grinding machine was used to grind these small pieces and then pulverized as shown in Fig. 15.2a, b.

Pine needles as additives were used to create a water-based drilling mud based on the API standard, using 22.5 gm of treated bentonite, 3.5×10^{-4} m^3 of water, and NaOH (sodium hydroxide and barite) as a pH control agent and densifier. Taking four samples with weights of 5, 10, 15, and 20 gm and mixing them in a Hamilton Beach Commercial high-speed mixer (Model 550) and a mixer cup. As per the mixing order shown in Table 15.1**,** the mud sample was prepared and mixed at low speed for 10 min, so that a homogeneous mixture was prepared and to prevent mud loss.

The first step is to create a CTRL MUD with no Pine Needle. In the following steps, with varying weights of pine needles, mud is to be prepared in the order listed in the table, followed by the time in seconds. The 8 entities are used to formulate a sample with pine needles weighing from 0.005 to 0.02 kg. Pine needle mud is the mud obtained from these 0.005 to 0.020 kg pine needle samples. Before conducting further testing, the prepared drilling mud samples were kept at room temperature for 24 h. Table 15.2 shows the API specifications for conventional fluid loss control additives.

Table 15.1 Order of mixing of materials

Order	Materials	Amount	Mixing time (s)	Purpose
1	Water	3.5×10^{-4} m^3		Base liquid
2	Bentonite	0.0225 kg	900	Viscosifier
3	NaOH	0.0225 kg	600	pH control
4	Pine Needle (PN1)	0.005 kg	600	Fluid loss control agent
	Pine Needle (PN2)	0.010 kg	600	Fluid loss control agent
	Pine Needle (PN3)	0.015 kg	600	Fluid loss control agent
	Pine Needle (PN4)	0.020 kg	600	Fluid loss control agent
5	Barite	0.01 kg	600	Densifier

Table 15.2 API Specifications for fluid loss additives

Parameter	API specification
API filter loss	For Carboxy methyl cellulose (CMC) 1×10^{-5} m^3 and for polyanionic cellulose (PAC) 2.3×10^{-5} m^3
Filter cake thickness (Drilling formulas, 2016)	< 2 mm

The LPLT filter press dead weight assembly experiment for filtrate volume estimation was carried out in accordance with API RP 13B-1, Recommended Practice for Field Testing Water-Based Drilling Fluids as shown in Fig. 15.3. In a static filtration test, a low pressure low temperature (LPLT) filter press with a width of 9 in, depth of 9 in, and height of 20 inches is used to measure fluid loss and filter cake characteristics of drilling fluids. The cell is filled to within 6 mm of the top with mud sample. The cell body's bottom is lined with filter paper. When all of the connections are tightened, a dead weight pressure gauge delivers 100 psi to the cell's top. After 30 min, the filtrate is collected in a graduated cylinder which was placed beneath the cell. Disassemble the filter cell by removing the filter cake and filter paper from the base cap. Remove any excess drilling fluid from the cake and use a scale to measure and record the filter cake thickness to the nearest millimeter. The filtrate collected and the cake deposited is shown in Fig. 15.4a, b.

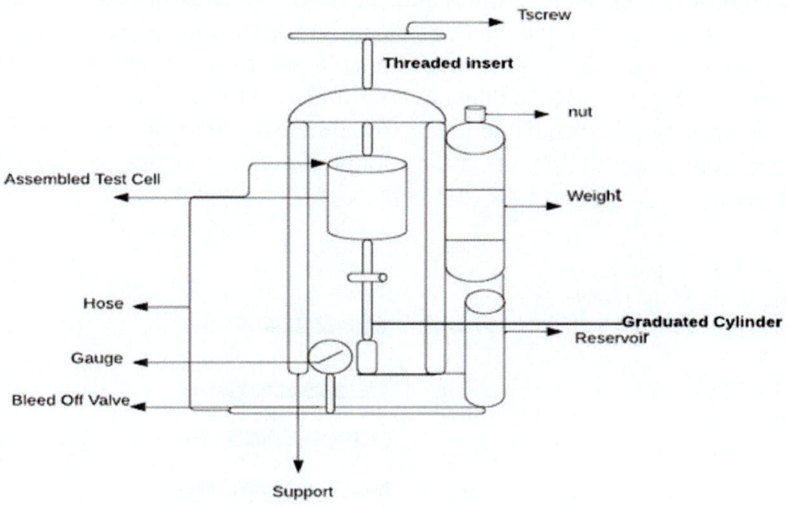

Fig. 15.3 Layout of Filter press deadweight assembly

Fig. 15.4 **a** Filtrate volume **b** Filter cake

15.3 Results and Discussions

Figures 15.5 and 15.6 present the filter loss volume and mud cake thickness obtained from the drilling fluid samples. CTRL mud has a filter loss of 32 ml, and as the concentration of pine needle powder increased from 0.005 kg to 0.02 kg, the filtrate loss decreased from 18.8 ml to 16 ml. This depicts that as the content of the local additive in the drilling fluid samples increases, there was a significant decrease in the filter loss volume from the drilling mud samples. As the additive content increases, the cellulose content in the drilling fluid increases, which attributes to the low filtrate volume. From Fig. 15.4**,** the filter cake deposited on the filter paper is soft rather than dry, which shows that the pine needle mud would avoid differential pipe sticking. Figure 15.6 depicts the cake thickness (millimetres) for the formulated water-based mud at varying pine needle contents. The thinness of the cake was measured and ranged from 1 mm to 3.2 mm as the additive content increases, the cake thickness increases.

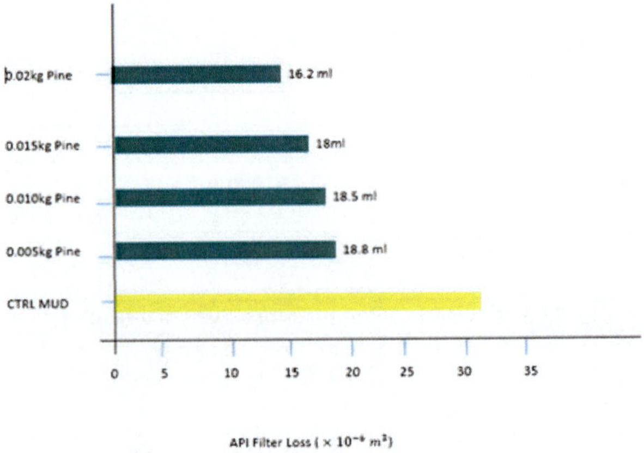

Fig. 15.5 Filter loss volume for the drilling fluid samples at varying content of pine needle

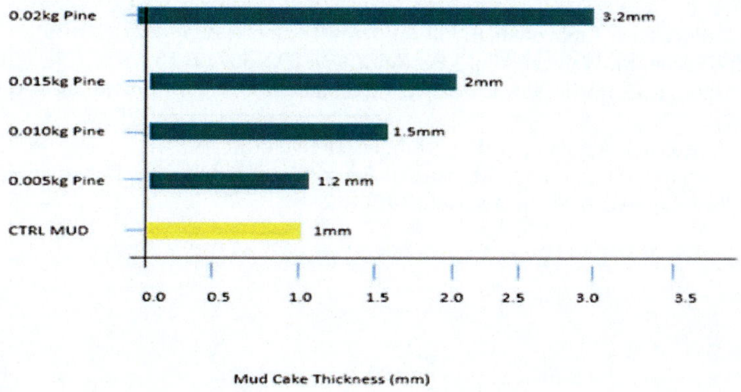

Fig. 15.6 Mud Cake thickness for the drilling fluid samples at varying content of pine needle

15.4 Conclusions

This study observed the impact of using pine needle powder as a fluid loss and loss circulation agent in water-based muds. Because they are renewable, environmentally friendly, and abundant in nature, pine needle powder promises to be an encouraging fluid loss control additive for water-based muds. As the amount of pine needle powder in the drilling mud increases, the filtrate loss control properties are more apparent. The experimental results investigated that as the weight of the pine needle powder in the water-based mud increases, then the filtrate volume decreases and the filter cake thickness increases, which shows the inverse proportionality between the filtrate volume and the cake thickness.

References

1. J.M. Neff, S. McKelvie, R.C. Ayers, J. Hazard Mater. Adv. (2022)
2. A. Agi, J.O. Oseh, A. Gbadamosi, C.K. Fung, R. Junín, M. Zaidi Jafar, J. Pet .Res. (2022)
3. H. Jia, P. Huang, Q. Wang, Y. Han, S. Wang, F. Zhang, W. Pan, K. Lv, Fuel **244**, (2019)
4. R. Caenn, H.C. Darley, G.R. Gray, Gulf Professional publishing (2011)
5. R. Borges, B. Oechsler, B.R. Olivia, L.D.R. Andrade, L.A. Calcada, C. Scheid, V. Calado, Powder Technol. **386**, (2021)
6. International association of drilling contactor (IADC), TX, 2015/IADC UBO/MPD glossary. I.D. Manual Global Standards (2011)
7. M. Al-Saba, K. Ahmadi, K.O. Al-Hadramy, M.A. Dushaishi, A.T.T. Al-hameedi, H. Alkinani, Environment, and social responsibility. Soc. Petrol. Eng. (2018)
8. F. Dias, R. Souza, E. Lucas, Fuel **140** (2015)
9. S.P.B. Almagro, C. Frates, J. Garand, A. Meyer, Oilfield Rev. Autumn **26**(3) (2014)
10. A. Azizi, A., Mohd. S.N. Ibrahim, K.H.K. Hamid, A.S.N. Ghazali, T.A.T. Mohammad, Int. J. Sci. Eng. **5**, (2013)
11. I.L. Egun, A.M. Abah, Discov. J. **3**(9) (2013)
12. A.N. Okon, F.D. Udoh, P.G. Bassey, *Nigerian Annual International Conference and Exhibition*, Lagos (2014)

13. A.L. Anawe Paul, Efeovbokhan. Int. J .Energy Technol. Policy **4**(2), (2014)
14. M.E. Hossain, M. Wajheeuddin, J.Pet.Sci. **13**(2), 292–303 (2016)
15. C.Y. Onuh, K.C. Igwilo, P.A.L. Anawe, O. Daramola, O. Ogunwomoju, Int.J.Appl.Eng. **12**(8) (2017)
16. O.E. Agwu, J.U. Akpabio, G.W. Archibong, Heliyon (2019)
17. Forest Survey of India (FSI), Ministry of Environment and Forests, Dehradun, India
18. http://www.ureda.uk.gov.in Accessed 19 July 21

Chapter 16
Generation of Green Hydrogen Using Semiconductor-Based Nanomaterials

Ajay Mittal and Rajeshwar Mahajan

Abstract Photocatalytic water splitting using semiconductors-based nanomaterials has recently attracted attention from scientists and businesses for its potential to produce hydrogen in a sustainable manner. This article discusses the mechanism of photocatalytic water splitting and production of green hydrogen using semiconductor-based nanomaterials, different approaches for structural modification for better efficiency and the challenges in the selection of best semiconducting material. According to the study, oxide-based semiconductors are preferred to other semiconductors because they are abundant, affordable, and stable, but their low efficiency limits their commercial application. The fundamental difficulty can be overcome by combining binary, ternary, or quaternary photoelectrode materials. This research will contribute to the creation of stable tertiary or quaternary photoelectrode materials, which will enhance the efficiency of green hydrogen production.

Keywords Green hydrogen · Semiconductor · Water splitting · Photoelectrolysis

16.1 Introduction

In recent years, green hydrogen's potential as a renewable energy source has risen to a higher level of attention because of its low environmental impact [1, 2]. Green hydrogen can be produced in a sustainable manner through photocatalytic water splitting using nanomaterials based on semiconductors [3, 4]. As shown in Fig. 16.1, lots of semiconducting materials like TiO_2, Fe_2O_3, WO_3, CdS, Ta_2O_5, $BiVO_4$, ZnO, Cu_2O and many more have been the focus of extensive study over the past decade because of their suitable band gap, stability, and efficient response to solar irradiation in the creation of hydrogen gas but lack mass scale production due to low efficiency due to their low reactivity to visible light, low reduction potential, and high charge carrier recombination rate [5, 6].

A. Mittal (✉) · R. Mahajan
Energy Cluster, School of Engineering, University of Petroleum & Energy Studies, Dehradun, Uttarakhand 248007, India
e-mail: ajay.mittal@ddn.upes.ac.in

A. K. Sahu et al. (eds.), *Energy Storage and Conservation*, Springer Proceedings in Energy, https://doi.org/10.1007/978-981-99-2870-5_18

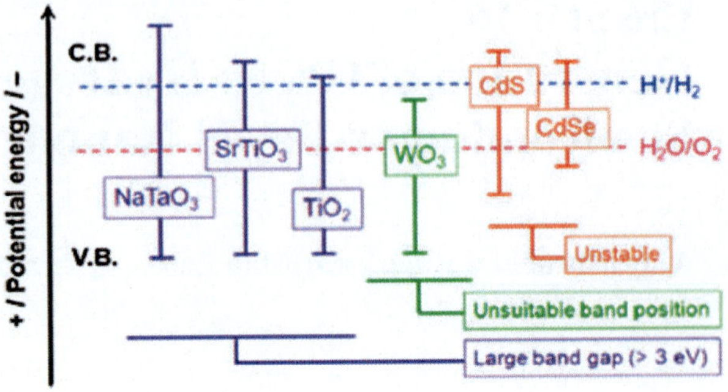

Fig. 16.1 Band structure of most used semiconductors for photocatalytic water splitting [7]

Semiconductor materials with band gaps that do not match the water splitting redox potentials are ruled out by the redox potential criterion unless the structure is adjusted. TiO_2, ZnO, Cu_2O have attracted the most attention from researchers interested in structural modification because of its low toxicity, excellent stability, and near-visible light photocatalytic activity [8, 9]. This article discusses the mechanism of photocatalytic water splitting and production of green hydrogen using semiconductors-based nanomaterials, different approaches for structural modification for better efficiency and the challenges in the selection of best semiconducting material. Additionally, this article facilitates the development of hybrid oxide–semiconductor materials, thereby facilitating future research that is more informed and fruitful.

16.2 Methods of Hydrogen Production

Hydrogen can be produced either from fossil fuels or renewable source of energy like biomass, solar energy, wind, or hydropower. The production of hydrogen from fossil fuels or biomass is not environmental-friendly and energy intensive. The splitting of water using solar energy, wind energy, or hydropower are the greener methods that can be utilized to produce hydrogen from renewable sources. The most promising technique for producing green hydrogen is water splitting using solar energy, which is also fortunate because it is less constrained by geographical location than hydropower and wind energy [10].

In the past decade, researchers utilized various techniques which water splitting solar energy for production of green hydrogen illustrated in Fig. 16.2. The photoelectrochemical (PEC) and photocatalysis water splitting technologies are the most promising methods for the creation of hydrogen. This is since both methods employ the limitless energy source of solar light and do not generate any CO_2 [6, 11, 12].

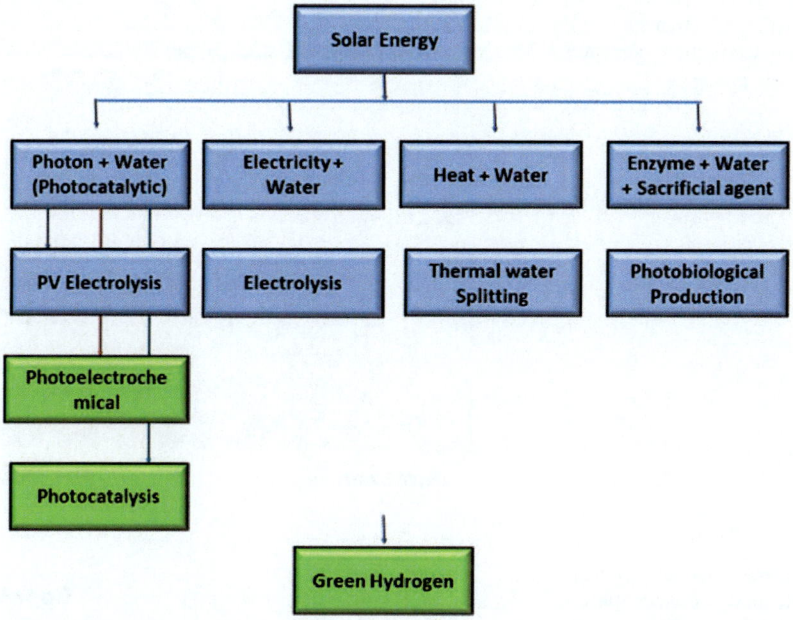

Fig. 16.2 Hydrogen generation techniques using solar energy (Created by adaptation of Fig. 16.1 from [8])

16.3 Technological Pathway of Photo-Electrochemical or Photocatalysis Water Splitting and Production of Hydrogen

Photocatalytic water splitting is photoelectrolysis of water, i.e., the absorption of solar energy and the electrolysis of water are brought together in both the photoelectrochemical and photocatalytic processes [13]. In a photoelectrochemical cell, water is reduced at the photocathode and oxidized at the photoanode, respectively [2, 11, 14–16] as shown in Fig. 16.3. In photocatalysis, reduction and oxidation reaction take place at photocatalyst and cocatalyst, respectively [11, 12, 14, 17, 18] as shown in Fig. 16.4. The reactions that take occur at the anode and cathode during water splitting are described below:

$$Anode : 2p^{+} + H_2O \rightarrow 0.5O_2 + 2H^{+}$$

$$Cathode : 2e^{-} + 2H^{+} \rightarrow H_2$$

In the process of photo electrolysis, solar irradiation with an energy equal to or greater than the band gap of the photoanode (photocatalyst) causes electrons in the valence band to be excited into the conduction band, leaving holes in the valence band. These photogenerated electrons and holes, respectively, create an oxidation

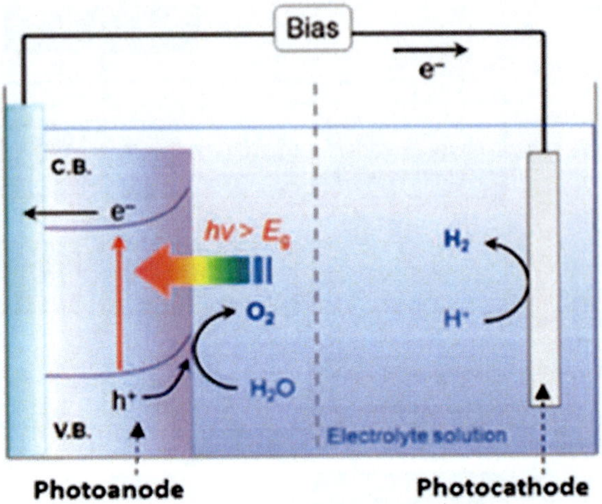

Fig. 16.3 Mechanism of photo-electrochemical water splitting process [7]

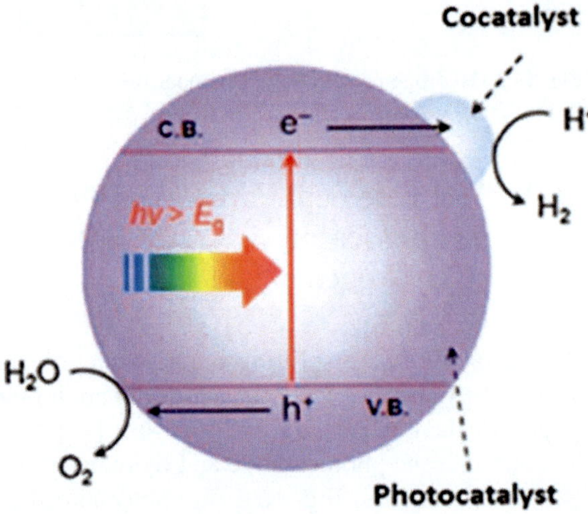

Fig. 16.4 Mechanism of photocatalysis water splitting process [7]

process at the photoanode (cocatalyst) and a reduction reaction at the photocathode (photocatalyst). To achieve water splitting, the bottom of the conduction band needs to be more negative than the reduction potential of H + to H2, which is equal to 0; conversely, the top of the valence band needs to be more positive than the oxidation potential of H_2O to O_2, which is equal to 1.23. Both conditions must be met. Therefore, the minimum photon energy that is thermodynamically necessary to drive the reaction is 1.23 eV [14, 16]. Additionally, for the water splitting process to be extremely efficient, semiconductor materials should have a bandgap that is greater than 3 eV [11].

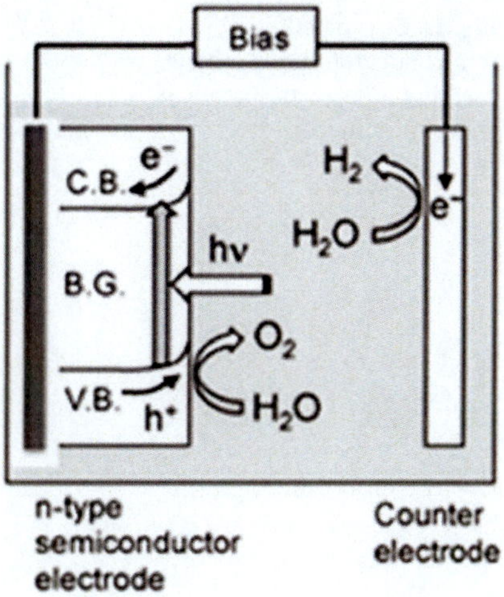

Fig. 16.5 n-type semiconductor photoanode used in photoelectrochemical water splitting [2]

Three types of photoelectrochemical frameworks are utilized by researchers that uses n-type semiconductor as photoanode (Fig. 16.5), p-type semiconductors as photocathode (Fig. 16.6), and a tandem system uses p-type semiconductors as photocathode and n-type semiconductor as photoanode (Fig. 16.7) are being developed for water splitting and production of hydrogen. n-type oxide semiconductors like TiO_2, $BiVO_4$, Fe_2O_3 are highly efficient and have frequently been used as photoanodes for the oxidation of water since they are resistant to photo corrosion and tend to be very affordable. p-type oxide semiconductors like Cu_2O are not efficient due to self-destruction [18]. Tandem system like p-GaP/n-TiO_2 [8] developed by researcher was inefficient but tandem system has promising potential for efficient photocatalytic water splitting but need structural modifications [8].

16.4 Semiconductor-Based Nanomaterials for Production of Green Hydrogen

In the past decade, chalcogenide, carbide, nitride and oxide semiconductors are extensively research for water splitting and production of hydrogen. Chalcogenide, carbide, nitride semiconductors found limited attention due to photo corrosion and chemical instability in water solution [8]. Metal oxide semiconductors become the preferred choice over other semiconductor materials for water splitting and production of hydrogen, due to their earth abundance and stability in water solution but the main challenge lies to meet the criteria suitable for effective water splitting like band-edge potentials suitable for overall water splitting, band-gap energy lower than 3 eV, stability in the photocatalytic reaction [7, 8]. Many oxide–semiconductor materials have been the subject of extensive study, but those whose band gap structures do not match the water splitting redox potentials are ruled out based on this

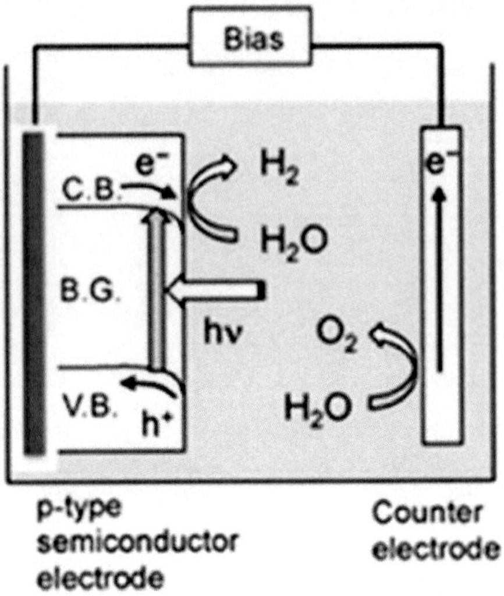

Fig. 16.6 p-type semiconductor photocathode used in photoelectrochemical water splitting [2]

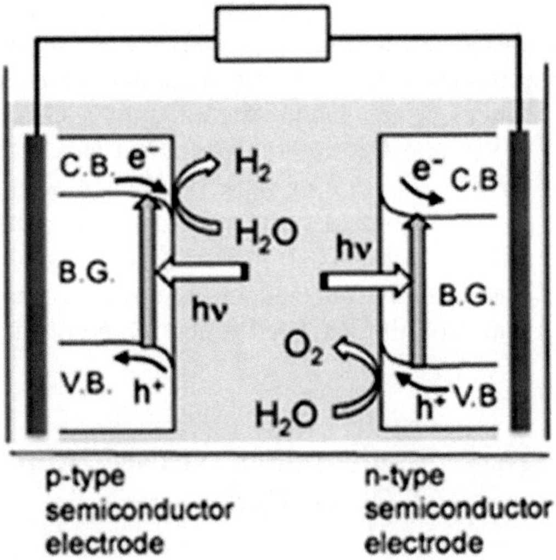

Fig. 16.7 Tandem system used in photoelectrochemical water splitting [2]

criterion until the structures are adjusted [11]. Thus, the focus of current studies is on developing improved methods for modifying oxide catalysts. For photocatalytic water splitting, researchers are investigating five ways for improving the hydrogen production efficiency of an oxide–semiconductor material. These include changes to the form and structure of crystals, doping a semiconductor with metallic and/ or nonmetallic ions, the sensitivity produced by quantum dots in semiconductors, the development of stable heterojunctions or solid solutions, and the insertion of a cocatalyst into a process [11]. Developing effective composite photocatalysts for solar water splitting using metal–organic framework (MOF) materials like Ti-based

MOF (MIL-125-NH_2) achieved optical bandgap shifts towards visible light region (2.6 eV/475 nm) [16], Ti-MOF/TiO_2 composite semiconductors attains an improved photocurrent density by 50% in comparison to p-TiO_2 [16], the photocurrent density achieved by Ar-TiO_2 is over 500 times higher than that of p-TiO_2 [11]. Despite several years spent doing research, scientists have not been successful in identifying a single binary metal-oxide semiconductor that is superior to all the others. In this regard, it is vital to take into consideration semiconductor materials that have three or four components and are based on metal oxides. It is predicted that by combining three or four components in a variety of different ways and subsequently adjusting the bandgap structure, it will be possible to find the optimum photoelectrode material with outstanding photocatalytic efficacy for splitting water.

16.5 Challenges in Generation of Nanomaterials for Photocatalytic Water Splitting

16.5.1 Low Photoconversion Efficiency

Solar-to-hydrogen conversion efficiency using photocatalytic method for all the reported semiconductor-based nanomaterials is low and limit their use for commercial application. Also, there is a need to have a comprehensive method to calculate solar-to-hydrogen conversion efficiency [8, 19, 20].

16.5.2 Photocorrosion or Corrosion Stability

Most of the semiconductors such as CdS and SiC are not corrosion resistant and not suitable for water splitting [21] and limit their applicability. Proper method needs to be devised to calculate corrosion stability to avoid using semiconductors, which leads to photo corrosion [11, 14].

16.5.3 Chemical and Thermal Stability

Chemical and thermal stability of semiconductor materials are also crucial for improved photoconversion efficiency and commercial application.

16.5.4 Photostability

High photostability helps in semiconductor material resistant to change under the influence of solar energy and leads to increase in photoconversion efficiency [8, 14, 22].

16.5.5 Low Cost

Cost of the semiconductor material should be low for mass-scale production and widespread application [22, 23].

16.5.6 Non-toxic

Semiconductor material should be environmentally friendly and possess no harm to human beings [10, 13].

16.6 Conclusion and Future Perspective

Due to their natural abundance, high and tunable bandgaps, stability in aqueous conditions, low cost, and ease of manufacturing, semiconductor-based nanomaterials have shown remarkable potential to produce green hydrogen through photocatalysis and photoelectrical water splitting methods. The main challenge lies in achieving optimum properties of semiconductor like effective absorption of a wide spectrum, photochemical stability, low cost, low recombination possibility, efficient charge carrier, which is only possible through combination of semiconductor-based hybrid nanomaterials. Recent research in the production of green hydrogen through two-component metal oxide–semiconductor materials has shown increased efficiency of hydrogen production but lack commercial application. Even with current innovations, market viability still requires further enhancements in efficiency, durability, and cost. More research into developing three- and four-component photocatalytic semiconductor materials is needed to attain high efficiency, stability, and cheap cost in photocatalytic water splitting for green hydrogen production.

Acknowledgements I would like to thank the University of Petroleum & Energy Studies for providing all the necessary assistance for the completion of this project.

References

1. M.B. Tahir, G. Nabi, N.R. Khalid, Enhanced photocatalytic performance of visible-light active graphene-WO3 nanostructures for hydrogen production. Mater. Sci. Semicond. Process. **84**, 36–41 (2018). https://doi.org/10.1016/J.MSSP.2018.05.006
2. M.B. Tahir et al., Semiconductor based nanomaterials for harvesting green hydrogen energy under solar light irradiation. Int. J. Environ. Anal. Chem. **101**(14), 2255–2271 (2021). https://doi.org/10.1080/03067319.2019.1700970
3. J. Xie et al., Simple preparation of WO3–ZnO composites with UV–Vis photocatalytic activity and energy storage ability. Ceram. Int. **40**(8), 12519–12524 (2014). https://doi.org/10.1016/J.CERAMINT.2014.04.106
4. Q. Xue, Y. Liu, Q. Zhou, M. Utsumi, Z. Zhang, N. Sugiura, Photocatalytic degradation of geosmin by Pd nanoparticle modified WO3 catalyst under simulated solar light. Chem. Eng. J. **283**, 614–621 (2016). https://doi.org/10.1016/J.CEJ.2015.08.016
5. G. Xi, et al., Synthesis of multiple-shell WO3 hollow spheres by a binary carbonaceous template route and their applications in visible-light photocatalysis. Chem. A Eur. J. **18**(44), 13949–13953 (2012). https://doi.org/10.1002/CHEM.201202312
6. S. Zhang, L. Song, Preparation of visible-light-active carbon and nitrogen codoped titanium dioxide photocatalysts with the assistance of aniline. Catal. Commun. **10**(13), 1725–1729 (2009). https://doi.org/10.1016/J.CATCOM.2009.05.017
7. K. Maeda, Photocatalytic water splitting using semiconductor particles: history and recent developments. J. Photochem. Photobiol. C Photochem. Rev. **12**(4), 237–268 (2011). https://doi.org/10.1016/J.JPHOTOCHEMREV.2011.07.001
8. S.K. Saraswat, D.D. Rodene, R.B. Gupta, Recent advancements in semiconductor materials for photoelectrochemical water splitting for hydrogen production using visible light. Renew. Sustain. Energy Rev. **89**, 228–248 (2018). https://doi.org/10.1016/J.RSER.2018.03.063
9. Y. Ma, X. Wang, Y. Jia, X. Chen, H. Han, C. Li, Titanium dioxide-based nanomaterials for photocatalytic fuel generations. Chem. Rev. **114**(19), 9987–10043 (2014). https://doi.org/10.1021/CR500008U/SUPPL_FILE/CR500008U_SI_001.PDF
10. T. Iqbal, A. Riaz, M. Zafar, An overview of photocatalytic water splitting on semiconductor oxides for hydrogen production. (2021). https://doi.org/10.1080/03067319.2021.1981304
11. S. Grushevskaya, I. Belyanskaya, O. Kozaderov, Approaches for modifying oxide-semiconductor materials to increase the efficiency of photocatalytic water splitting. Materials (Basel) **15**(14), 4915 (2022). https://doi.org/10.3390/MA15144915
12. M.S. Reza, N.B.H. Ahmad, S. Afroze, J. Taweekun, M. Sharifpur, A.K. Azad, Hydrogen production from water splitting through photocatalytic activity of carbon-based materials, in *Chemical Engineering and Technology* (John Wiley & Sons Ltd., 2022). https://doi.org/10.1002/ceat.202100513
13. A.M. Abdalla, S. Hossain, O.B. Nisfindy, A.T. Azad, M. Dawood, A.K. Azad, Hydrogen production, storage, transportation and key challenges with applications: a review. Energy Convers Manag **165**, 602–627 (2018). https://doi.org/10.1016/J.ENCONMAN.2018.03.088
14. J.M. Yu, et al., High-performance and stable photoelectrochemical water splitting cell with organic-photoactive-layer-based photoanode. Nat. Commun. **111**(1), 1–9 (2020). https://doi.org/10.1038/s41467-020-19329-0
15. T. Favet, T. Cottineau, V. Keller, M.A. El Khakani, Comparative study of the photocatalytic effects of pulsed laser deposited CoO and NiO nanoparticles onto TiO2 nanotubes for the photoelectrochemical water splitting. Sol. Energy Mater. Sol. Cells **217**, (2020). https://doi.org/10.1016/J.SOLMAT.2020.110703
16. X. Li, Z. Wang, L. Wang, Metal–organic framework-based materials for solar water splitting. Small Sci. **1**(5), 2000074 (2021). https://doi.org/10.1002/SMSC.202000074
17. N. Lakshmana Reddy, et al., Nanostructured semiconducting materials for efficient hydrogen generation. Environ. Chem. Lett. **16**(3). (Springer Verlag, 2018), pp. 765–796. https://doi.org/10.1007/s10311-018-0722-y

18. R. Abe, Recent progress on photocatalytic and photoelectrochemical water splitting under visible light irradiation. J. Photochem. Photobiol. C Photochem. Rev. **11**(4), 179–209 (2010). https://doi.org/10.1016/J.JPHOTOCHEMREV.2011.02.003
19. T. Hisatomi, J. Kubota, K. Domen, Recent advances in semiconductors for photocatalytic and photoelectrochemical water splitting. Chem. Soc. Rev. **43**(22), 7520–7535 (2014). https://doi.org/10.1039/C3CS60378D
20. J.W. Ager, M.R. Shaner, K.A. Walczak, I.D. Sharp, S. Ardo, Experimental demonstrations of spontaneous, solar-driven photoelectrochemical water splitting. Energy Environ. Sci. **8**(10), 2811–2824 (2015). https://doi.org/10.1039/C5EE00457H
21. B. Radhakrishnan, M. Thirumarimurugan, R. Azarudeen, Recent developments on photocatalytic water splitting using semiconductors for hydrogen production. Int. J. Adv. Res., 2017. www.ijariit.com. Accessed 07 Aug 2022
22. G. Wang, Y. Ling, H. Wang, L. Xihong, Y. Li, Chemically modified nanostructures for photoelectrochemical water splitting. J. Photochem. Photobiol. C Photochem. Rev. **19**(1), 35–51 (2014). https://doi.org/10.1016/J.JPHOTOCHEMREV.2013.10.006
23. S.C. Wang, F.Q. Tang, L.Z. Wang, Visible light responsive metal oxide photoanodes for photoelectrochemical water splitting: a comprehensive review on rational materials design. J. Inorg. Mater. **33**(2), 173 (2018). https://doi.org/10.15541/JIM20170352

Chapter 17
Studies on Indanthrene Blue Dye Degradation Using Iron II Oxide Nanoparticles

Mathivanan Varatharajan, Murugesan Kumarasamy, and Siddharth Sampathkumar

Abstract In this manuscript, an efficient technique for successful degradation of Indanthrene blue dye using Iron II oxide nanoparticles has been reported. Iron II nanoparticles were prepared by using sodium hydroxide synthesis. Degradation pattern of Indanthrene blue dye was identified by using Ultraviolet–visible (UV–vis) spectroscopy. The rate of degradation was found to increase in acidic environment. For an initial concentration of Indanthrene blue dye of 75 mg/l (pH = 3.5) and Iron II nanoparticle addition of 1 g/l, the degradation efficiency of Indanthrene blue dye reached up to 96.8% within 6 min. The degradation constant was found to be up to 1.974 min^{-1}. Comparison studies were done regarding rate of decolorization for disperse orange 11 and alizarin dyes with indanthrene blue dye. Rate of decolorization was found to be greater for dyes with lower molecular mass. Studies were conducted by using Cobalt-based nanoparticles as dye degradation agent and the efficiency of it was evaluated with Fe II-based nanoparticles. It was found that the dye effectiveness of Fe II nanoparticle was greater than Co-based nanoparticles. The use of Fe II nanoparticle was found to be effective and it can be recycled as it is magnetic.

Keywords Dye degradation · Fe II oxide nanoparticles · Indanthrene blue dye · Disperse orange 11 · Alizarin

---

M. Varatharajan (✉) · M. Kumarasamy
Department of Environmental Science, Periyar University, Tamil Nadu, Salem 636001, India
e-mail: mathivanan999@gmail.com

S. Sampathkumar
Department of Mechanical Engineering, PSN College of Engineering and Technology, Tirunelveli, Tamil Nadu, India

A. K. Sahu et al. (eds.), *Energy Storage and Conservation*, Springer Proceedings in Energy, https://doi.org/10.1007/978-981-99-2870-5_19

17.1 Introduction

Different kinds of dyes are being used in different industries and sectors for incorporating color to cloths, leathers, etc. Indanthrene blue dye is being used in many textile and textile-based industries [1]. The wastage and effluents from textile industries affect the environment [2]. The toxic nature of the dyes affects biodiversity and pollutes the environment [3]. Hence, dye degradation is being studied and much research is being conducted to reduce the dye and degrade it from textile effluents before it is dumped into the environment [4]. Different processes are being used for degrading dyes. A few chemical and biological dye degradation methods are microbial process [5], photocatalyst method [6] and reductive dye removal [7]. In certain cases a combination of two different methods are being used for effective removal of the dyes and degrade it. For industrial purpose, reduction methods are cheap and are being extensively used [8]. For reducing industrial dyes, nanoparticles are being used. Ag-based [9] and Ni-based nanoparticles [10] have been used for successful dye degradation. For some cases, catalytic dye degradation methods have been developed.

In this manuscript, a quick degradation process has been developed for dye degradation by using Fe-based nanoparticles, which were synthesized by using sodium hydroxide synthesis. The synthesized Fe II nanoparticles were made to degrade Indanthrene blue dye. The dye degradation effectiveness was compared with two other dyes such as disperse orange 11 and alizarin dyes. In addition to that, kinetic studies were conducted by comparing the effectiveness of Fe II nanoparticles with Co-based nanoparticles.

17.2 Materials and Methods

17.2.1 Chemicals Used

In this investigation, Fe II oxide powder, sodium borohydride, hydrochloric acid, and sodium hydroxide were used. The synthesis equipments and glass beakers were cleaned using hydrochloric acid to eliminate any kind of contamination. After cleaning the equipments with hydrochloric acid, they were dried. Three dyes such as indanthrene blue, disperse orange 11, and alizarin dyes were purchased. 99.9% pure Co-based nanoparticles were purchased and used for comparing the degrading effectiveness of Fe II oxide-based nanoparticles.

17.2.2 Synthesis of Fe II Oxide-Based Nanoparticles

The Fe II oxide-based nanoparticles were synthesized by using wet synthesis method [11]. In 20 ml of deionized water, 0.5557 g of Fe II oxide was mixed at room temperature. Using magnetic stirrer, it was mixed for 5 minutes and 0.92 g of sodium borohydride powder was added to it. Then it was stirred continuously. Immediately after dissolving the sodium borohydride powder, it was placed in a freezing environment. It was subjected to vigorous stirring. Then it was centrifuged for half an hour to coagulate. Then Fe II oxide-based nanoparticles were separated by using neodymium magnet and then it was washed in acetone. Then it was dried in vacuum drying equipment. They were collected as powder and they were weighed. After weighing, they were added in quantitative manner for conducting dye degradation experiments.

17.2.3 Degradation of Indanthrene Blue, Disperse Orange II, and Alizarin

The synthesized Fe II oxide-based nanoparticles were made to degrade the three dyes such as indanthrene blue, disperse orange II, and alizarin dye. At ambient conditions, indanthrene blue dye molecular weight was 442.4. For comparing the rate of degradation, two different dyes were used for comparison. One dye was chosen to have a higher molecular weight of 450.27 for disperse orange II and another dye was chosen as a lower molecular weight dye (240.1) for alizarin. For degrading indanthrene blue dye at different conditions, experiments were conducted. For conducting degradation studies, centrifuge tube of 15 ml capacity was used. The centrifuge tube was fixed in a rotator and at low rpm it was revolved to conduct the degradation experiments. At 45 mg/l of concentration, indanthrene blue dye was added in the centrifuge tube. The pH of the solution was adjusted by using 2M conc. hydrochloric acid solution and 2 M conc. sodium hydroxide solution. After adding the indanthrene dye, Fe II oxide nanoparticles were added to it. After adding the Fe II oxide nanoparticles, the reaction was identified at different time intervals. Till 2 hrs, the reaction was monitored by using Ultraviolet Visible spectroscopy. At the specific time interval, 75 μL of the solution was extracted and subjected to UV-Vis spectroscopic analysis. Before subjecting the specimen to UV-Vis spectroscopic analysis, it was centrifuged at high speeds for 30 seconds to halt the degradation process. 75 μL dye solution was diluted to 1.5 ml by using deionized water [12]. The interval of samples was adjusted as to obtain proper kinetics curves. Depending on the initial concentration of the dyes, pH value of the dyes and type of nanoparticle used, the kinetics was adjusted.

17.2.4 Characterization and Equipment Used

The microstructural aspects of the Fe II oxide-based nanoparticles were evaluated by using Field Electron Scanning Electron Microscopy. The FE-SEM equipment was attached with energy dispersive X ray spectrometer (EDS). EDS analysis was conducted to identify the elemental composition of the Fe II oxide nanoparticles. Fourier Transform Infrared Ray (F-TIR) Spectroscopic analysis was conducted on the synthesized Fe II nanoparticles, procured Cobalt-based nanoparticles, indanthrene blue, disperse orange II and alizarin dyes. UV-Vis spectroscopic analysis was conducted by using Fischer Scientific UV-Vis spectrometer.

17.3 Results and Discussion

17.3.1 Fe II Oxide Nanoparticles

Microstructural evaluation of Fe II nanoparticles has been conducted by using FE-SEM imaging. Figure 17.1a shows FE SEM image of Fe II oxide nanoparticles. On evaluating the FE SEM image of Fe II oxide nanoparticles, their average sizes were found to be 90–115 nm. The Fe II nanoparticles exhibited round shape and structure. It was found to exhibit abrupt edges in the ends.

Figure 17.1b shows UV-Vis spectra of Fe II oxide. By evaluating the absorbance peaks, at 162.3 nm wavelength, the absorbance peaks were found to be maximum [13]. This UV-Vis spectroscopic evaluation was used to identify the presence of the desired nanoparticles. Figure 17.1c shows F-TIR image of Fe-II oxide nanoparticles. FT-IR peaks of Fe II oxide nanoparticles exhibited absorbance peaks at 3439 cm^{-1}, 2139 cm^{-1} and 1463 cm^{-1} [13]. Figure 17.1d shows EDS spectrum of Fe II oxide nanoparticles, indicating the presence of Fe. The EDS spectrum indicates that the Fe nanoparticles are pure [13].

17.3.2 Evaluating Degradation Efficiency

Indanthrene blue dye is an organic compound made using potassium hydroxide treated 2-aminoanthraquinone. On reducing the pH value, the absorption band of indanthrene dye indicates a red shift. It indicates delocalization of the dye molecules. Indanthrene blue dye treated with Fe II oxide nanoparticles is shown in Fig. 17.2a. Untreated due indanthrene blue dye is shown in Fig. 17.2b.

Figure 17.3 shows the (time dependent) UV-Vis absorption spectra of indanthrene blue dye. From the time dependent UV Vis absorption spectra, it was observed that the reaction and degradation have started immediately up on adding Fe II oxide nanoparticles in the indanthrene blue dye medium. At minimum time, the UV-Vis

Fig. 17.1 **a** FE SEM image of Fe II oxide nanoparticles, **b** UV–Vis spectra of Fe II oxide nanoparticles, **c** F-TIR spectra of Fe II oxide nanoparticles, **d** EDS spectra of Fe II oxide nanoparticles

(a) FE-SEM image of Fe II oxide nanoparticles

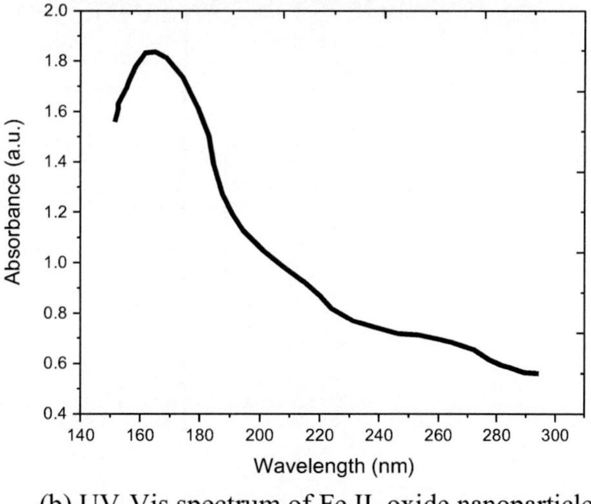

(b) UV-Vis spectrum of Fe II oxide nanoparticles

curve represents the indanthrene blue dye before adding Fe II oxide nanoparticles. On adding Fe II oxide nanoparticles at pH = 3, a sudden reduction in the absorption wavelength was observed. The wavelength in the absorption spectra was reduced by 300–45 nm. This reduction in absorption wavelength is a clear indication that degradation of indanthrene dye has taken place. Decolorization of the solution starts at a few minutes and the entire process was completed within 5 min.. Absorption peaks at 515 nm were observed. The intensity of variations reduced on prolonging the time duration. The new peaks which were formed is a clear indication that decolorization has occurred due to degradation of the dye and not by just physical adsorption alone.

Fig. 17.1 (continued)

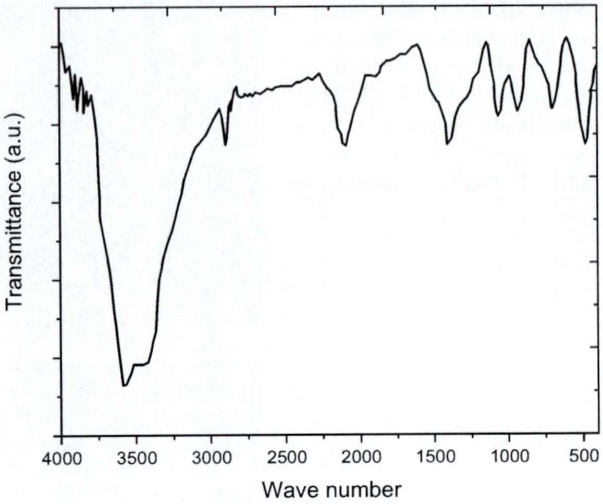

(c) F-TIR image of Fe II oxide nanoparticles

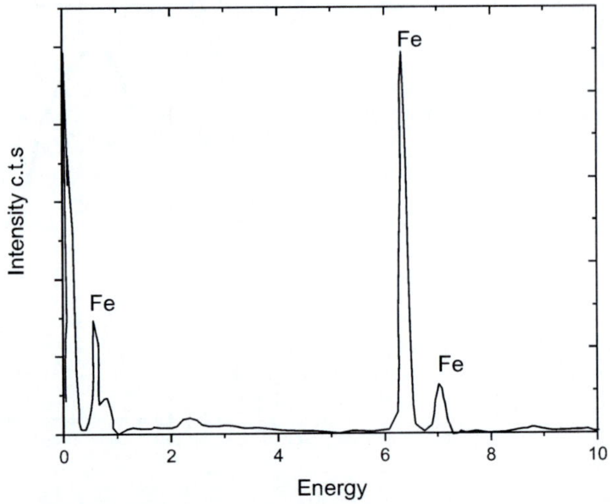

(d) EDS spectra of Fe II oxide nanoparticles

17.3.3 Degradation Mechanisms

The degradation of dyes by decolorization experiments using Fe II oxide nanoparticles was performed on two more dues such as disperse orange II dye and alizarin dye.

Variations in absorption peaks in UV-Vis spectrum of dye degradation studies of disperse orange II are shown in Fig. 17.4. Similarly, the variations in absorption peaks in UV-Vis spectrum of dye degradation studies of alizarin are shown in Fig. 17.5. On using Fe II oxide nanoparticles for degrading disperse orange II dye and alizarin dye, decolorization was witnessed. Decolorization of disperse orange II was found to be

(a) Fe II oxide nanaoparticle treated indanthrene blue dye

(b) Untreated indanthrene blue dye

Fig. 17.2 **a** Fe II oxide nanoparticle treated indanthrene blue dye, **b** Untreated indanthrene blue dye

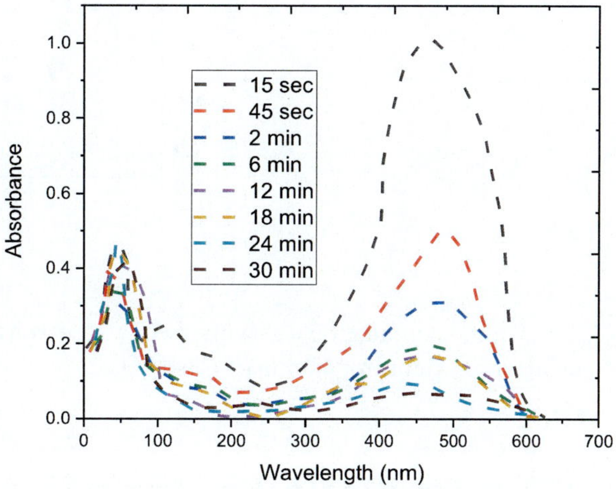

Fig. 17.3 Time-dependent UV Vis results for indanthrene blue dye, degraded using Fe II oxide nanoparticles

faster as it exhibits a simpler chemical structure. Dyes with larger molecules tend to delay, during degradation experiments. On the basis of the experiments conducted, the degradation mechanism involved absorption of disperse orange II dye and alizarin

Fig. 17.4 Time-dependent UV Vis spectrum of disperse orange II degraded by using Fe II oxide nanoparticles

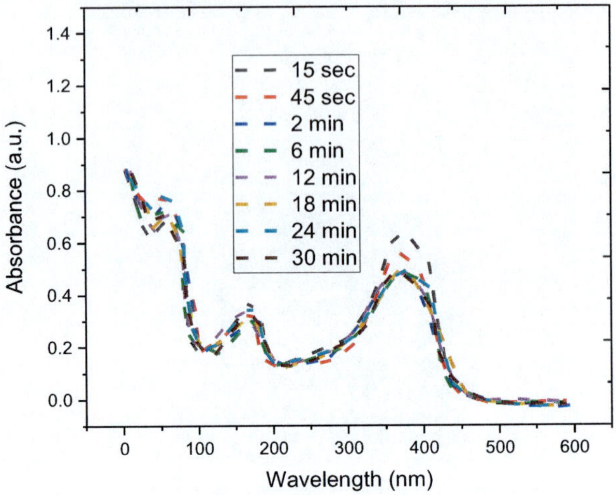

Fig. 17.5 Time-dependent UV–Vis degradation of alizarin using Fe II oxide nanoparticles

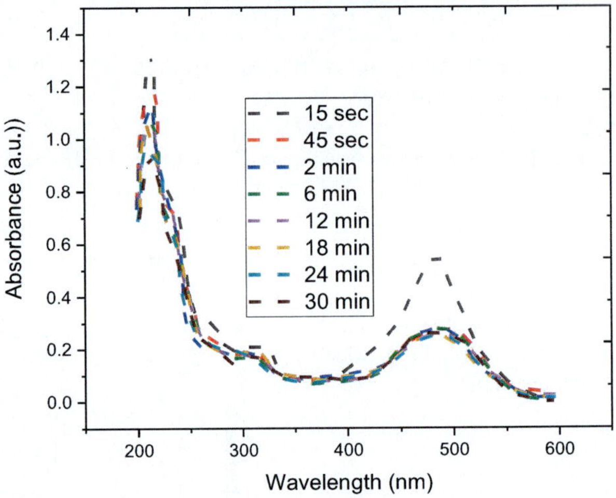

dye. Similar degradation was witnessed by research investigations on dye degradation using zero valent iron nanoparticles [14].

17.3.4 Kinetic Study Experiments

Three experiments were conducted using indanthrene blue dye, disperse orange II and alizarin dye for kinetic studies. By evaluating the maximum absorbance peak [15] of indanthrene blue dye, disperse orange II and alizarin dye the concentration of the dyes were identified. Kinetic study was conducted for duration of 100 min. The kinetic study graphs are shown in Fig. 17.6. Initially, certain instabilities were witnessed.

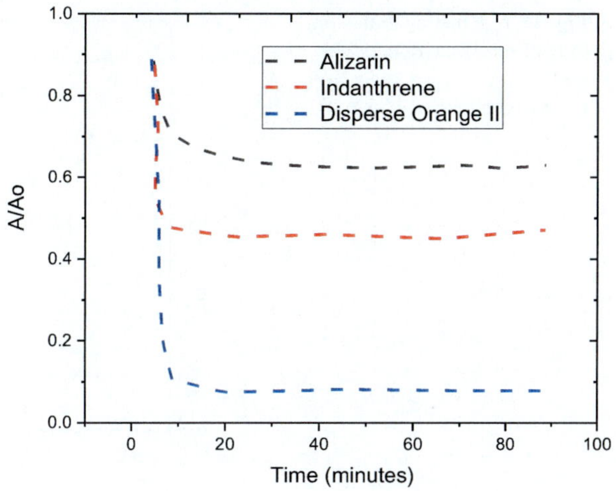

Fig. 17.6 Kinetic studies on indanthrene blue dye, disperse orange II and alizarin dye

The rate of reactive chemical degradation of disperse orange II dye was more that of the other two. The fitting curves indicated that decolorization of disperse II orange was more than the other two dyes. For same quantity of Fe II oxide nanoparticles, disperse II orange was found to degrade at a faster rate than the other two dyes. The degradation order is as follows Alizarin <Indanthrene blue> disperse orange II.

Other methods can also be used for the synthesis of Fe II oxide nanoparticles. For identifying the effect of Fe II oxide nanoparticles on dye degradation, Cobalt-based nanoparticles were used for degradation of indanthrene dye. Figure 17.7 shows the kinetic studies of indanthrene dye degradation by using Fe II oxide nanoparticles and Cobalt-based nanoparticles. On comparing the do colorization effectiveness of Fe II oxide based nanoparticles performed better than Cobalt based nanoparticles. The rate of degradation of indanthrene dye depends on the pH of the solution. When the pH of the solution was increased to 10, in the basic solution, the degradation efficiency drastically decreased. Acidic conditions were found to be favorable for fast degradation of indanthrene dye. At pH value 2.5, the effectiveness of Fe II oxide nanoparticles was better than the effectiveness of Cobalt based nanoparticles. The presence of nanoparticles poses a threat and hazard to the environment. As Fe II oxide nanoparticles are magnetic, they can be easily isolated by using magnetic separation. After separation of the Fe II oxide nanoparticles from the reaction solution, they are free to be disposed and they would be harmless to the environment.

17.4 Conclusions

In this study, Fe II oxide nanoparticles were successfully used for degradation of indanthrene blue dye. Indanthrene blue dye is a widely used coloring agent in textile industries. The wastes should be properly disposed for environmental safety

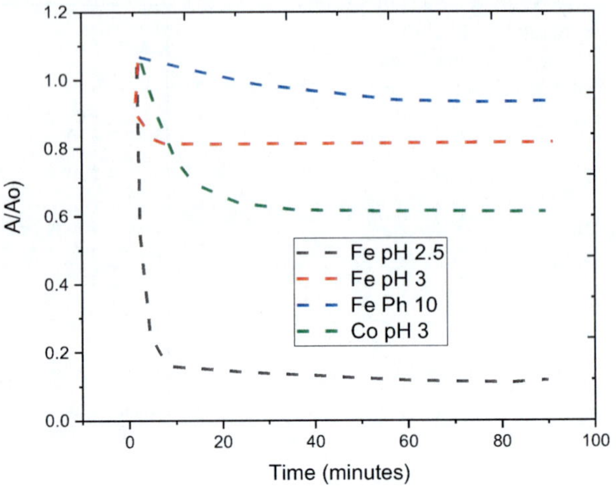

Fig. 17.7 Kinetic studies on indanthrene blue dye, using Fe II oxide nanoparticles and Co-based nanoparticles

concerns. Fe II oxide nanoparticles synthesized by using wet chemical synthesis were found to be effective in degrading indanthrene blue dye. For comparison, two more dyes such as disperse orange II and alizarin dye were degraded by using Fe II oxide nanoparticles. Using UV-Vis spectroscopy, time-dependent studies were conducted. Kinetic studies were conducted and the variation in degradation rates of indanthrene blue dye, alizarin and disperse orange II dye was identified. The effectiveness of the synthesized Fe II oxide nanoparticles was found to be better than Co-based nanoparticles in degrading indanthrene blue dye.

References

1. S. El Harfi, A. El Harfi, Classifications, properties and applications of textile dyes: a review. Appl. J. Environ. Eng. Sci. **3**(3), 00000–00003 (2017)
2. A. Lahkimi, M.A. Oturan, N. Oturan, M. Chaouch, Removal of textile dyes from water by the electro-Fenton process. Environ. Chem. Lett. **5**(1), 35–39 (2007)
3. M. Berradi, R. Hsissou, M. Khudhair, M. Assouag, O. Cherkaoui, A. El Bachiri, A. El Harfi, Textile finishing dyes and their impact on aquatic environs. Heliyon **5**(11), e02711 (2019)
4. A.E. Ghaly, R. Ananthashankar, M.V.V.R. Alhattab, V.V. Ramakrishnan, Production, characterization and treatment of textile effluents: a critical review. J. Chem. Eng. Process. Technol. **5**(1), 1–19 (2014)
5. S. Varjani, P. Rakholiya, H.Y. Ng, S. You, J.A. Teixeira, Microbial degradation of dyes: an overview. Biores. Technol. **314**, 123728 (2020)
6. N.L. Stock, J. Peller, K. Vinodgopal, P.V. Kamat, Combinative sonolysis and photocatalysis for textile dye degradation. Environ. Sci. Technol. **34**(9), 1747–1750 (2000)
7. C.I. Pearce, J.R. Lloyd, J.T. Guthrie, The removal of colour from textile wastewater using whole bacterial cells: a review. Dyes. Pigm. **58**(3), 179–196 (2003)
8. V. Katheresan, J. Kansedo, S.Y. Lau, Efficiency of various recent wastewater dye removal methods: a review. J. Environ. Chem. Eng. **6**(4), 4676–4697 (2018)

9. P. Kumar, M. Govindaraju, S. Senthamilselvi, K. Premkumar, Photocatalytic degradation of methyl orange dye using silver (Ag) nanoparticles synthesized from Ulva lactuca. Colloids Surf., B **103**, 658–661 (2013)
10. A.D. Bokare, R.C. Chikate, C.V. Rode, K.M. Paknikar, Effect of surface chemistry of Fe− Ni nanoparticles on mechanistic pathways of azo dye degradation. Environ. Sci. Technol. **41**(21), 7437–7443 (2007)
11. A. Molla, H. Choi, H. Sakong, J.H. Youk, Sulfur-source dependent wet mechanochemical synthesis of pyrrhotite nanoparticles and evaluation of their sonocatalytic dye degradability. Mater. Res. Bull. **145**, 111519 (2022)
12. E. Routoula, S.V. Patwardhan, Degradation of anthraquinone dyes from effluents: a review focusing on enzymatic dye degradation with industrial potential. Environ. Sci. Technol. **54**(2), 647–664 (2020)
13. L. Huang, F. Luo, Z. Chen, M. Megharaj, R. Naidu, Green synthesized conditions impacting on the reactivity of Fe NPs for the degradation of malachite green. Spectrochim. Acta Part A Mol. Biomol. Spectrosc. **137**, 154–159 (2015)
14. Z.H. Farooqi, R. Begum, K. Naseem, W. Wu, A. Irfan, Zero valent iron nanoparticles as sustainable nanocatalysts for reduction reactions. Catal. Rev. **64**(2), 286–355 (2022)
15. M. Chakraborty, D.C. Mukherjee, B.M. Mandal, Dispersion polymerization of aniline in different media: a uv− visible spectroscopic and kinetic study. Langmuir. **16**(6), 2482–2488. LNCS Homepage. http://www.springer.com/lncs. Accessed 21 November 2016.

Chapter 18
Synthesis of Fe Oxide Nanoparticles by Using Mangifera Indica and Its Effectiveness in Photocatalytic Degradation

Mathivanan Varatharajan, Murugesan Kumarasamy, and Siddharth Sampathkumar

Abstract In this study, Fe oxide nanoparticles were prepared by using Mangifera indica leaf extract. Using UV–Vis spectroscopy, the properties of the prepared Fe oxide nanoparticles were studied. The presence of Fe oxide nanoparticles was confirmed by studying the absorption peaks from UV–Vis spectrum. The prepared Fe oxide nanoparticles were subjected to microstructural investigations using scanning electron microscopy. Using dynamic light scattering experiments, the average size of the synthesized Fe oxide-based nanoparticles was found to be 115 nm. At −31.35 mV, stability of the prepared Fe oxide nanoparticles was identified using negative zeta potential disclosure experiments. Fourier transform infrared ray spectroscopic experiments were used to identify the pattern of the prepared Mangifera indica synthesized Fe oxide nanoparticles. Using X-ray diffraction experiments, the crystallinity of the prepared Fe oxide nanoparticles was observed. In photocatalytic dye degradation experiments, Fe oxide nanoparticles in colloidal form degraded alizarin dye.

Keywords Mangifera indica · Fe oxide nanoparticles · Alizarin dye · Photocatalytic degradation · FTIR · X-RD

M. Varatharajan (✉) · M. Kumarasamy
Department of Environmental Science, Periyar University, Tamil Nadu, Salem 636001, India
e-mail: mathivanan999@gmail.com

S. Sampathkumar
Department of Mechanical Engineering, PSN College of Engineering and Technology, Tirunelveli, Tamil Nadu 627152, India

A. K. Sahu et al. (eds.), *Energy Storage and Conservation*, Springer Proceedings in Energy, https://doi.org/10.1007/978-981-99-2870-5_20

18.1 Introduction

In recent times, a lot of development is being witnessed in the field of biotechnology [1]. In specific, advancements in nanotechnology and its related fields have sky rocketed. Nanoparticle synthesis has developed by using advance techniques [2]. Metal nanoparticles are being used for different applications [3]. For improving the effectiveness of metal-based nanoparticles, different synthesis techniques [4] are being used. Even though metal-based nanoparticles can be effectively developed by using chemical process, these chemicals react with the metal-based particles to form toxic products. Green synthesis of metal-based nanoparticles is relatively simple and eco-friendly [5]. Even though different metals are used to prepare nanoparticles, nanoparticles synthesized from Fe metal are advantageous as it is cheap and it exhibits desirable biological characteristics [6]. In recent times, Fe-based nanoparticles were found to be effective in treating pharmaceutical waste and textile dye wastes [7]. However, the effectiveness of Fe-based nanoparticles on photocatalytic degradation is not known.

A lot of pollutants are dumped into the environment from textile industries and dye tanneries. The presence of untreated dye affects the environment and creates imbalance in the ecosystem [9]. Natural resources such as rivers, ponds and lakes are heavily polluted when untreated dye wastages are dumped in it. As the chemical stability of these dye-based pollutants is higher, simple degradation techniques such as ultraviolet radiation are not effective. Dye decolorization can be done using photocatalytic degradation [10]. Semiconductor materials are being used in photocatalytic degradation mechanisms and a significant improvement in the rate of decolorization was observed [11]. Recent findings have shown that metal-based nanoparticles can be successfully used for degradation of dyes. Green synthesis of metal-based nanoparticle synthesis was found to be easy and profitable than physical and chemical nanoparticle synthesis methods. In this study, an attempt has been made to prepare Fe oxide-based nanoparticles and its effectiveness as a photocatalytic dye degradation agent.

18.2 Materials and Methods

18.2.1 Synthesis of Fe Oxide Nanoparticles by Using Mangifera Indica Extract

In this investigation, Mangifera indica extract was prepared from Mangifera indica leaves to be used for synthesis of Fe oxide nanoparticles. The leaves were washed in water and dried in shade for 2 weeks. Then it was powdered and 20 g of the powder was mixed in 200 ml of deionized waster and stirred using a magnetic stirrer. The temperature of the mangifera extract was increased to 65 °C by using an electrical heater. Throughout the heating process, the extract was continuously stirred. The solution was filtered using nanofilters, and the extract was stored at very low temperature in freeze condition till further use. 120 ml of 5 mM Ferrous oxide heptahydrate solution was added to the mangifera extract solution, and it was continuously stirred for the synthesis process to occur. The solution was allowed to stay at room temperature till color change occurred. Then Fe oxide nanoparticles were separated, filtered and used for the experiments.

18.2.2 Characterization of Mangifera Indica Synthesized Fe Oxide Nanoparticles

For evaluating the Fe oxide nanoparticles prepared by using Mangifera indica green synthesis, ultraviolet visual spectroscopy (Make—Shimadzu) (UV–Vis) evaluation was done. Fourier transform infrared ray spectroscopy was used for identifying the properties of the Mangifera indica synthesized Fe oxide nanoparticles. At 4 cm^{-1} resolution, experimental evaluation was conducted. The zeta potential distribution of mangifera indica synthesized Fe oxide nanoparticles, Zetazier equipment was used. Using JOEL make Scanning Electron Microscope, microstructural evaluation of Fe oxide nanoparticles was observed. X-RD spectra of Fe oxide nanoparticles were evaluated by using Maverick make X-ray Spectroscopic equipment. Using Cu target, scanning was done from 20 to 80 degree two theta, using a step size of 0.002.

18.2.3 Photocatalytic Degradation of Alizarin Dye Using Mangifera Indica Synthesized Fe Oxide Nanoparticles

For identifying the effectiveness of the Mangifera indica synthesized Fe oxide nanoparticles, alizarin dye was used. At 10 mg/L, alizarin dye was diluted and 15 mg of Fe oxide nanoparticles was added to it and stirred. The setup was conducted under bright sunlight. Sunlight was used as the major light source. At specific time intervals, a small quantity of the solution was taken and diluted, stirred and centrifuged at 10,000 rpm for 15 min to stop the degradation activity. The degradation wastes were evaluated by using UV Vis spectroscopy.

18.3 Results and Discussion

The initiation and dye degradation process can be identified by observing the color change, during photocatalytic dye degradation process. UV–Vis spectrum of Mangifera indica synthesized Fe oxide nanoparticles is shown in Fig. 18.1. From

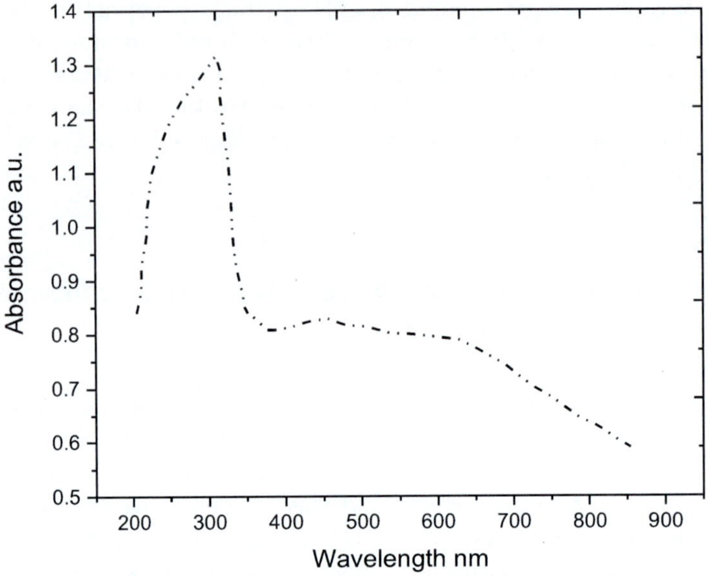

Fig. 18.1 UV–Vis spectra of Mangifera indica synthesized Fe oxide nanoparticles

Fig. 18.1, it was observed that at 3.4.3 nm, absorbance peak was at maximum, indicating the presence of Fe oxide nanoparticles.

The effect of bioactive molecules of Mangifera indica extract on Fe oxide nanoparticles was observed using FTIR evaluation. FTIR spectrum of Mangifera indica synthesized Fe oxide nanoparticles is shown in Fig. 18.2. FTIR evaluation was conducted from 400 to 4000 cm^{-1}. Peaks were identified at 3521, 2967, 2215 and 1423 cm^{-1}. For evaluating the surface charge and size of the Mangifera indica synthesized Fe oxide nanoparticles, Dynamic Light Scattering (DLC) evaluation was done.

The variations in volume percentage at different sizes of the Mangifera indica synthesized Fe oxide nanoparticles are shown in Fig. 18.3. The average sizes of the nanoparticles were found to be around 115 nm.

Zeta potential was found to be −31.35 mV. The variations in zeta potential from −250 to +250 m V are shown in Fig. 18.4. This indicates that the nanoparticles are repellent and aggregation is prevented. Under SEM evaluation, Mangifera indica synthesized Fe oxide nanoparticles exhibit irregular agglomeration and pear shape (Fig. 18.5).

Energy Dispersive X-ray Spectroscopic analysis was conducted on Mangifera indica synthesized Fe oxide nanoparticles, and it has been shown in Fig. 18.6. X-ray diffraction analysis on Mangifera indica synthesized Fe oxide nanoparticles was conducted, and the spectra are shown in Fig. 18.7. Peakss at 33.22 and 66.39 two theta values indicate Fe particles. The obtained X-RD analysis was compared with JCPDS data and the evaluation was done.

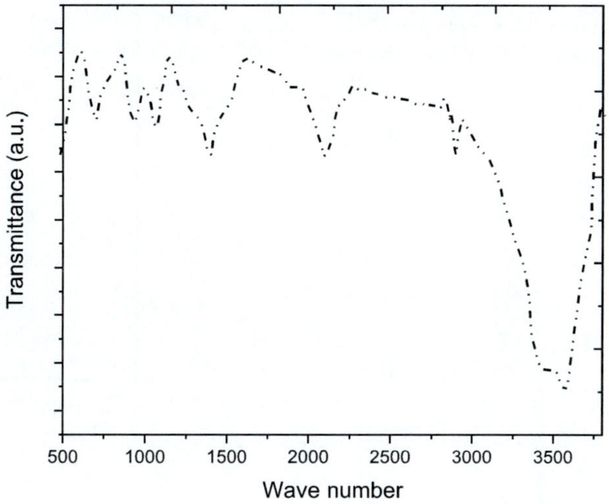

Fig. 18.2 FTIR spectra of Mangifera indica synthesized Fe oxide nanoparticles

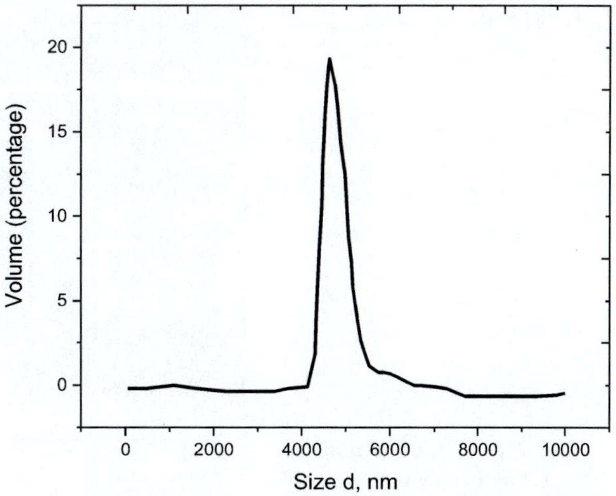

Fig. 18.3 Dynamic light scattering variations in Mangifera indica synthesized Fe oxide nanoparticles

The green synthesized Mangifera indica synthesized Fe oxide nanoparticles were made to degrade alizarin dye. The decrease in absorbance peaks indicates that degradation has occurred. Initially, absorbance caused decoloration and afterwards photocatalytic degradation was observed. The peak intensity decreased on increasing the time from 15 to 60 min. The variations in absorbance peaks have been shown in Fig. 18.8. In Fig. 18.9, the reduction of alizarin dye at 561 nm at different time intervals is shown. The color degradation was attributed to the variations in size, shape and morphology of the green synthesized Mangifera indica-treated Fe oxide nanoparticles.

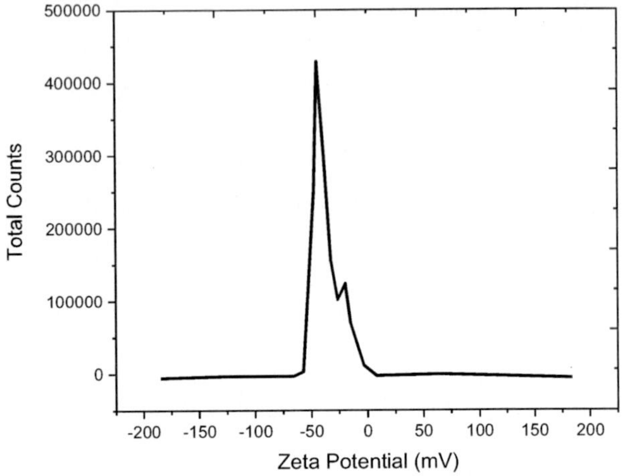

Fig. 18.4 Zeta potential variations of Mangifera indica synthesized Fe oxide nanoparticles

Fig. 18.5 SEM image of Mangifera indica synthesized Fe oxide nanoparticles

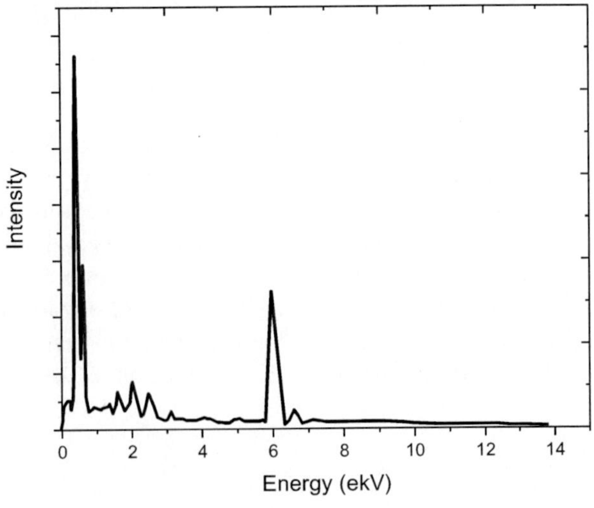

Fig. 18.6 EDAX spectrum of MAngifera indica synthesized Fe oxide nanoparticles

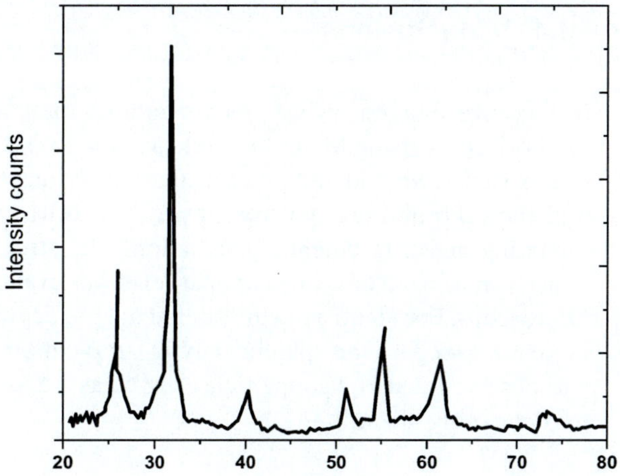

Fig. 18.7 XRD spectrum of Mangifera indica synthesized Fe oxide nanoparticles

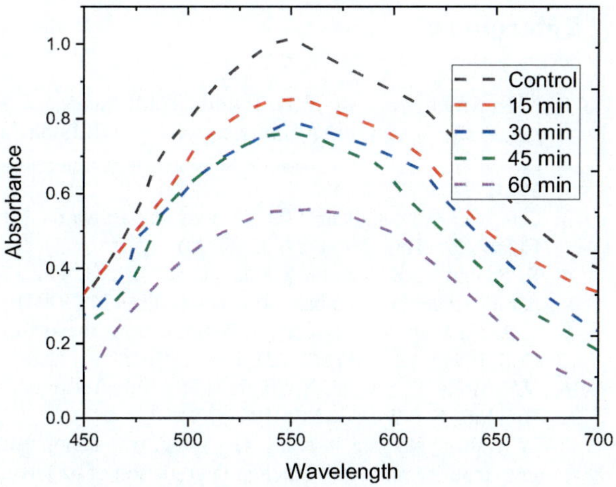

Fig. 18.8 Photocatalytic dye degradation of alizarin using Mangifera indica synthesized Fe oxide nanoparticles

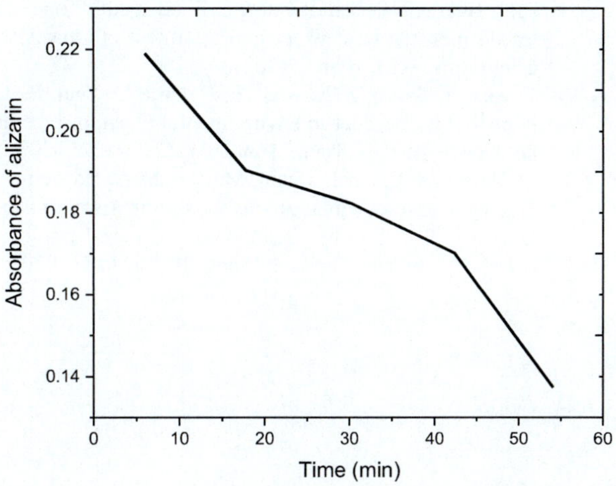

Fig. 18.9 Absorbance of alizarin with time

18.4 Conclusions

In this investigation, using green synthesis method, Fe oxide nanoparticles were synthesized by using Mangifera indica extract. The properties of the synthesized Fe nanoparticles were identified by using ultra-violet visual ray spectroscopy, Fourier transform infrared ray spectroscopy, X-ray diffraction spectroscopy, dynamic light scattering and zeta potential evaluation. The structure of the prepared Mangifera indica synthesized Fe oxide nanoparticles was evaluated by using Scanning Electron Microscopy. For identifying the photocatalytic degradation ability of the synthesized Fe oxide nanoparticles, photocatalytic degradation experiments were conducted. It was observed that Fe nanoparticles served as a good photocatalytic degrading agent for decoloring alizarin.

References

1. M.K. Smith, H.L. Ko, S.D. Hamill, G.M. Sanewski, M.W. Graham, Biotechnology, in *The pineapple: botany, production and uses.* (CABI Publishing, Wallingford UK, 2003), pp.57–68
2. A. Curtis, C. Wilkinson, Nantotechniques and approaches in biotechnology. Trends Biotechnol. **19**(3), 97–101 (2001)
3. C.R. Rao, G.U. Kulkarni, P.J. Thomas, P.P. Edwards, Metal nanoparticles and their assemblies. Chem. Soc. Rev. **29**(1), 27–35 (2000)
4. S. Panigrahi, S. Kundu, S. Ghosh, S. Nath, T. Pal, General method of synthesis for metal nanoparticles. J. Nanopart. Res. **6**(4), 411–414 (2004)
5. K.B. Narayanan, N. Sakthivel, Biological synthesis of metal nanoparticles by microbes. Adv. Coll. Interface. Sci. **156**(1–2), 1–13 (2010)
6. V.V. Mody, R. Siwale, A. Singh, H.R. Mody, Introduction to metallic nanoparticles. J. Pharm. Bioallied Sci. **2**(4), 282 (2010)
7. W. Wang, Y. Cheng, T. Kong, G. Cheng, Iron nanoparticles decoration onto three-dimensional graphene for rapid and efficient degradation of azo dye. J. Hazard. Mater. **299**, 50–58 (2015)
8. R. Kishor, D. Purchase, G.D. Saratale, R.G. Saratale, L.F.R. Ferreira, M. Bilal, R. Chandra, R.N. Bharagava, Ecotoxicological and health concerns of persistent coloring pollutants of textile industry wastewater and treatment approaches for environmental safety. J. Environ. Chem. Eng. **9**(2), 105012 (2021)
9. T. Akar, I. Tosun, Z. Kaynak, E. Ozkara, O. Yeni, E.N. Sahin, S.T. Akar, An attractive agro-industrial by-product in environmental cleanup: Dye biosorption potential of untreated olive pomace. J. Hazard. Mater. **166**(2–3), 1217–1225 (2009)
10. M. Alizadeh, E. Ghahramani, M. Zarrabi, S. Hashemi, Efficient de-colorization of methylene blue by electro-coagulation method: comparison of iron and aluminum electrode (2015)
11. D. Vaya, P.K. Surolia, Semiconductor based photocatalytic degradation of pesticides: an overview. Environ. Technol. Innov. **20**, 101128 (2020)

Chapter 19
Studying the Effectiveness of Multicriteria Decision-Making Tool Such as Response Surface Methodology for Dissimilar Friction Stir Spot Welding

Siddharth Sampathkumar

Abstract Optimization techniques in manufacturing and automotive sectors are being widely used for speeding up research activities as well as reducing the expenditure and time duration involved in it. In this study, a very prudent hybrid optimization technique known as response surface methodology has been used for obtaining better mechanical and tensile properties of dissimilar Al and Cu friction stir spot joints. Using a central composite design matrix, the process parameters were appropriately selected to fabricate 20 joints. The appropriate parameter selection for obtaining maximum tensile strength and minimum interface hardness was obtained by evaluating the contour and surface plots of the optimized model developed. It was found that the use of response surface methodology significantly contributed in prediction of the desired output from the model developed with high precision and accuracy.

Keywords Hybrid optimization technique · Response surface methodology · Contour plot · Surface plot

19.1 Introduction

Optimization techniques have gained much importance in research and research-based activities due to the enormous advantages and help they offer for quick calculation and evaluation of the developed experimental models. A hybrid methodology is a calculation that consolidates two or more different sequences that tackle the same issue, either selection of the one more contingent upon information or exchanging during the course of the calculation span. Response surface methodology (RSM), an empirical model development tool using a collection of statistical and mathematical techniques is widely used as an advance optimization method [1]. Khuri and

S. Sampathkumar (✉)
Department of Mechanical Engineering, PSN College of Engineering and Technology, Tirunelveli, Tamil Nadu 627152, India
e-mail: sksmsiddharth1@yahoo.in

A. K. Sahu et al. (eds.), *Energy Storage and Conservation*, Springer Proceedings in Energy, https://doi.org/10.1007/978-981-99-2870-5_21

Mukhopadhyay [2] applied hybrid RSM with genetic algorithm to optimize oil trans-esterification on nanoscale mesoporous agglomerates. Using RSM, the consequences of the ratios on the fuel synthesis were evaluated and a quadratic function was developed for data analysis using central composite design (CCD) [2]. Athi Sankar et al. [3] identified the signals extended due to vibrations to their corresponding material properties by including particle swarm methodology into the hybrid RSM technique. Mohammad et al. [4] used RSM to evaluate the parameters that determine the main characteristics of injection molding process. The important parametric features were incorporated with an effective simulated algorithm to ensure better results. Namura et al. [5], integrated the radial basis function networks, for the sole purpose of increasing the model's robustness, with Kriging model, using RSM and obtained highly accurate models. Saibaba and King [6] combined the positive aspects of both genetic algorithm and RSM to effectively create an efficient model for the process of removal of die and also optimized it. Yang and Tseng [7] successfully solved using RSM in an ink-marking case study by incorporating advance goal programming concepts. Vairamani et al. [8] used RSM strategy in welding process. Process parameters of dissimilar combinations were successfully optimized for obtaining excellent weld quality [8]. Yang et al. [9] displayed a methodology for anticipation of the acoustic gushing in a thermo acoustic heat system incorporating decreased projection point and other dimensional parameters. RSM was used for contemplate the relations, and a model was created to anticipate the streaming. Sodeifian et al. [10] investigated the process of extracting a vital oil utilizing a supercritical system. The investigations and procedure enhancement were done with RSM [10]. Mamourian et al. [11] conducted 2D examination of the radiation heat exchange combined with investigation of consolidated turbulent blended heat transfer using a two-stage blend model. Numerical re-enactments were conducted to contemplate the impacts of the important parameters using RSM [11]. To accomplish the most extreme evacuation rate of Triadimefon and the base creation of its relating metabolite, Kong et al. [12] used RSM in fermentation process. The preparing variables were additionally assessed amid the advanced blending handle and were observed amid the malting processing, heating and heat reduction stages. For assessment, RSM strategies were applied.

In this investigation using response surface methodology, dissimilar joints of Aluminum and Copper using friction stir spot welding were optimized. Evaluation of the obtained results and the data evaluation was conducted using Design expert software.

19.2 Materials and Methods

19.2.1 Response Surface Methodology

RSM is an effective tool used for investigation of the connections between a few logical variables and one or more reaction variable. The principle ideology of RSM is utilization of a group of outlined examinations for getting an ideal response, for which a second-order polynomial was suggested by [13]. A simple approach for assessment of a first-degree polynomial model is to utilize a factorial analysis or a fragmentary factorial configuration. This is presumed to be adequate to figure out which illustrative variables affect the reaction variables of interest. When it is clear that the lone critical informative variables are left, then a more confounded outline, for example, a central-based composite configuration is actualized for assessment of a higher degree polynomial model, which still accounts for best-case scenario estimation. Nevertheless, the second-degree model can be utilized to enhance, augment, minimize or accomplish a particular focused requirement [13]. RSM utilizes measurable models, and in this way, researchers are to be understood that even the best factual model is an estimate to reality. By and by, in reality, both the generated models and the parameter qualities are not known and they are subjected to various vulnerability on top of obliviousness. Box's configuration diminished the expenses of experimentation so that a quadratic model could be fit [14]. RSM is transforming into zones including the utilization of summed up direct models, and ideal outlines for these zones are very difficult to actualize by the client. Other RSM strategies that will appreciate use by experts later on includes techniques involving both non- and semi-parametric systems [15].

19.2.2 Base Materials and Welding Process

The base materials used for this investigation were Aluminum alloy Al 5052 and Commercial Copper alloy, both of 1.25 mm thickness. Both the materials were sized to length 75 mm and breadth 30 mm as shown in Fig. 19.1a. For weld joints, overlap was made to be 30 mm (as per ASTM).

Friction stir spot welding (FSSW) is a linear version of its predecessor, FSW is a solid-state material joining technique. It is a three-step method involving piercing of the sheets with a non-consumable tool, a period of dwell and withdrawal [16]. It is illustrated in Fig. 19.1b.

The non-consumable tool used was made up of H 13 material, with straight cylindrical shoulder geometry of shoulder diameter of 14 mm, pin diameter 5 mm, pin height 1.5 mm. Important process parameters of FSSW welding process were identified to be the rotational speed of the tool (R) in rpm, duration of plunge of tool at that position (T) in seconds, extent of piercing of the tool onto the sheets (D) in seconds. A modified vertical milling CNC machine was used for conducting the

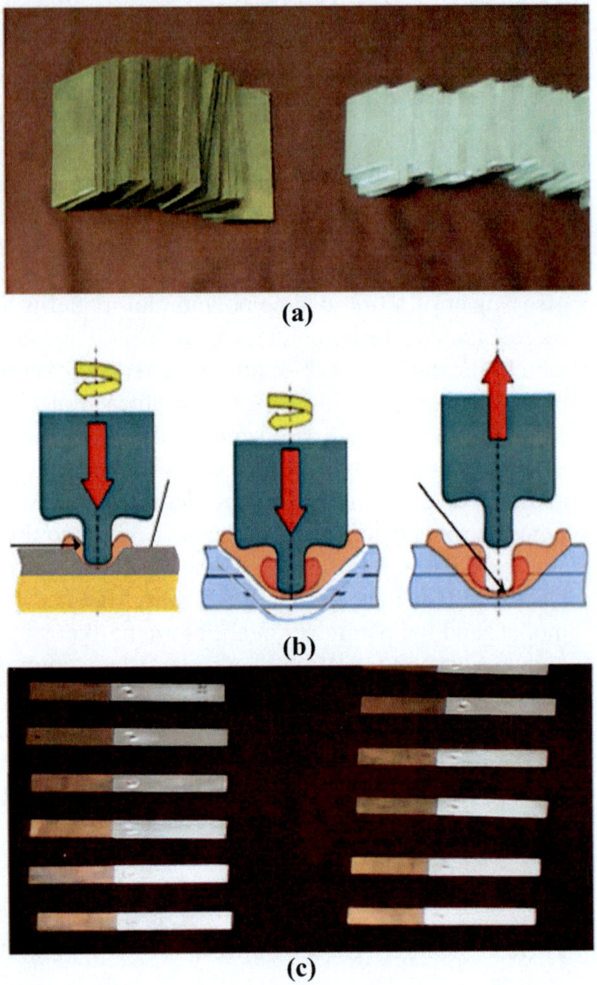

Fig. 19.1 **a** Base materials, **b** FSSW process, **c** Fabricated joints

experiments. The feasible limits of the process parameters for the selected dissimilar Al/Cu samples were evaluated by trial and error process and iterative techniques to obtain a reasonable and good quality joint. The limits are given in Table 19.1.

Table 19.1 Feasible limits

Parameters	Level				
	−1.682	−1.0	0	+ 1.0	+ 1.682
Tool rotational speed (R)	980	1085	1225	1425	1485
Plunge depth (D)	1.15	1.25	1.35	1.45	1.55
Dwell time (T)	10	12	15	17	19

A three-factor five-level central composite design matrix was developed. According to the procedure given by [17], with 20 coded conditions, 6, 6 star and center points.

+ 1.682 was taken as upper limit and −1.682—lower limit. For finding intermediate values, the following relationship was used [17].

$$H_i = 1.682\,[2H - (H_{max} + H_{min})]\,/\,(H_{max} + H_{min}) \tag{19.1}$$

From H_{min} to H_{max}, Z assumes any value of the variable. The developed design matrix is given in Table 19.2.

After welding, the specimens were subjected to tensile tests using Universal testing machine, and the interface hardness was measured using Vickers hardness testing machine.

As per ASTM standards, the loading for shearing the specimens was given with a uniform incremental rate of 1.5 kN per minute. At three different regions in the joint interface, the microhardness values were observed, and the average of the taken values is presumed to give the most appropriate value of microhardness. The joints

Table 19.2 Design model matrix

No	Coded factor value				
Run	R	T	D	TSFL(kN)	IH(HV)
1	1	−1	1	3671	100
2	1	1	1	3137	118
3	0	0	0	3654	94
4	0	0	0	3550	113
5	0	1.68	0	2809	124
6	0	−1.68	0	3688	87
7	−1.68	0	0	3309	117
8	1	1	1	3430	118
9	1	-1	-1	3119	120
10	0	0	1.68	3050	122
11	0	0	0	3137	118
12	0	0	0	3654	94
13	−1	−1	1	3430	118
14	−1	1	−1	3533	108
15	0	0	−1.68	3585	100
16	0	0	0	4016	65
17	−1	1	1	3602	100
18	1	1	−1	3637	97
19	−1	−1	−1	2891	123
20	1.68	0	0	2309	138

are required to have a high tensile shear failure load value (TSFL) and minimum interface hardness (IH) to ensure good quality. The values of TSFL and IH are recorded in Table 19.2. The joints that have been fabricated according the process parameter values are given in the given design table, are shown in Fig. 19.1c.

19.2.3 Establishing Empirical Relationships

An adaptable DOE technique was utilized. A fold over was created and the run criteria were determined. Using half normal plotting, the robustness was validated for the given information. Plan assessment techniques were used to verify the aliases. If necessary another set of runs are made with the element levels at the inverse setting of the first design plan.

The observed output values, tensile properties and the weld interface hardness were attributed to the three identified FSSW process parameters. By using the methodology stated by [18], their relationship is indicated as.

$$\text{TSFL}\,(Tensile\ force\ N) = f\{Speed,\ Time\,of\ plunge,\ Depth\} \tag{19.2}$$

$$Microhardness = f\{\,Speed,\ Time\,of\ plunge,\ Depth\,\} \tag{19.3}$$

The response surface Y two input characteristics are represented by second order polynomial regression equation [19]

$$Y = a_0 + \sum a_i x_i + \sum a_{ii} x_{i2} + \sum a_{ij} x_i x_j \tag{19.4}$$

$$\begin{aligned}\text{TSFL/IH} = \{&\text{a}_0 + \text{a}_1\text{R} + \text{a}_2\text{T} + \text{a}_3\text{D} + \text{a}_{12}\text{RT} + \text{a}_{13}\text{RD} \\ &+ \text{a}_{23}\text{TD} + \text{a}_{11}\text{R}^2 + \text{a}_{22}\text{T}^2 + \text{a}_{33}\text{D}^2\}\end{aligned} \tag{19.5}$$

a_0 is termed to be the average of response. The regression coefficients are made to be dependent and fluctuate with linear term, interaction and squared terms of factors are a_1, a_2, a_3,…,a_{nn} [19]. Design expert software was used for the evaluation of the coefficients. Student's t test and p values were used for determination of the model's significance. ANNOVA can be utilized as an exploratory methodology to evaluate the variations. ANNOVA analysis was conducted for TSFL and IH models and the results are shown in Table 19.3 and Table 19.4. ANOVA gives a factual test of regardless of whether the method for a few gatherings is equivalent, and in this way sums up the t-test to more than two sets. The structure of the added substance model permits answer for the added substance coefficients by straightforward polynomial mathematical analysis. "Prob > F" values clearly indicate that the terms of the model are significant at 95% confidence level as it is less than 0.0500. The values greater than 0.10 indicate that the model terms are not significant. These coefficients obtained

Table 19.3 Analysis of variance for Al and Cu FSSW joint tensile strength

Source	Sum of squares (SS) $\times 10^3$	Degree of freedom (df)	Mean square (MS) $\times 10^3$	F ratio	p-value Prob > F	Note
Model	3832	9	3832	623	< 0.0001	Significant
R	3124	1	3124	368	< 0.0001	
T	1961	1	1961	421	< 0.0001	
D	8436	1	8436	1146	< 0.0001	
RT	6330	1	6330	859	< 0.0001	
RD	2975	1	2975	69	< 0.0001	
TD	2854	1	2854	745	< 0.0001	
R^2	2186	1	2186	421	< 0.0001	
T^2	3167	1	3167	621	< 0.0001	
D^2	3821	1	3821	741	< 0.0001	
Residual	4963	10	694			
Lack of fit	1964	5	341	0.27	0.8541	Not significant
Std. Dev		21.24		R^2	0.9846	
Mean		3175		Adj	0.9942	
C.V. %		0.69		Pred	0.9865	
PRESS		142.21		Adeq precision	99.36	

adds to the formation of TSFL and IH relationship, which were formed empirically. Tensile shear failure load (TSFL) of the dissimilar Al Cu joints,

$$\begin{aligned} \text{TSFL} = \{ &3651 + 127\text{R} + 118\text{T} + 235\text{D} - 260\text{RT} + 62\text{RD} - 221\text{TD} \\ &- 110\text{R}^2 - 146\text{T}^2 - 162\text{D}^2 \} \end{aligned} \tag{19.6}$$

Interface hardness of the dissimilar Al- Cu joints,

$$IH = \left\{95 - 5R + 0.60T - 7D + 9RT - 6RD + 10TD + 6R^2 + 7T^2 + 6D^2\right\} \tag{19.7}$$

.

Table 19.4 ANNOVA test results for IH

Source	Sum of squares (SS)	Degree of freedom (df)	Mean square (MS)	F ratio	p-value Prob > F	Note
Model	4782	9	531	12	0.0003	Significant
R	352	1	352	8	0.0168	
T	2	1	5	0.12	0.7397	
D	844	1	844	19	0.0013	
RT	78	1	780	18	0.0017	
RD	325	1	325	7	0.0204	
TD	903	1	903	21	0.0010	
R^2	580	1	580	13	0.0043	
T^2	908	1	908	21	0.0010	
D^2	376	1	376	8	0.0143	
Residual	429	10	42			
Lack of fit	309	5	61	2	0.1591	Not significant
Std. Dev		6.55		R^2	0.9176	
Mean		108		Adj	0.8435	
C.V. %		6		Pred	0.4840	
Press		2689		Adeq precision	14.437	

19.3 Results and Discussion

RSM is an effective tool used for investigation of the connections between a few logical variables. Using design expert software, RSM evaluation was conducted. The element ranges characterize both the area of interest and operability, so the CCD model is evaluated and the results were obtained. The standard error (SE) of the design model developed is shown in Fig. 19.2. In the event that the SE of a few individual amounts is known then the SE of some capacity of the amounts can be effortlessly ascertained most of the time.

The correlation between the predicted and actual values of TSFL and IH is shown in Fig. 19.3. It could be observed from Fig. 19.3 that the predicted and actual values have a high level of correlation. For the predicted TSFL values, the correlation was excellent from 2300 to 3650 N. Beyond and above that values, the correlation was sparingly repetitive. For the IH model, from a microhardness value of 103.5 HV to 122.75, the prediction was very close to the actual values. Beyond and above the determined values, the model was not much feasible.

RSM was used to optimize the dissimilar joints. Two components at once can be envisioned in a shape plot, the others must be set to regularly at their mid ranges. Generally, a polynomial of a specific level (not more than 2 or 3) with flexible

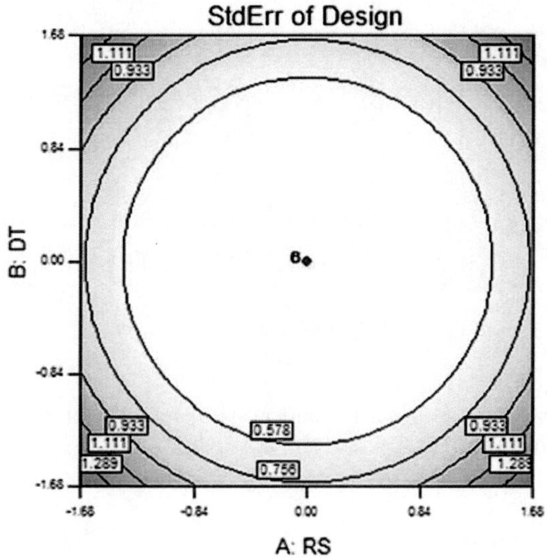

Fig. 19.2 Standard error of the design model developed

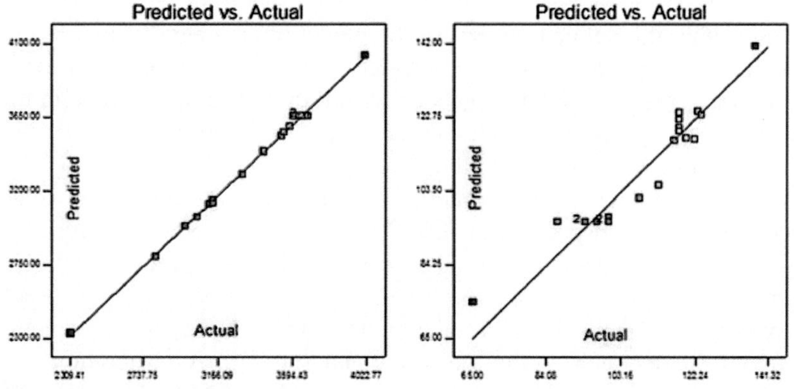

Fig. 19.3 Predicted versus actual values of TSFL and IH model

parameters is utilized. The parameters are evaluated exactly from the trial plan. The representation conceivable is through the "projections" of the hyperspace into a few two-dimensional spaces, which are actually the two-dimensional form plots that were created. RSM is used to optimize the independent variables such as tool rotational speed (R), dwell time (T) and plunge depth (D) have been observed to change with the variables that are dependent, which are tensile shear failure load (TSFL) or interface hardness (IH). The representation of RSM in the qualitative form for clearly indicating the factors that are independent is given by

$$A = \Phi(x_1, x_2, \ldots\ldots x_k) \pm er \tag{19.8}$$

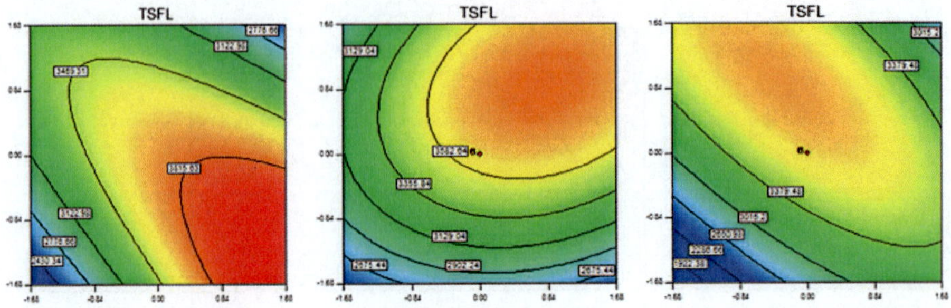

(a) Tool rotational speed vs dwell time

(b) Tool rotational speed vs plunge depth

(c) Dwell time vs plunge depth

Fig. 19.4 TSFL contour plots. **a** Tool rotational speed versus dwell time. **b** Tool rotational speed versus plunge depth. **c** Dwell time versus plunge depth

A is the response and x_1, x_2,......x_k are the quantitative factors. Φ—response surface or response function. The errors that originate during the experimental process are recorded and evaluated by the residual error function. A characteristic surface is responded for a given set of independent variables.

When the value of Φ cannot be determined exactly using mathematical formulations, within the defined regions of the model, it is generally approximated with the help of a polynomial. When the degree of the polynomial becomes greater and greater, the level of correlation gets better but there is a distinct rise in the overall expenditure of the cost of the entire research activity. In this present investigation, the formation and development of the empirical procedures were done using RSM. These were developed in the form of multiple regression equations, which could clearly describe the quality aspects of the FSSW joints. The variables that are clearly independent were viewed as surface and upon which the appropriate model developed is fitted.

By utilizing the distinct and unique circle types of shapes, the variation of the important factors such as the tool rotational speed in rpm, duration of dwell in seconds and the extent of piercing of the tool is plotted in contour plots. The region that consists of the optimum values is clearly indicated and can be seen visually. Simple and small contours are a characteristic for the models of first-order type and as the number of orders increases the degree of complexity also increases.

A stationary point is selected and it is required to evaluate the responses within the region around the stationary point. The purpose of evaluation has to be determined whether it is for increasing to the maximum value or decreasing to the minimum value. Study of the contours guides in evaluation of the point that is given in the study area. Using design expert software, the contour plots were generated. The geometry of the surface determines the optimum values that could be identified with a high level of accuracy [19]. The contour plots were generated considering the process parameters in the mid-range and plotting them in x and y axis. The responses of TSFL and IH values were made to take z axis. The contour plots of TSFL model are indicated in Fig. 19.4, where Fig. 19.4a shows contour plots of Tool rotational speed versus

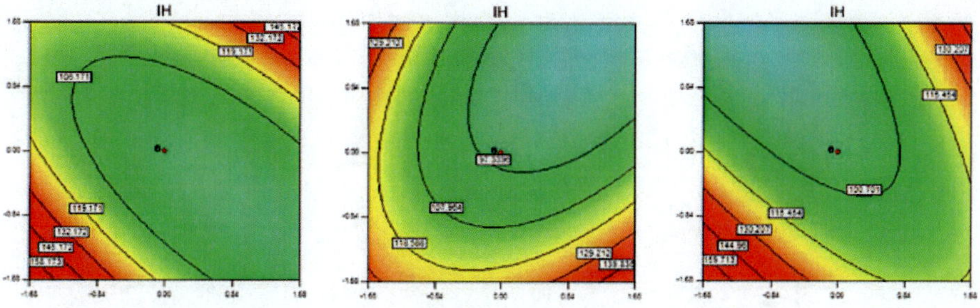

(a) Tool rotational speed vs dwell time (b) Tool rotational speed vs plunge depth (c) Dwell time vs plunge depth

Fig. 19.5 IH contour plots. **a** Tool rotational speed versus dwell time. **b** Tool rotational speed versus plunge depth. **c** Dwell time versus plunge depth

dwell time, Fig. 19.4b TSFL contour plots—Tool rotational speed versus plunge depth, Fig. 19.4c TSFL contour plots—dwell time versus plunge depth.

The contour plots for IH model developed are given in Fig. 19.5a IH contour plots—tool rotational speed versus dwell time, Fig. 19.5b IH contour plots—tool rotational speed versus plunge depth, Fig. 19.5c IH contour plots—dwell time versus plunge depth. The surface plots developed for TSFL model are given in Fig. 19.6a. TSFL surface plot—Tool rotational speed versus dwell time, Fig. 19.6b. TSFL surface plot—Tool rotational speed versus plunge depth, Fig. 19.6c. TSFL surface plot—dwell time versus plunge depth. The best conditions are appeared to be at middle estimations of all the elements. The surface plots developed for IH model are given in Fig. 19.7a. IH surface plot—tool rotation speed versus dwell time, Fig. 19.7b. IH surface plot—tool rotation speed versus plunge depth Fig. 19.7c. IH surface plot—dwell time versus plunge depth. The cube representation of the developed TSFL model is shown in Fig. 19.8 and that for IH model is shown in Fig. 19.9. Post-analysis of the developed plots it was observed that the maximum tensile shear failure load value was found to be 3840 N and the minimum interface hardness value obtained was 97 HV. Process parameters for obtaining high tensile strength and minimal harness were found, tool rotational speed of 1350 rpm, dwell time of 14.5 s, plunge depth of 1.35 mm. For confirming the predicted mode, validation experiments were conducted with the obtained process parameter values, and the results obtained were within 0.1% error, which determines that the developed model is of high accuracy.

19.4 Conclusions

Thus, a new optimization technique—Response Surface Methodology has been successfully used for developing empirical models for the prediction of the process parameter selection for maximum tensile properties and minimum hardness for the

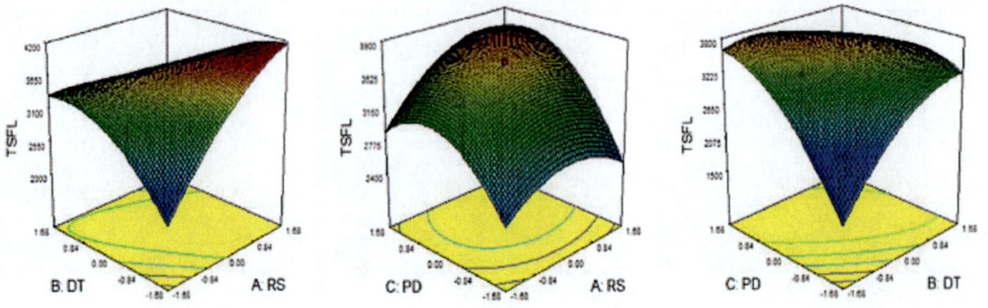

(a) Tool rotational speed vs dwell time

(b) Tool rotational speed vs plunge depth

(c) Dwell time vs plunge depth

Fig. 19.6 TSFL surface plots. **a** Tool rotational speed versus dwell time. **b** Tool rotational speed versus plunge depth. **c** Dwell time versus plunge depth

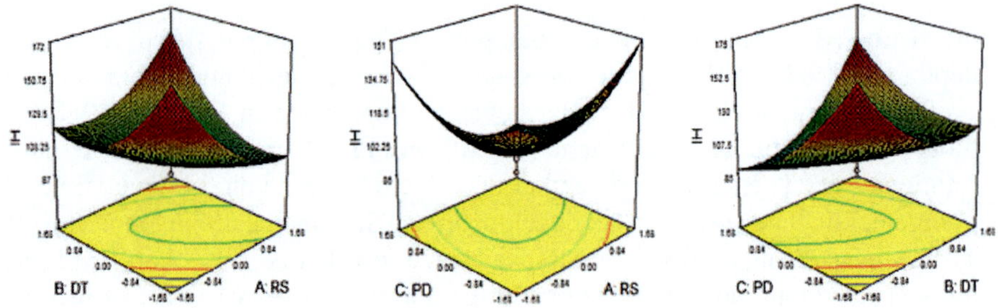

(a) Tool rotational speed vs dwell time

(b) Tool rotational speed vs plunge depth

(c) Dwell time vs plunge depth

Fig. 19.7 IH surface plots. **a** Tool rotational speed versus dwell time. **b** Tool rotational speed versus plunge depth. **c** Dwell time versus plunge depth

Fig. 19.8 TSFL cube model

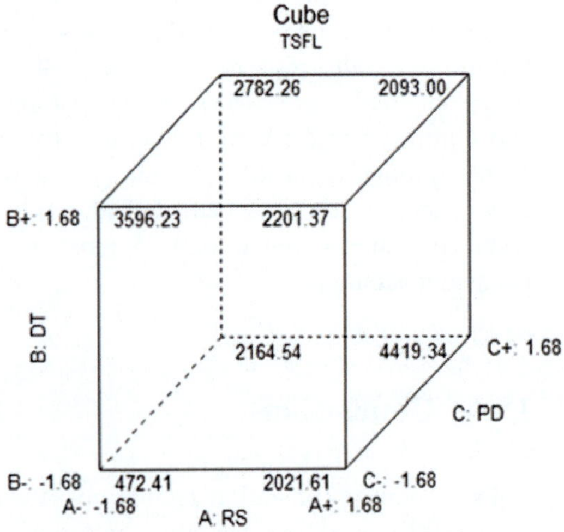

Fig. 19.9 IH cube model

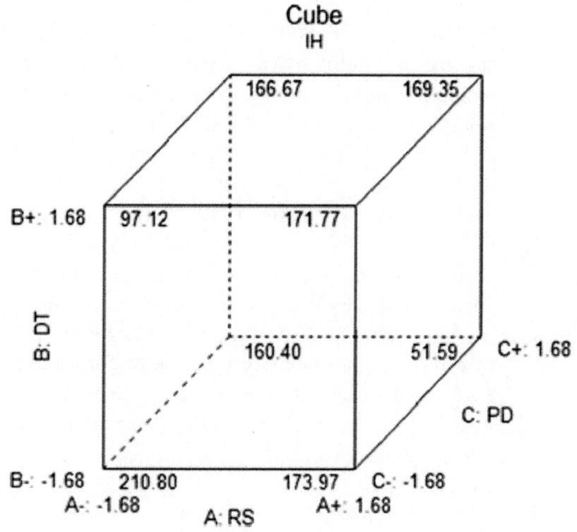

friction stir spot welded Al/Cu dissimilar joints. A central composite design matrix was chosen for the developmental model with 20 encoded values, using ANNOVA the significance of the model was analyzed with a confidence level of 95%. Contour and surface plots were generated to obtain the maximum TSFL value to be 3840 N and minimum hardness to be 97 HV.

References

1. Khuri, S. Mukhopadhyay, Response surface methodology. WIREs Comp Stat. **2**, 128–149 (2010)
2. M. Karimi, A. Keyhani, A. Akram, M. Rahman, B. Jenkins, P. Stroeve, Hybrid response surface methodology-genetic algorithm optimization of ultrasound-assisted transesterification of waste oil catalysed by immobilized lipase on mesoporous silica/iron oxide magnetic core-shell nanoparticles. Environ Technol. **34**, 2201–2211 (2013)
3. P. Athi Sankar, R. Machavaram, K. Shankar, System identification of a composite plate using hybrid response surface methodology and particle swarm optimization in time domain. Measurement. **55**, 499–511 (2014)
4. M.A. Mohammad, M. Eshan, J.J. Mostafa, A hybrid response surface methodology and simulated annealing algorithm. in *2011 International Conference on Computer Communication and Management Proc .of CSIT*, vol. 5 (IACSIT Press, Singapore, 2011)
5. N. Namura, K. Shimoyama, S. Jeong, S. Obayashi, Kriging/RBF-hybrid response surface method for highly nonlinear functions. in 2011 *IEEE Congress of Evolutionary Computation (CEC)* (New Orleans, LA 2011), pp. 2534−2541
6. N. Saibaba, P. King, Modelling and optimization of dye removal process using hybrid response surface methodology and genetic algorithm approach. J Fundam Renew. Energy Appl **4**, 126 (2013)
7. T. Yang, L. Tseng, Solving a multi-objective simulation model using a hybrid response surface method and lexicographical goal programming approach-a case study on integrated circuit ink-marking machines. Part Spec. Issue: Process. OR. **53**(2), 211–221 (2002)

8. G. Vairamani, T. Senthilkumar, S. Malarvizhi, V. Balasubramanian, Application of response surface methodology to maximize tensile strength and minimize interface hardness of friction welded dissimilar joints of austenitic stainless steel and copper alloy. Trans. Nonferrous Met. Soc. China **23**, 2250–2259 (2013)
9. P. Yang, Y.-W. Liu, G.-Y. Zhong, Prediction and parametric analysis of acoustic streaming in a thermoacoustic stirling heat engine with a jet pump using response surface methodology. Appl. Therm. Eng. **103**, 1004–1013 (2016)
10. G. Sodeifian, N.S. Ardestani, S.A. Sajadian, S. Ghorbandoost, Application of supercritical carbon dioxide to extract essential oil from cleome coluteoides boiss: experimental, response surface and grey wolf optimization methodology. J. Supercrit. Fluids. **114**, 55−63
11. M. Mamourian, K.M. Shirvan, S. Mirzakhanlari, Two phase simulation and sensitivity analysis of effective parameters on turbulent combined heat transfer and pressure drop in a solar heat exchanger filled with nanofluid by response surface methodology. Energy. **109**, 49−61
12. Z. Kong, M. Li, J. Chen, Y. Bao, B. Fan, F. Francis, X. Dai, Processing factors of triadimefon and triadimenol in barley brewing based on response surface methodology. Food Control **64**, 81–86 (2016)
13. G.E.P. Box, K.B. Wilson, On the experimental attainment of optimum conditions (with discussion). J. R. Stat. Soc. Ser. **B13**(1), 1–45 (1951)
14. M. Soltani, J. Soltani, Determination of optimal combination of applied water and nitrogen for potato yield using response surface methodology (RSM). J. Biosci. Biotechnol. Res. Commun. **9**(1), 46–54 (2016)
15. R.H. Myers, Response surface methodology-current status and future directions, J. Qual. Technol. **31**(1), 30–44 (1999)
16. H. Badarinarayan, Fundamentals of friction stir spot welding. (PhD thesis, Missouri University of Science and Technology, United State, 2009)
17. D.C. Montgomery, *Design and analysis of experiments*, 4th edn. (John Wiley & Sons, New York, 2001)
18. R. Paventhan, P.R. Lakshminarayanan, V. Balasubramanian, Prediction and optimization of friction welding parameters for joining aluminium alloy and stainless steel. Trans. Nonferrous Metals Soc. China **21**, 1480–1485 (2011)
19. Miller, J.E. Freund, R. Johnson, Probability and statistics for engineers, Vol. 5. (Prentice Hall, New Delhi, 1996)

Chapter 20
Surface Morphology Studies on Al-Steel Dissimilar Friction Stir Spot Joints

Siddharth Sampathkumar

Abstract Friction stir spot welding, a linear variant of friction stir welding has been widely used in aerospace and automotive industries for spot joining of dissimilar materials. Due to its environment-friendly nature and its consumable free material joining process, it has gained wide recognition in joining materials such as aluminum and steel. In this investigation, friction stir spot welding process was used to join aluminum alloy Al 5083 and SPCC steel plates. The effect of important friction stir spot welding process parameters such as tool rotational speed, plunge depth and dwell time on joint strengths was studied. The effect of stir on the plate deformation and the changes induced in grain structure were evaluated.

Keywords Friction stir spot welding · Dissimilar joints · Micro-structure

20.1 Introduction

Dissimilar material joining poses much difficulty owing to the difference in mechanical and metallurgical properties of the dissimilar materials. Conventional fusion welding process for joining dissimilar materials with difference in melting point, thermal coefficient of expansion and density causes lots of issues. Friction stir spot welding (FSSW), which is a solid-state material joining process, can be used for joining such dissimilar combinations easily [1].

It is a three-step process (Fig. 20.1) involving (a) *plunging*—a high-speed non-consumable tool is forced into the sheets, (b) *stirring*—for a specific time duration, frictional heat is induced in the weld zone and (c) *drawing out*—withdrawal from the location to allow the spot to cool [2]. A typical cross-section of a FSSW joint (Fig. 20.2) consists of four distinct zones such as (i) parent material (PM)—the area which is unaffected by heat, (ii) heat affected zone (HAZ) —region where the grains

S. Sampathkumar (✉)
Department of Mechanical Engineering, PSN College of Engineering and Technology, Tirunelveli, Tamil Nadu 627152, India
e-mail: sksmsiddharth1@yahoo.in

A. K. Sahu et al. (eds.), *Energy Storage and Conservation*, Springer Proceedings in Energy, https://doi.org/10.1007/978-981-99-2870-5_22

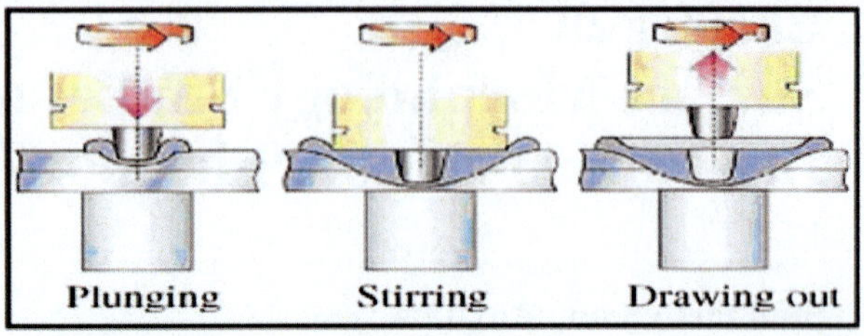

Fig. 20.1 Three-step friction stir spot welding process—plunging, stirring and drawing out [2]

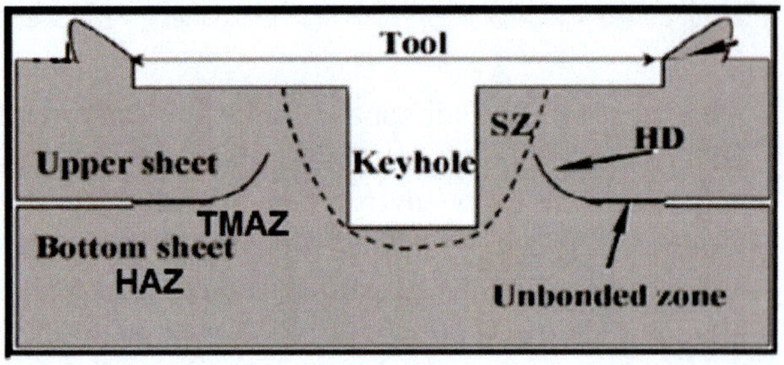

Fig. 20.2 Schematic representation of FSSW cross-section [3]

have undergone a distinct change but no plastic deformation occurs, (iii) thermo-mechanically affected zone (TMAZ)—the region near weld zone where significant grain modifications due to effect of heat and presence of plastic deformation is seen, (iv) Stir zone (SZ)—region closest to weld spot which has undergone completer recrystallization [3].

Investigations in dissimilar joints of aluminum and steel using FSSW techniques have been reported in literatures. Chen et al. [4], produced high-quality dissimilar joints of 1 mm thick aluminum AL 6111-T4 andDC04 low carbon steel sheet using FSSW process within a weld time of 1 min. Sun et al. [5], joined Aluminum 6061 alloy with mild steel using FSSW technique and observed no intermetallic compounds, which is desirable for high-quality joints. It was also observed that the pin length had less effect on the joint properties, which is important for extension of tool life. Bozzi et al. [6], joined aluminum AA6016 with galvanized IF-steel using FSSW process and found that the thickness of intermetallic compounds increased with increase in tool rotational speed and penetration depth. Figner et al. [7] joined aluminum AA5754-H111 and HX340 LAD steel with FSSW technique and observed that weld strength can be significantly improved by selecting proper FSSW process parameters such as spindle speed and dwell time. From the literature survey, conducted study on the grain structure on dissimilar joints of aluminum and steel was found to be very less or nil. In this investigation, the grain structure and changes in the grain morphology due to the effect of heat input from the rotating non-consumable tool have been studied.

Table 20.1 Nominal chemical composition of the base materials (wt. %)

Al 5083	Si	Mg	Mn	Fe	Cr	Cu	Zn	Ti	Al
	0.4	4.0–4.9	0.4–1.0	0.4	0.25	0.1	0.25	0.15	Balance
SPCC	C	Mn	P	S	Si	Cu	Zn	Ti	Fe
	0.25	0.6	0.035	0.04	0.10	0.2	–	0.05	Balance

Table 20.2 Important mechanical properties of the base materials

Material	Yield Strength (MPa)	Tensile strength (MPa)	Elongation (%)	Hardness (HV)
Al 5083	145	290	22	87
SPCC	185	340	41	185

20.2 Materials and Methods

In this investigation, aluminum alloy 5083—H111 of 1.5 mm thickness (as per ASM standards) and G3141 SPCC steel plate of 1.5 mm thickness (as per JIS standards) were used. The plates were sized to a dimension of 100 mm by 30 mm. An overlap of 30 mm was given for the joints. The chemical composition of the base materials is given in Table 20.1. The important mechanical properties of the base materials are given in Table 20.2.

The non-consumable rotating tool was made with H13 tool steel material. A cylindrical straight tool nomenclature was selected. The tool shoulder diameter was chosen as 16 mm, with cylindrical probe of 6 mm diameter and 1.25 mm length. Overall length of tool was 125 mm. The sized and surface-cleaned plates are shown in Fig. 20.3a. The prepared tool is shown in Fig. 20.3b. The FSSW process setup with the plates fixed in clamped fixture and the tool in the rotating spindle is shown in Fig. 20.3c. The vertical milling CNC machine used for performing the FSSW experiments is shown in Fig. 20.3d. The important process parameters that influence the quality of FSSW joints are tool rotational speed in rpm, dwell time in sec, plunge depth in mm [8]. By conducting trial experiments the feasible limits of the process parameters were selected. The joint visual appearance, depth of keyhole, cracks on the macro surface, flash formation were the criteria employed for finding the feasible limits. Joints were fabricated with the process parameters given in Table 20.3.

20.3 Results and Discussion

Three sets of samples were welded for each criterion and the strength of the joints was tested for mechanical properties. Schematic representation of the specimens prepared for tensile tests is shown in Fig. 20.4. The tensile tests were conducted using a 100 kN electro-mechanically controlled universal testing machine.

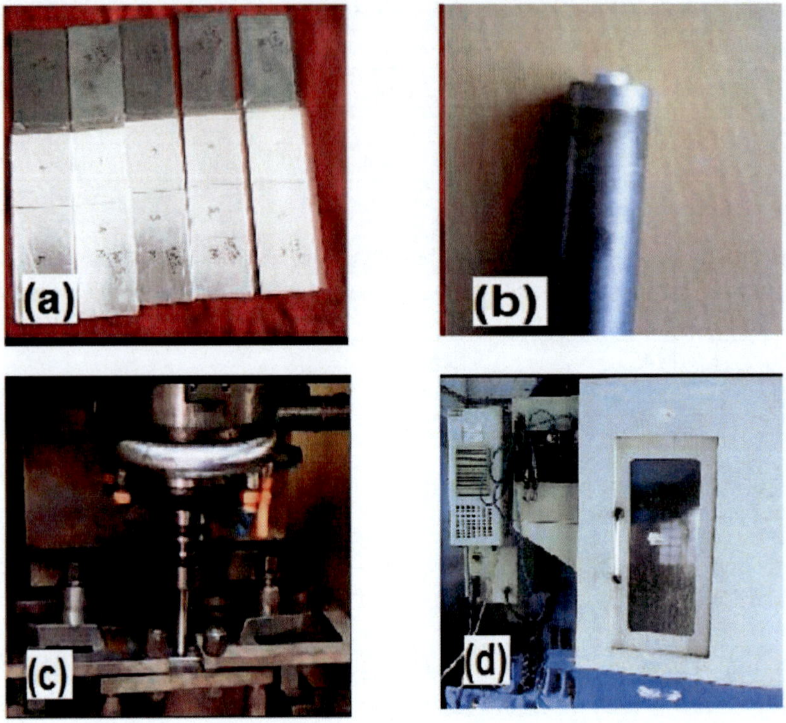

Fig. 20.3 **a** Lap configuration of Al/SPCC plates, **b** FSSW tool, **c** FSSW process setup and **d** equipment used for FSSW experiments

Table 20.3 Friction stir spot welding parameters

Experiment Number	Rotational speed rpm	Dwell time sec	Plunge depth mm	Feed rate mm/min
1	950	15	1.7	15
2	1175	14	1.6	15
3	1300	12	1.4	15
4	1475	9	1.3	15
5	1600	7	1.2	15

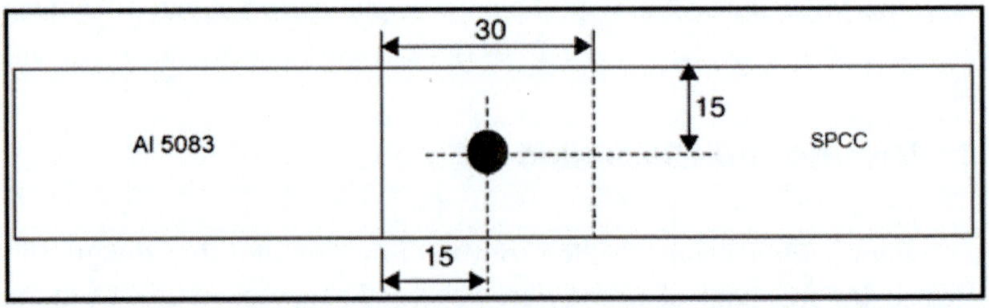

Fig. 20.4 Tensile test specimens

Table 20.4 Results of tensile and micro-hardness experiments with macrostructure

Experiment No	TSFL (kN)	Micro-hardness (HV)
1	8.9	113
2	11.3	127
3	12.6	124
4	11.7	131
5	8.6	129

All dimensions are in millimeters.

As per ASTM specifications, the specimens were loaded at a rate of 1.5 kN/min till fracture of the specimens. The hardness at the interface of the joints was tested using Vickers micro-hardness testing machine. The values of tensile shear failure load and the micro-hardness at the joint interface are given in Table 20.4. The macrostructure of the joints with keyhole dimensions and flash height is also shown in Table 20.5.

Comparison of tensile shear failure load and micro-hardness of the joints is given in Fig. 20.5. For a certain extent, the tensile shear failure load value increases. Beyond certain speed of rotation, excessive heat causes reduction in bond strength. The micro-hardness distribution gives an indication regarding the effect of stir of the tool. Scanning electron microscopy was used to investigate the microstructure of the joints. The unaffected parent material is the region more than 16 mm away from the weld center. This region is not at all affected by the heat generated by the welding process. The microstructure of the third specimen is taken for consideration.

The microstructure of the parent material (Fig. 20.6a) of SPCC rolled steel plate indicates ferrite. It is fine-grained and equiaxed. Heat affected zone reveals (HAZ) (Fig. 20.6b) ferrite matrix consisting of bainite [9]. Similar microstructural investigations were done using AISI 1005 steel by W. Zhang [10].

In those, experiments, coarse, fine and partially transformed grain regions were found. As the proximity towards weld center is increased, the thermo-mechanically affected zone (Fig. 20.6c) corresponds to α-ferrite to ¥-austenite toα–ferrite. The microstructure of parent Al 5083 base material (Fig. 20.7a) consists of equiaxed secondary phase particles. The effect of rolling on the grains is more pronounced. The heat affected zone (Fig. 20.7b) consists of inter-granular precipitates dispersed in trans-granular precipitates. The thermo-mechanically affected zone (Fig. 20.7c) undergoes distinct change in microstructure. Deformed pancake grains get elongated owing to thermal stresses due to frictional stir.

The nugget zone or stir zone is the region directly under the tool pin. Maximum heat and mechanical stir deform the grains. The temperature is raised significantly and as the joints re-crystallize, the grain structure gets changed. Pan cake-shaped Al structures are visible in Fig. 20.8a (at 973 kx magnification). Erratic and chaotic mix due to the mechanical trans-location of the material is seen in Fig. 20.8b. Certain regions are smoothly distributed with a fine layer of Al over SPCC (Fig. 20.8c). Finely fused layer of Al is visible. The nugget center zone is completely re-crystallized and a nearly steady boundary separating the two materials is seen in Fig. 20.8d.

Table 20.5 Macrostructure of the joints fabricated at different process parameter values

Experiment No	TSFL In kN	Micro hardness HV	Macrostructure	Few of the welded joints
1	8.9	113	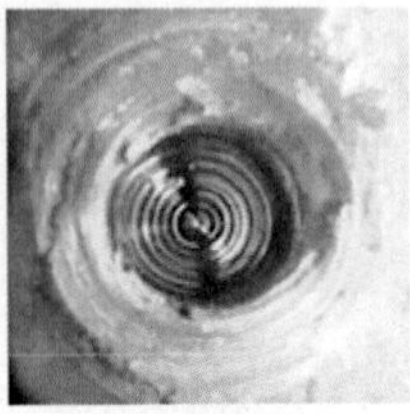	
2	11.3	127		
3	12.6	124	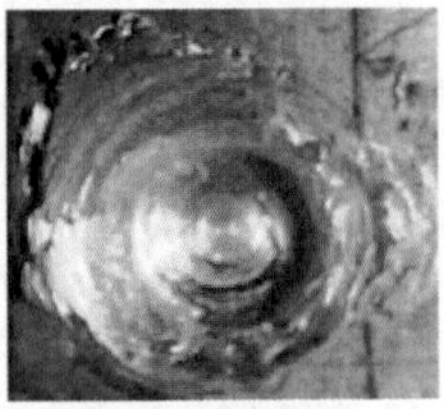	
4	11.7	131	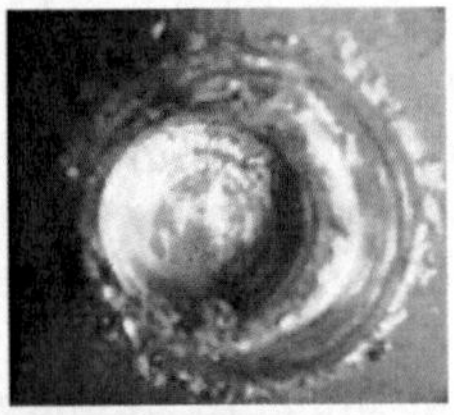	
5	8.6	129	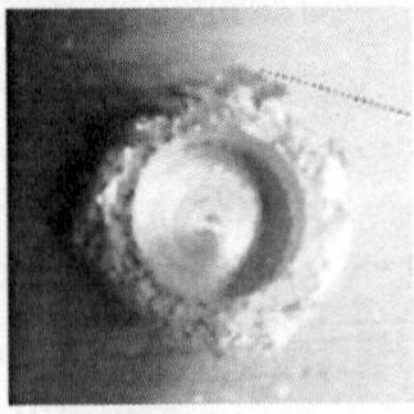	

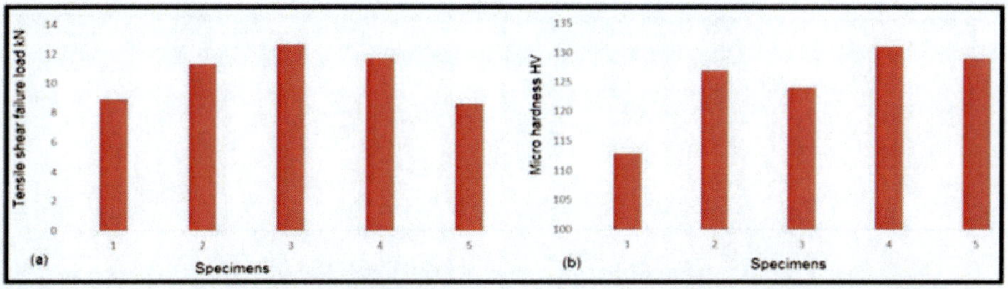

Fig. 20.5 **a** Comparison of tensile shear failure loads (kN), **b** Comparison of micro-hardness

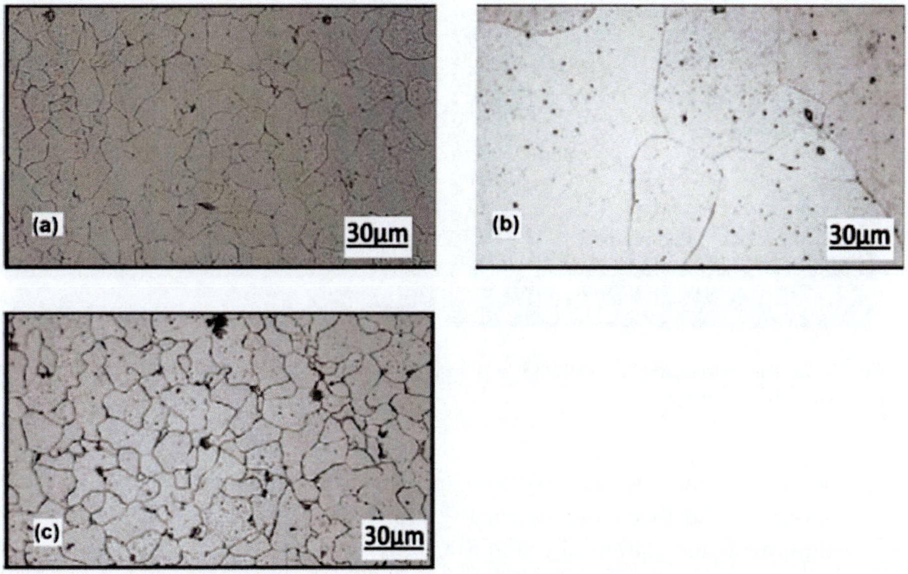

Fig. 20.6 Microstructures of SPCC **a** Parent material, **b** Heat affected zone, **c** Thermo-mechanically affected zone

20.4 Conclusions

Thus using friction stir spot welding technique, dissimilar joints of Al 5083—H111 and SPCC were made, and the following observations were made:

(a) The effect of the important process parameters such as tool rotational speed in rpm, dwell time in seconds, plunge depth in mm on the quality was studied.
(b) The variations in tensile shear failure load and micro-hardness at the interface for the joints were studied.
(c) The grain morphology and the changes induced by the plunge and the heat induced by the non-consumable tool at various zones were studied.

Fig. 20.7 Microstructures of Al5083 **a** Parent material, **b** Heat affected zone, **c** Thermo-mechanically affected zone

(d) As the proximity towards weld center increases the grains were more thermally affected. The thermo-mechanically affected zones of Al 5083 were heavily deformed and phase changes in SPCC were observed.

(e) The nugget zone was completely re-crystallized with fine distribution of Al over SPCC and chaotic in certain regions.

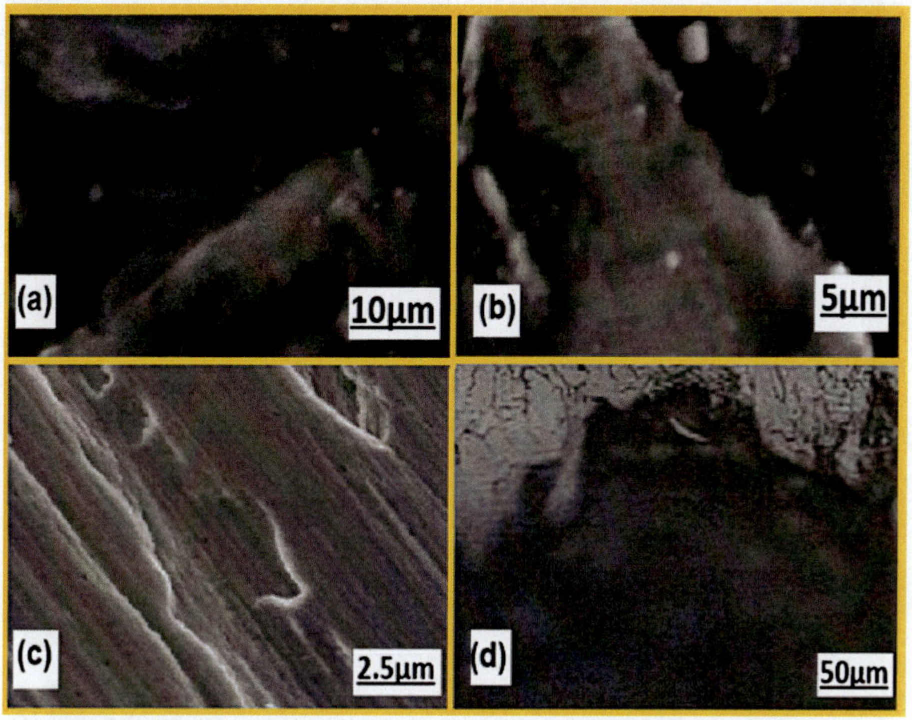

Fig. 20.8 Region directly under pin: **a** Elongated grain structure, **b** Chaotic mix of Al and SPCC, **c** Finely fused layer of both the alloys, **d** Fine-grained with sparsely distributed Al

References

1. M.P. Mubiayi, E.T. Akinlabi, Friction stir spot welding of dissimilar materials-an overview. in *Proceedings of the World Congress on Engineering and Computer Science 2014 Vol II WCECS 2014* (San Francisco, USA, 2014), pp 22–24
2. M. Merzoug, M. Mazari, L. Berrahal, A. Imad, Parametric studies of the process of friction spot stir welding of aluminium 6060–T5 alloys. Mater. Des. **31,** 3023−3028
3. R.Z. Xu, D.R. Ni, Q. Yang, C.Z. Liu, Z.Y. Ma, Pin-less friction stir spot welding of Mg–3Al–1Zn alloy with Zn interlayer. J. Mater. Sci. Technol. **32**(1), 76–88 (2016)
4. Y.C. Chen, A. Gholinia, P.B. Prangnell, Interface structure and bonding in abrasion circle friction stir spot welding: a novel approach for rapid welding aluminium alloy to steel automotive sheet. Mater. Chem. Phys. **134**, 459–463 (2012)
5. Y.F. Sun, H. Fujii, N. Takaki, Y. Okitsu, Microstructure and mechanical properties of dissimilar Al alloy/steel joints prepared by a flat spot friction stir welding technique. Mater. Des. **47**, 350–357 (2013)
6. S. Bozzi, A.L. Helbert-Etter, T. Baudin, B. Criqui, J.G. Kerbiguet, Intermetallic compounds in Al 6016/IF-steel friction stir spot welds. Mater. Sci. Eng., A **527**, 4505–4509 (2010)
7. G. Figner, R. Vallant, T. Weinberger, H. Schrottner, H. Pasic, N. Enzinger, Friction stir spot welds between aluminium and steel automotive sheets: influence of welding parameters on mechanical properties and microstructure. Weld. World **53**, R13–R23 (2009)
8. H. Badarinarayan, Fundamentals of friction stir spot welding. (PhD thesis, Missouri University of Science and Technology, United State, 2009)

9. M. Okayasu, Y. Ohkura, T. Sakamoto, S. Takeuchi, H. Ohfuji, T. Shiraishi, Mechanical properties of SPCC low carbon steel joints prepared by metal inert gas welding. Mater. Sci. & Eng. AA **560**, 643–652 (2013)
10. W. Zhang, J.W. Elmer, T. DebRoy, Modeling and real time mapping of phases during GTA welding of 1005 steel. Mater. Sci. & Eng. **333**(1–2), 320–335

Chapter 21
An Investigation into the Use of Online Assessments and Its Impact on Students' Learning Outcomes: Evidence from Universities of United Arab Emirates

Sarwat Jahan, Shadi Hijazi, Zafarullah Khan, and Mohammed Ismail Iqbal

Abstract Assessment of knowledge is regarded as one of the important components in achieving the desired learning goals by students. Besides the importance of quantifying the knowledge of students, online assessments play various other roles like stimulating the learning process and assisting in the development of skills required. This paper focuses on the impact of online assessments and evaluation methods on learning outcomes of students. For this purpose, the research design selected was of quantitative nature. Due to the quantitative nature of study as well as the formulation of hypotheses for the study, deductive approach has been selected as the main research approach. The quantitative data have been collected with the help of a survey questionnaire comprising of five-point Likert scale. The population of the study are the students in UAE pursuing higher education, i.e. after passing their schools. From the population, a random sample of a total of 350 students has been selected irrespective of any bias and on the basis of purposive sampling technique. The findings of this study have revealed that online assessment has significant positive impact on learning outcomes, and online assessments have enabled students to demonstrate the level of new information they gained during the course. Furthermore, this investigation infers that online assessment exercises help students to understand and analyze

S. Jahan (✉)
School of Business, City University Ajman, Ajman, United Arab Emirates
e-mail: sarwat_usman@hotmail.com

S. Hijazi
Continuing Education and Talent Development Center, General Directorate of Residency and Foreigners Affairs, Dubai, UAE

Z. Khan
Faculty of Business and Accountancy, Lincoln University College, Petaling Jaya, Malaysia

M. I. Iqbal
Faculty of Engineering Department, University of Technology and Applied Sciences, Nizwa, Oman

A. K. Sahu et al. (eds.), *Energy Storage and Conservation*, Springer Proceedings in Energy, https://doi.org/10.1007/978-981-99-2870-5_23

the topic after the lecture, as well as the online assessment tasks, encourage students to demonstrate comparative skills and the assessment methods are relevant to the course material and learning outcomes.

21.1 Introduction

To identify the progress of cognitive and interactive processes in online learning, it is necessary to have a sensible and valid performance evaluation strategy. These must be combined with a series of tools to detect complex changes in the knowledge constructions of the students. These tools can be used to regulate the knowledge construction in students [33]. There are three stages in the assessment of online learning (the same in case of traditional learning for that matter). The first is the initial assessment which is aimed to gain reliable idea of the level of knowledge and skills of the students. The second is the formative assessment which is carried out during the teaching course, at established intervals as per the course design and planning. The third is the summative assessment which is carried out at the end of the course [1].

Given the characteristics of online education, there are communication limitations inherent in the nature of the internet as medium, which in turn distinguish online assessment from traditional face to face assessment. The underlying purpose of assessment is to evaluate performance and provide feedback to students [2]. Student assessment-focused studies stated that the attributes of an online assessment include reliability [3], validity [3], objectivity [4] and authenticity [4]. Reliability refers to the confidence that an instrument generates to reflect the student's level of achievement [3]. The validity refers to the instrument measuring what is really intended, and not something else [3]. The objectivity refers to the neutrality with which students are graded [4]. Authenticity relates to the contents of the assessment with the practical reality which reflects the skills and competences that students gain to get ready for practical world [4]. Literature review indicates that although much has been written about online assessment, however, there are few studies analyzing assessment with these factors.

In the research on online learning, Batu et al. (2018) analyzed many studies in which the effectiveness of online courses is analyzed and reported that fundamentally following assessment strategies are used. The first category is the knowledge questionnaires that are conducted at the end of the courses. These are critiqued for their assurance of reliability, validity or authenticity, although they present a certain guarantee of objectivity, inherent to the online medium. The second assessment methods are the interviews or using indices/inventories that probed student satisfaction in the course (implying that satisfaction and learning are equal). This article mentions [5] who differentiates both concepts and argued that two constructs are not comparable, since the satisfaction variable confuses motivational and instructional effects. This study agreed with [6] to posit that the variety of online learning assessment resources should include objective tests, essays, projects, rubrics, concept mapping, and/or evidence-based assessment. A possible classification of the types of online assessment: (a) score-based assessment such as learning needs analysis; (b) elaborative assessment such as essays, concept maps, projects, etc.; and (c) collaborative assessment, in which students are assessed in group work such as peer review [6].

According to [7], although assessment is considered a fundamental part of the online learning process, and even when a great potential of technology is recognized to create effective assessment systems, there is no relevant development of technological applications to carry it out. Research in the evaluation of online learning is scarce, and among other things, this is why the knowledge that is had in this regard is unsatisfactory (Ho, et al., 2018). Therefore, this study is based on the rationale that online assessments must have aforementioned features to fulfill their purpose, i.e. to regulate or influence students' learning. In other words, this research aims to explore the relationship between online assessment and students' learning outcomes. This study has adopted the conceptual framework based on cognitive task analysis model to design a questionnaire, which is used to understand students' perception about online assessments and its impact on learning outcomes.

21.2 Literature Review

21.2.1 Online Assessment

A critical step in the development of student-centered systems is having quality information about the student's competencies in the educational domain. The technology includes a set of tools with the potential to optimize the evaluation of learning in online environments. In fact, in the online educational modality, assessment should be a continuous activity and the backbone of learning [8]. Willcox et al. [9] argued that in the online education much attention has been paid to the relevance of the contents, the instructional design, the optimal use of technological resources, therefore the meticulous consideration of the assessment of learning has been grossly ignored. Ultimately, assessment is considered to be a weak point or limitation in online learning [7]. Hence, there is a need to improve online assessment with support from theoretical models and empirical evidence. This is the basic purpose of this research work.

The technological infrastructure could allow the flexible and comprehensive recording of performance data and the monitoring of students' progress in their process of cognitive development, i.e. knowledge construction and skills acquisition. Despite such potential, the evaluation of online learning or online assessment is something that needs to be improved, something on which we must reflect in greater depth, which must be developed in a more creative way and return to it by investing more pedagogical and technological resources [10].

According to [11] although assessment is considered a fundamental part of the online learning process, and even when a great potential of technology is recognized to create effective assessment systems, there is no relevant development of technological applications to carry it out. Research in the evaluation of online learning is scarce, and among other things, this is why the knowledge that is had in this regard is unsatisfactory.

21.2.2 Cognitive Task Analysis (CTA)

Fives and Barnes [12] argued that CTA developed complex learning assessment models based on the analysis of the cognitive operations and skills underlying the tasks performed in a given educational domain. An educational domain is a delimited thematic area composed of a set of knowledge that is object of assessment.

To carry out these analyses, in the first instance, it is necessary to identify the competences that make up the educational domain. Once this macrostructure has been identified, a Cognitive Task Analysis is carried out to identify the micro components that constitute each competence. The procedure analyzes the task through steps in which the knowledge, skills and dispositions associated with each step are identified, in a progressive sequence of greater detail and precision [13]. In this process, it is important to identify the cognitive complexity (e.g., understanding, application, problem-solving) and the type of knowledge it involves (e.g., conceptual, structural, causal) [14].

Based on both analyses, a structure is built that identifies fine-grained components. In this way, a universe of measurement is organized that has made it possible to design representative assessment situations of critical competences and their components, in an articulated assessment scheme [12]. The dimensions to perform these classifications are:

A continuum of complexity in terms of the cognitive skills required in the domain includes three categories [15]: (a) understanding of the topics, which includes the recognition of information, classification, and organizing; (b) application of knowledge and skill, including operations such as analysis, extrapolation, inference, and comparison; and (c) problem-solving, which includes operations such as error correction, action planning, evaluation and decision-making.

The complexity dimension in terms of the mental models built by the students. Cognitive skills represent categories of operations that the student can apply in the domain; mental models reflect the integration of knowledge that allow you to explain the reality of phenomena. The inclusion of mental models as a diagnostic axis responds to evidence that students, depending on their level of expertise, could only conceptually describe the domain, or could have a highly structured knowledge about it, for which they could explain it [13]. The mental models that are included in this model are three: (1) conceptual, they answer the question: "what is this?", They describe the meaning of the phenomenon or theme, and the interrelation of the elements that compose it; (2) structural, which answer the question: "how is this structured?", And describe how the conceptual field in question is organized; and (3) causal, which answer the question: "how does this work?", and which describe how

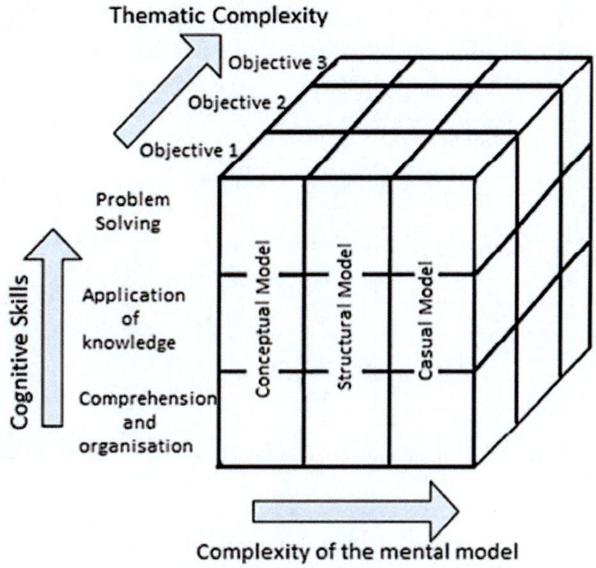

Fig. 21.1 Model of the cognitive analysis of tasks of a domain

the principles affect each other and help to interpret processes, give explanations of events and make predictions (ibid).

The dimension of the complexity of the learning topics, determined by the content units as a universe to be evaluated applying different mental models with different levels of cognitive complexity. This aspect has a logical order from the simple to the complex, depending on the objectives of the units of the chosen learning program [16].

Figure 21.1 shows the structure that would result from integrating the three relevant dimensions of the analysis. In it, each space of the cube includes a category of content that together make up the universe of the educational domain analyzed. This means that assessment items, resources and learning activities can be generated from each box.

21.2.3 Conceptual Model of Cognitive Task Analysis in Online Assessment

Based on the logic of Cognitive Task Analysis, following model is proposed to describe the application of these techniques in online assessment. Figure 21.2 illustrates the proposed model. It includes on the left the modeling of the educational domain, which leads to the instructional design. This makes it possible to propose what is the route to follow to meet the learning objectives and outcomes. The instructional design is a set of decisions about the activities, content and assessments that the course will have, based on the objectives.

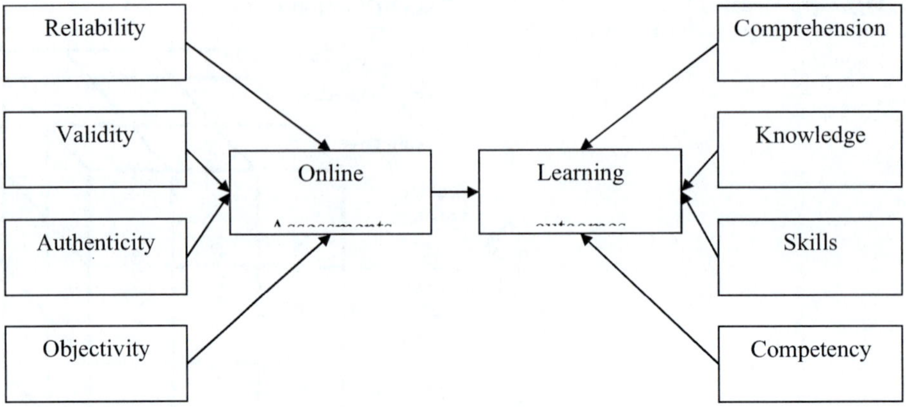

Fig. 21.2 Application model of cognitive task analysis to online assessment

It was previously pointed out that assessment is a weak point of online education, although it is the backbone of learning in these contexts. There is not a significant body of knowledge derived from research in this field. Online assessments can be systematized by applying Cognitive Task Analysis as suggested in the model, so that initial/diagnostic assessment to allow obtaining valid, reliable, objective and authentic assessment results. Since the online environment is especially suitable for this type of assessment, which can include objective reagents, exercises, cases, problems, etc., derived from a design of the assessment that starts from the Cognitive Analysis of Tasks.

Formative assessments represent the most extensive work in online education. They include a wide variety of activities related to the topics of instructional design and related to the fulfillment of course objectives. Some of the formative assessments can be automatic, since interactive exercises can be included based on the levels required for the topics to be reviewed in the course [17]. Other formative evaluations can be elaborative such as essays, project reports, etc., its complexity would also be derived from the levels detected in the Cognitive Task Analysis.

Finally, summative evaluations can also be constructed based on Cognitive Task Analysis. It is suggested that the questionnaire-based assessments used as initial diagnosis be applied again at the end as a post-test; however, equally a special evaluation could also be constructed, depending on the needs of the course in question [18].

21.3 Methodology

In order to apply the conceptual model presented in previous section, deductive research approach was selected. The researcher started by developing a conceptual model and developed tentative hypotheses followed by collection of data from specific settings [19], i.e. higher education students in the UAE. Therefore, this study has attempted to test CTA theory within the specific context of the UAE higher education system and hence deductive reasoning-based research approach has been conducted.

Since the study aims to assess the impact of online assessment on learning outcomes, therefore quantitative data analysis is essential. Hence, the research method applied in the research process is quantitative method. The quantitative method is distinguished from others because it uses quantitative data only [20]. Furthermore, selection of quantitative data is also suitable for this study because it provides high level of reliability and validity as compared to qualitative data. In addition, quantitative methods were preferred over mixed method because the former provides higher time and cost efficiency [21]. Furthermore, the study is based on cross-sectional time horizon in which data were collected from a sample of higher education students in the UAE at a specific point in time. The cross-sectional design suits this study as compared to longitudinal design because the former facilitates small-scale research with better time and cost efficiency as compared to the latter where multiple data collection at different time points are essential requirements [22].

For data collection, this study developed a self-administered questionnaire. The self-administered questionnaire was designed to apply the conceptual framework designed above. The research used closed ended statements to operationalize the variables. The Likert five-point scale system was used to obtain ratings with respect to responses of students to the statements. In order to establish the reliability of questionnaire, this study assessed internal consistency using Cronbach's alpha technique [23], the results of which are reported in following section.

With respect to sampling technique, the researcher used random sampling method. A sample of 350 students was selected from five Universities of UAE including City University College of Ajman, Al Ghurair University Dubai, Capital College Dubai, Manipal University Dubai and Skyline University College Sharjah. Initially, the researcher designed the questionnaire and sent request for participation to 500 students; however, only 396 students responded back. However, during the data collection period, it was realized that 46 questionnaires were incomplete and hence were excluded from research process.

21.4 Results

Based on the data collected through questionnaire, this study explored the relationship between online assessments and learning outcomes using correlation and regression techniques. Online assessments were assessed in terms of their reliability, validity, authenticity, and objectivity, while learning outcomes were operationalized in terms of understanding, knowledge, skills, and problem-solving. Following table provides correlations for the aforementioned variables.

The correlation between reliability and learning outcomes is ($r = 0.9$, $p = 0.000$) from which it can be inferred that there is strong positive impact of online assessment on learning outcomes as the former has level of reliability to assess learning outcomes. In other words, from perception of students, the online assessments are able to check students' achievement and performance. Furthermore, validity as a feature of online assessment also shows very strong and positive correlation with learning outcomes ($r = 0.899$, $p = 0.000$). These results shows that online assessment are able to check learning outcomes with high level of validity. In other words, online assessments serve their purpose and check what they are aimed to check. Thirdly, the correlation between authenticity and learning outcomes is ($r = 0.898$, $p = 0.000$), which leads to the assertion that online assessments are able to check skills and competences of students. This result also reinforces positive relationship between online assessment and learning outcomes. Finally, the correlation between objectivity and leaning outcomes is ($r = 0.914$, $p = 0.000$), which indicates that online assessment shows high level of objectivity which means that they are able to test students fairly, equally, and objectively.

Correlations

		Reliability	Validity	Authenticity	Objectivity	Learning outcomes
Reliability	Pearson correlation Sig. (2-tailed) N	1 350	0.906 0.000 350	0.907 0.000 350	0.892 0.000 350	0.900 0.000 350
Validity	Pearson correlation Sig. (2-tailed) N	0.906 0.000 350	1 350	0.333 0.000 350	0.903 0.000 350	0.899 0.000 350
Authenticity	Pearson correlation Sig. (2-tailed) N	0.907 0.000 350	0.383 0.000 350	1 350	0.898 0.000 350	0.898 0.000 350
Objectivity	Pearson correlation Sig. (2-tailed) N	0.392 0.000 350	0.903 0.000 350	0.398 0.000 350	1 350	0.914 0.000 350
Learning outcomes	Pearson correlation Sig. (2-tailed) N	0.900 0.000 350	0.399 0.000 350	0.398 0.000 350	0.914 0.000 350	1 350

**Correlation is significant at the 0.01 level (2-tailed)

Despite the fact that correlation coefficients provide valuable and clear insights about relationship between online assessment and learning outcomes, yet the inherent limitation in the coefficients requires further analyses. The main limitation is that correlation only reflects nature and strength of change occurring on learning outcomes due to change in online assessment and does not gauge the impact of the former on the latter [24]. Hence, regression model was constructed to overcome these limitations.

The regression model below uses learning outcomes as dependent variable while online assessment indicators as independent variable which are reliability, validity, authenticity and objectivity. The coefficient of determination in the model is 0.884. On the one hand, it shows that the model is sound and that online assessments explain 88.4% of change or variability in learning outcomes. Therefore, it can be concluded that online assessments have strong impact on learning outcomes. Furthermore, the significance level in the model is 0.000 < 0.05 implying that the impact of online assessment is significant. Finally, the beta values indicate the predicted change in learning outcomes provided that there is a unit change in online assessment. The model predicts that there is likely to be a 0.205 unit increase in learning outcomes if reliability of the online assessment increase by 1 unit ($\beta = 0.205$). Furthermore, the model predicts that there is likely to be a 0.208 unit increase in learning outcomes if validity of the online assessment increase by 1 unit ($\beta = 0.208$). Furthermore, the model predicts that there is likely to be a 0.206 unit increase in learning outcomes if authenticity of the online assessment increase by 1 unit ($\beta = 0.206$). Finally, the model predicts that there is likely to be a 0.205 unit increase in learning outcomes if objectivity of the online assessment increase by 1 unit ($\beta = 0.361$).

Model Summary

Model	R	R square	Adjusted R Square	Std. error of the Estimate
1	940a	0.884	0.883	0.40761

[a]Predictors: [Constant), Objectivity[1], Reliability, Authenticity, Validity

ANOVflª

Model	Sum of squares	df	Mean square	F	Sig
1 Regression	437.119	4	109.280	657.734	0.000[b]
Residual	57.320	345	0.166		
Total	494.439	349			

Dependent Variable: Learning outcomes
Predictors: [Constant), Objectivity, Reliability, Authenticity, Validity

Coefficients[3]

Model	Unstandardized coefficients		Standardized coefficients	t	Sig
	B	Std. error	Beta		
1 [Constant)	0.062	0.074	0.204	0.826	0.409
Reliability	0.205	0.053	0.210	3.895	0.000
Validity	0.208	0.050	0.209	4.167	0.000
Authenticity	0.206	0.049	0.355	4.198	0.000
Objectivity	0.361	0.051		7.080	0.000

[a]Dependent Variable: Learning outcomes

Lastly, in order to further enhance the methodological rigor of the results above, this study calculated the Cronbach's alpha coefficient to check the internal consistency of questionnaire. The internal consistency refers the degree to which items in the questionnaire are able to collect what they are aimed for. The coefficient reported is 0.955 > 0.7 and indicates that there is very high level of internal consistency and hence reflects high level of reliability in the questionnaire.

Case Processing Summary

	N	%
Cases Valid	350	100.0
Excluded[3]	0	0.0
Total	350	100.0

[a]Listwise deletion based on all variables in the procedure

Reliability Statistics

Cronbach's alpha	N of items
0.955	**18**

21.4.1 Discussion

The model presented adapts a proven methodology in the area of the evaluation of complex learning in face-to-face higher education, to incorporate it into virtual environments as a mechanism that can give structure and order to content, materials, interactions and assessments in an online learning environment. In this way, the competencies and their components, which represent the learning goals/outcomes of educational programs, are the central axis of work. Students interact with objects that represent the levels of complexity expected of their performance (Batu et al. 2018).

Once the contents have been analyzed and the competences of the educational domain have been identified, the Cognitive Task Analysis leads to a powerful assessment scheme that allows diagnosing the structuring of students' knowledge throughout online courses, taking into account dimensions such as training of mental models as cases of knowledge complexity, as well as skill levels, as cases of cognitive complexity, throughout the dimension of the subject of the course [25]. The creation of this model of the domain of knowledge of the courses leads to two fundamental results for the development of research: on the one hand, the construction of valid and reliable assessment instruments, and on the other hand, the prescription of activities and instructional materials that can promote learning [26].

This study has performed the modeling of these structures, which can be useful in e-learning research, as it helps determine the influence of variables (see regression results in previous section). In sum, this conceptual model could be applied to carry out research on various methods of assessment and promotion of online learning, due to high level of reliability it offers in the sense of delimiting the educational domain by means of a valid and sensitive taxonomy [27]. Online education is considered a flexible, effective and viable option to cover the high educational demand of modern world, but if we do not generate research that explains the effects of online assessments and learning promotion interventions, we could be heading towards a scenario characterized by poor results, which is part of what we no longer want in the education system of a country that aims to become developed economy [28].

21.5 Conclusions

The aim of this study was to evaluate the impact of online assessment on students' learning outcomes. The study collected data from higher education students in UAE using survey questionnaire. With high level of internal consistency (Cronbach's $\alpha = 0.0.955$), this study reports that online assessment has significant positive impact on learning outcomes. Furthermore, Additionally, in the light of the data procured, this inquiry endorses that students think online assessments that they have taken are designed to check how much they have learned in the course. Moreover, from the evidence accumulated, this investigation understands that online assessments have enabled students to demonstrate the level of new information they gained during the course. Furthermore, based on the analysis rendered, this investigation infers that online assessments are relevant with the course and topics covered in the course. Furthermore, in the light of the discussions shown, this study explains that assessment exercises help students to understand and analyze the topic after the lecture. Moreover, in view of the analysis obtained, this study exhibits that assessment tasks encourage students to demonstrate comparative skills. In addition, in the light of the facts yielded, this inquiry recognizes that assessment methods are relevant to the course material and learning outcomes. In addition, in view of the illustrations collected, this study imparts that many of the assessment tasks are focused on real-world settings. Furthermore, taking into account the data accumulated, this research accounts that assessment tasks require students to show and apply the lecture and book knowledge in practical settings. In addition, taking into account the assessments contributed, this work demonstrates that assessment and evaluations require to solve problems and to apply theoretical knowledge. Furthermore, in view of the statistics found, this inquisition highlights that students think online assessments are fair and transparent. Furthermore, in the light of the empirical information produced, this research corroborates that students believe that all students are tested equally through online assessments. Additionally, in the light of the empirical data gathered, this research reveals that students believe that marking schemes, learning outcomes and assessment tasks are suitable for the education level (they are not excessive burden). Moreover, considering the assessments secured, this inquiry explicates that online assessments have the ability to properly check the knowledge and skills of students. In addition, using the arguments discovered, this research declares that online assessments test analytical and problem-solving skills of students. Moreover, in view of the empirical evidence accumulated, this investigation surmises that online assessments can be used to evaluate the performance of students during the course.

21.6 Future Research Opportunities

One of the main implications of this study is for future researchers to use the conceptual framework defined above and assess the impact of online assessment methods on learning outcomes such as using the framework above to check the reliability, validity, authenticity and objectivity of essays/reports/case studies, etc. Teachers and school administrations can use this framework to assess the impact of their assessment methods and identify opportunities for improvement. It is also important to emphasize on the fact that the evidence reported above involves students' perceptions and experiences only which is a limitation in generalizability. Future researchers are encouraged to conduct same research with other actors particularly teachers and assess its applicability from teachers' perspective.

Appendix a—Survey Questionnaire

Demographics

Gender

- Male
- Female

Education Format

- Online
- Traditional
- Both
- Others

Education Discipline

- Medical
- Business
- Engineering
- Art
- Vocational Education

Scale Strongly Agree (5), Agree (4), Neutral (3), Disagree (2), Strongly Disagree (1)

Independent variable—reliability	
I think online assessments that I have taken are designed to check how much I have learned in the course	(5)(4)(3)(2)(1)
Online assessments have enabled me to demonstrate the level of new information I gained during the course	
Online assessments are relevant with the course and topics covered in the course	
Independent variable—validity	
Assessment exercises help me to understand and analyze the topic after the lecture	
Assessment tasks encourage me to demonstrate comparative skills	
Assessment methods are relevant to the course material and learning outcomes	
Independent variable—authenticity	
Many of the assessment tasks are focused on business world settings	
Assessment tasks require me to show and apply the lecture and book knowledge in practical settings	
Assessment and evaluations require to solve problems and apply theoretical knowledge	
Independent variable—objectivity	
I think online assessments are fair and transparent	
I believe that all students are tested equally through online assessments	

(continued)

(continued)

Independent variable—reliability	
I think online assessments that I have taken are designed to check how much I have learned in the course	(5)(4)(3)(2)(1)
I believe that marking schemes, learning outcomes, and assessment tasks are suitable for my education level (they are not excessive burden)	
Dependent variable—learning outcomes	
Online assessments have the ability to properly check the knowledge and skills of students	
Online assessments test analytical and problem-solving skills of students	
Online assessments can be used to evaluate the performance of students during the course	

Note Online assessments include essays, case study reports, individual and collective presentations, research projects among others

References

1. N. Lander, P.J. Morgan, J. Salmon, S.W. Logan, L.M. Barnett, The reliability and validity of an authentic motor skill assessment tool for early adolescent girls in an Australian school setting. J. Sci. Med. Sport **20**(6), 590–594 (2017)
2. N. Haber, T.N. Mitchell, Using formative and summative assessment to evaluate library instruction in an online first year writing course. J. Libr. Inf. Serv. Dist. Learn. **11**(3–4), 300–313 (2017)
3. Z. Liu, T. Li, H. Diao, Analysis on the reliability and validity of teachers' self-designed English listening test. J. Lang. Teach. Res. **11**(5), 801–808 (2020)
4. S. Alaidarous, T.A. Mohamed, E. Masuadi, S. Wali, A. AlMalki, Saudi internal medicine residents' perceptions of the objective structured clinical examination as a formative assessment tool. Health Professions Educ. **2**(2), 121–129 (2016)
5. R. Ammigan, E. Jones, Improving the student experience: learning from a comparative study of international student satisfaction. J. Stud. Int. Educ. **22**(4), 283–301 (2018)
6. L.F. Khairil, S.E. Mokshein, 21st century assessment: online assessment. Int. J. Acad. Res. Bus. Soc. Sci. **8**(1), 659–672 (2018)
7. A. Okada, D. Whitelock, W. Holmes, C. Edwards, e-Authentication for online assessment: a mixed-method study. Br. J. Edu. Technol. **50**(2), 861–875 (2019)
8. C. Edwards, D. Whitelock, A. Okada, W. Holmes, Trust in online authentication tools for online assessment in both formal and informal contexts (2018)
9. K.E. Willcox, S. Sarma, P.H. Lippel, Online education: a catalyst for higher education reforms (2016). Final Report. Recuperado de https://goo.gl/Vfrzne
10. C. Hewson, J.P. Charlton, An investigation of the validity of course-based online assessment methods: the role of computer-related attitudes and assessment mode preferences. J. Comput. Assist. Learn. **35**(1), 51–60 (2019)
11. J.E. Lee, M. Recker, M. Yuan, The validity and instructional value of a rubric for evaluating online course quality: an empirical study. Online Learn. **24**(1), 245 (2020)
12. H. Fives, N. Barnes, Navigating the complex cognitive task of classroom assessment. Teach. Teach. Educ. **92**, 103063 (2020)
13. C. Mills, C. Chapparo, Use of perceive, recall, plan, perform stage two cognitive task analysis for students with autism and intellectual disability: The impact of a sensory activity schedule. J. Occupational Therapy, Schools, Early Intervention **10**(3), 232–253 (2017)

14. K.R. Koedinger, E.A. McLaughlin, Closing the loop with quantitative cognitive task analysis. Int. Educ. Data Mining Soc. (2016)
15. K.S. Steelman, K.L. Tislar, L.C. Ureel, C. Wallace, Eliciting best practices in digital literacy tutoring: a cognitive task analysis approach, in *International Conference on Human Aspects of IT for the Aged Population* (Springer, Cham, 2017), pp. 447–460
16. L. Elias, Cognitive task analysis-an innovative methodology for learning needs analysis. Training Dev. **43**(5), 16 (2016)
17. D.D. Dixson, F.C. Worrell, Formative and summative assessment in the classroom. Theory into Practice **55**(2), 153–159 (2016)
18. J. Dolin, P. Black, W. Harlen, A. Tiberghien, Exploring relations between formative and summative assessment, in *Transforming Assessment* (Springer, Cham, 2018), pp. 53–80
19. A. Bryman, *Social research methods* (Oxford University Press, 2016)
20. D. Silverman (Ed.). *Qualitative Research* (Sage, 2016)
21. S.J. Taylor, R. Bogdan, M. DeVault, *Introduction to Qualitative Research Methods: A Guidebook and Resource* (Wiley, 2015)
22. N. Walliman, *Social Research Methods: The Essentials* (Sage, 2015)
23. P. Eriksson, A. Kovalainen, *Qualitative Methods in Business Research: A Practical Guide to Social Research* (Sage, 2015)
24. M.A.H. Rizwan, Correlation and regression, in *Machine Learning and Big Data: Concepts, Algorithms, Tools and Applications* (2020), p. 53
25. B. Ogange, J. Agak, K. Okelo, P. Kiprotich, Student perceptions of the effectiveness of formative assessment in an online learning environment. Open Praxis **10**(1), 29–39 (2018)
26. P. Joyce, The effectiveness of online and paper-based formative assessment in the learning of English as a second language. PASAA: J. Lang. Teach. Learn. Thailand **55**, 126–146 (2018)
27. A. Felix, Using pre/post-testing to evaluate the effectiveness of online language programs. J. Second Lang. Teach. Res. **4**(1), 176–193 (2016)
28. J. Olson, N. Boyer, Effectiveness of online leadership education (2019)
29. K. Harish Professor, K. Srivvastava, Study on effect of meditation on individual's performance at workplace. Int. J. Manage. (IJM), **12**(1), 379–389 (2021). https://doi.org/10.34218/IJM.12.1.2021.033
30. M.K.R. NC, O. Isaac, K.S. Harish, M.I. Iqbal, K. Srivvastava, Study on effect of meditation on individual's performance at workplace. Int. J. Manage. (IJM), **12**(1), 379–389 (2021). https://doi.org/10.34218/IJM.12.1.2021.033
31. M. Ismail Iqbal, O. Isaac, I. Al Rajawy, S. Khuthbuddin, A. Ameen, Hazard identification and risk assessment with controls (Hirac) in oil industry-a proposed approach. Mater. Today: Proc. **44**(January), 4898–4902 (2020).https://doi.org/10.1016/j.matpr.2020.11.800
32. Y. Xiong, H.K. Suen, Assessment approaches in massive open online courses: possibilities, challenges and future directions. Int. Rev. Educ. **64**(2), 241–263 (2018)

Chapter 22
Impact of SiO_2 and Al_2O_3 Nanoparticles on Electroless Ni–P–B Corrosion Resistance Enhancement on AZ91D Magnesium Alloy: A Preliminary Study

Motilal Lakavat, Amiya Bhaumik, Suman Gandi, Sadi Reddy Parne, and Mohammed Ismail Iqbal

Abstract In recent years, nanoparticles have been the focus of a great deal of research due to their potential applications in a variety of fields. One such application is in the field of corrosion resistance enhancement. In this preliminary study, the impact of SiO_2 and Al_2O_3 nanoparticles to electroless Ni–P–B coated on the corrosion resistance of AZ91D magnesium alloy was investigated. The FESEM images showed that a thick coating of nanoparticles was formed on the surface of the alloy after coating. The XRD results indicated that the coatings were mainly composed of SiO_2 and Al_2O_3 nanoparticles. The corrosion resistance of AZ91D magnesium alloy was enhanced by Ni–P–B–SiO_2–Al_2O_3 ($j_{cor} = 0.82 \times 10^{-8}$ A cm^{-2}) coating compare to Ni–P–B–SiO_2 and Ni–P–B–Al_2O_3. The polarization curve showed that the corrosion current density was significantly reduced after the nanoparticle coating. The EIS results also demonstrated that the impedance of the coated samples was much higher than that of the uncoated samples. This study has found that the addition of SiO_2 and Al_2O_3 nanoparticles to AZ91D magnesium alloy significantly enhances corrosion resistance.

Keywords AZ91D magnesium alloy · Electroless coating · Corrosion resistance · Polarization · Impedance spectroscopy

M. Lakavat (✉) · A. Bhaumik (✉)
Department of Mechanical Engineering, Lincoln University College (LUC), 1440 Lincoln, Malaysia
e-mail: motilal@lincoln.edu.my

A. Bhaumik
e-mail: amiya@lincoln.edu.my

S. Gandi · S. R. Parne
Department of Applied Sciences, National Institute of Technology Goa, Ponda 403401, India

M. I. Iqbal
Faculty of Engineering Department, University of Technology and Applied Sciences, Nizwa, Oman

A. K. Sahu et al. (eds.), *Energy Storage and Conservation*, Springer Proceedings in Energy, https://doi.org/10.1007/978-981-99-2870-5_24

22.1 Introduction

The AZ91D magnesium alloy is a high strength material that is often used in the aerospace and automotive industries. This alloy contains a high amount of zinc and manganese, which gives it its high strength. The AZ91D alloy also has good corrosion resistance and weldability [1]. Electroless nickel-phosphorus (Ni–P) is a well-known corrosion-resistant coating that has been used for many years in a variety of engineering applications [2–4]. In recent years, the application of nanoparticles (NPs) to enhance the performance of electroless Ni–P coatings has been the subject of considerable research interest [5–8]. The characteristics of Ni–P coatings are very sensitive to changes in the NP concentration, temperature, time, and voltage [5, 6]. The optimum NPs concentration is usually determined on a per-coating basis, due to the complex nature of these coatings. NPs and temperature effects on morphology and mechanical characteristics behavior of the Ni–P nanocomposite coating were investigated by Rezaee Behzad et al. [5]. The results showed that with the increase of NPs concentration, the mechanical and wear properties of the Ni-P nanocomposite coating improved, while the electroless deposition coating exhibited better corrosion resistance. In addition, the dispersion stability of zinc ZnO NPs in an electroless bath with various surfactants was studied by Krishnakumar and Elansezhian [9]. The Zeta potential and absorption spectroscopy were used to characterize the dispersions. It was found that the addition of a surfactant could stabilize the ZnO NPs dispersion and reduce the size of ZnO NPs. The addition of anionic surfactant had the most significant impact on the size and stability of the dispersion.

Engineers have long been searching for coatings that can improve the tribological properties of AZ91D alloy engineering components. Recently, NiB and NiBP have shown promise in this area. These coatings are able to reduce friction and wear, which can lead to improved performance and longer component life [10]. Algul Hasan et al. studied the NiP, NiB, and NiBP electroless coatings [11]. The highest strength 8.76 GPa was achieved with NiBP coatings. The wear resistance of NiBP coatings was also better than that of NiB and NiP coatings. Further, the study of electroless Ni–P–B coatings on steel by Biswas Pritam et al. aimed to investigate the mechanical properties of these coatings [12]. The coatings were subjected to heat treatment, and it was found that the hardness of the heated ternary coating was higher than that of the as-plated coating (6 GPa versus 5 GPa). In addition, the friction coefficient of the heat-treated ternary coating was lower than that of the as-plated coating. This suggests that the coating treatment improved the mechanical properties of the alloys.

Considering the aforementioned findings, we concentrated our attention in the current work on the corrosion resistance of AZ91D magnesium alloy that was improved by electroless Ni–P coating. Recently, we demonstrated Ni-P-B electroless coating on AZ91D magnesium alloy [13]. This demonstrated that Ni–P–B can be effectively used as a corrosion-resistant coating for magnesium alloys. The Ni–P–B coating was found to be effective in preventing corrosion of the magnesium alloy, and also provided good adhesion to the substrate. In this paper, a comprehensive investigation was carried out to evaluate the impacts of Al_2O_3, SiO_2 NPs concentrations in the bath on the thickness of the coatings and their resistance to corrosion. This was accomplished through a systematic study of Ni–P–B (NPB), 5.5 g/L SiO_2 (NPB–Si), Ni–P–B 6 g/L Al_2O_3 (NPB–Al), and Ni–P–B 5.5 g/L SiO_2–6 g/L Al_2O_3 (NPB–Si–Al) hybrid nanocomposite coating on AZ91D magnesium alloy substrate.

22.2 Experimental Methods

In this study, magnesium alloy specimens were created using a wire cut electrical discharge machining (WEDM) process. This WEDM process uses a wire electrode to cut through the workpiece material, using electrical discharges. The WEDM process is well-suited for cutting complex shapes with tight tolerances. In this study, the WEDM process was used to create magnesium alloy specimens with precision and accuracy. The surface of the samples was polished with a series of emery papers up to 2500 grit. They were then degreased with acetone for 15 min (in an ultrasonic bath), rinsed with Deionized water, and acid pickled for two minutes. The cleaning procedure is repeated in stages prior to coating. After cleaning and dipping in a methanol wash are performed, followed by a rinse in a 40% HF solution, followed by a rinse in Deionized water.

In order to prepare an NPB, NPB-Si, NPB-Al, and NPB-Si-Al electroless bath for coating, we followed the procedures outlined in our prior research [13]. Nickel chloride ($NiCl_2 \cdot 6H_2O$) is used as the primary source of nickel in the coating bath used for electroless deposition. Sodium hypophosphite ($NaPO_2H_2 \cdot H_2O$) is used as a deduction catalyst. Sodium acetate $C_2H_3NaO_2$) is used as a complexing negotiator. Dimethylamine borane (DMAB) and sodium borohydride ($NaBH_4$) are both used as reducing agents. In addition, the additives nano Al_2O3 and SiO_2 were introduced in conjunction with the electroless NPB deposition. The temperature of the sample bath was kept at 90 degree Celsius and pH was adjusted to 9 during the entire process. Table 22.1 outlines the primary chemical components of electroless baths as well as the parameters used to prepare NPB, NPB-Si, NPB-Al, and NPB-Si-Al baths in order to coat AZ91D alloy.

The characterization of NPB, NPB-Si, NPB-Al, and NPB-Si-Al electroless coatings on AZ91D alloy was carried out in order to determine the effect of each coating on the overall corrosion resistance of the alloy when exposed to 3.5 weight percent NaCl. This was done in order to determine the effect that each coating has on the overall corrosion resistance of the alloy. X-ray diffraction, scanning electron microscopy, potentiodynamic polarization, and electrochemical impedance tests were used to characterize NPB, NPB-Si, NPB-Al, and NPB-Si-Al electroless coatings. The coating's crystallinity was determined using an X-ray diffractometer (X'Pert^3 MRD) and a monochromatic Cu K_α radiation source at 40 kV and 30 mA, with 0.05° resolution and a counting period of 2 s. Further, a high-resolution SEM (Zeiss GeminiSEM) is used to obtain detailed images of the structure and coating behavior of a specimen. In order to assess the corrosion behavior of coatings when

Table 22.1 Electroless bath with main chemical components and parameters

Chemical components and parameters	Quantity
$NiCl_2 \cdot 6H_2O$	15 g L^{-1}
$NaPO_2H_2 \cdot H_2O$	15 g L^{-1}
$C_2H_3NaO_2$	20 g L^{-1}
$NaBH_4$	10 g L^{-1}
DMAB	2 g L^{-1}
Thiourea	0.0001 g L^{-1}
SiO_2	5.5 g L^{-1}
Al_2O_3	6 g L^{-1}
Bath temperature	90 °C
pH	9

exposed to ambient conditions, potentiodynamic polarization, and electrochemical impedance spectroscopy were employed in a 3.5wt% of NaCl solution. This provided valuable information about the potential for corrosion and the ability of the coatings to resist it. The results showed that the coatings were effective in protecting the metal beneath them from corrosion.

22.3 Results and Discussions

Figure 22.1 shows the SEM images of four different types of coatings on AZ91D magnesium alloy. The first is a pure NPB coating (a), followed by NPB-Si (b), NPB-Al (c), and finally a hybrid NPB-Si-Al coating (d). All four coatings show good adhesion to the substrate and smooth surfaces. SEM images of coatings with different nodular structures are shown in Fig. 1a–d. Figure 1b shows that the nodular structure on the surface of the AZ91D alloy disappeared after deposition of the NPB-Si nanocomposite coating. The uniform SiO_2 distribution ensures that the coating is evenly applied to the surface of the alloy and the nodular structure provides a more durable and reliable coating. Further, Form Fig. 22.1b–d clearly indicates that the nodular structure on the surface of the coatings gradually contracted; meanwhile, the uniform and dense of the coatings increased [14]. The results indicate that the formation of nodules affects the micro-structure and morphology of coatings, which consequently affects their corrosion properties [15, 16]. Figure 1d demonstrates the homogenous density of nanoparticles in the coating corresponding to new growth zones, which resulted in grain refinement after the application of SiO_2 and Al_2O_3 nanoparticles (bright dense) on electroless NPB coating [17, 18].

Surface passivation is a process in which the surface of a metal is coated with a thin film of another material [19–21]. The purpose of this coating is to protect the metal from corrosion. Magnesium alloys are particularly susceptible to corrosion, so surface passivation is an important means of improving their resistance to this type

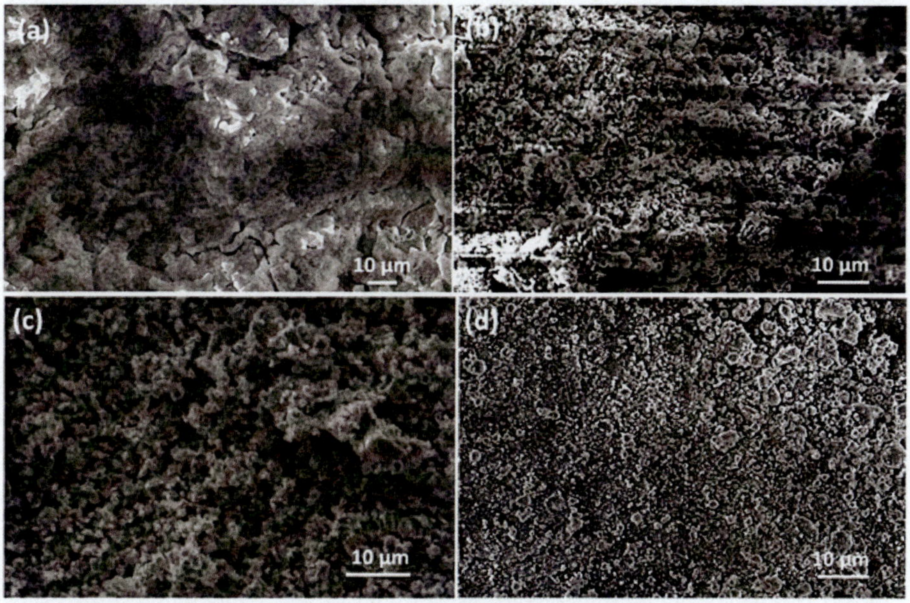

Fig. 22.1 SEM images of **a** NPB **b** NPB-Si **c** NPB-Al **d** NPB-Si-Al hybrid coatings

of degradation. Moreover, SEM image analysis showed that the surface passivation NPB-Si-Al coating improved the corrosion resistance of magnesium alloys. The results showed that the corrosion resistance was improved by the presence of the coating, which prevented the formation of corrosion products on the surface of the magnesium alloy.

The XRD profile of the NPB-Si-Al nanocomposite coating is shown in Fig. 22.2. The amorphous nature of the coating is clearly evident from the broad peak at ~44.49° by the nickel reflection. The broad peak (~40–50°) of the coating can be attributed to the presence of SiO_2 and Al_2O_3 nanoparticles in the NPB matrix. Several reports claim that the electroless amorphous coatings can enhance the corrosion resistance as demonstrated by XRD findings [22–25]. This is likely due to the fact that these coatings provide a barrier between the metal and the environment, preventing oxygen and water from coming into contact with the metal surface. Additionally, the amorphous nature of the coating helps to prevent localized corrosion by providing a more uniform coverage over the metal surface.

Figure 22.3 displays the polarization potentiodynamic curves of the nanocomposite coating that was applied to the AZ91D magnesium substrate and then immersed in 3.5% NaCl solution. The data that were obtained by sketching linear Tafel curves are presented in Table 22.2. The change in the corrosion potential values of the NPB coating from more negative to more positive potentials is depicted very clearly in Fig. 22.3, which displays the coating's relationship to the AZ91D magnesium substrate. Without any coating, a pure AZ91D magnesium substrate obtained the lowest corrosion potential (–1.53 V) and corrosion current density (1.55×10^{-4}

Fig. 22.2 XRD profile of NPB-Si-Al hybrid nanocomposite coating

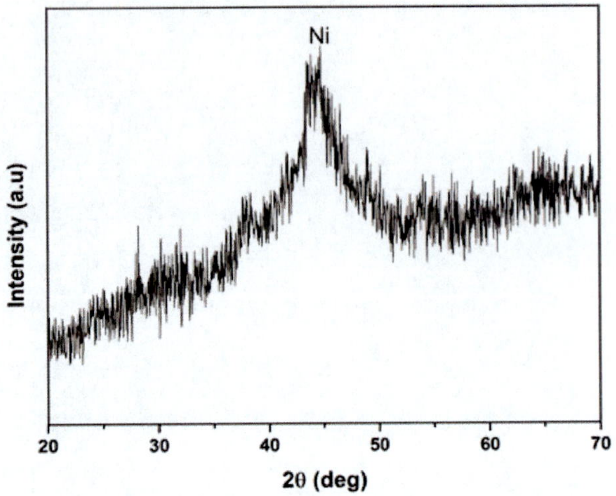

A.cm^{-2}). After electroless coating of NPB coating, the current density (1.1 × 10^{-6} A.cm^{-2}) reduced, while the corrosion potential (–0.365 V) increased [26–29]. This was caused by the incorporation of phosphorus and boron in the coated samples. The thickness and uniformity of the NPB coating on the AZ91D magnesium alloy is what causes the decrease in current density (Fig. 22.1). This coating stops the NaCl solution from reaching the substrate without causing any flaws or micropores in the material. The current density drops after the addition of the NPB-Si coating, going from 1.1 × 10^{-6} A.cm^{-2} to 7.99 × 10^{-7} A.cm^{-2}, while the corrosion potential rises (0.056 V). In contrast, the introduction of an NPB-Al coating caused the current density to drop to 8.49 × 10^{-7} A.cm^{-2}, while simultaneously increasing the corrosion potential to 0.35 V. The NPB-Si-Al hybrid coating exhibited the highest corrosion potential of 0.96 V and the lowest current density of 0.82 × 10^{-8} A.cm^{-2} during the test. The presence of nanoparticles with a composition of 5.5 g/L SiO_2 and 6 g/L Al_2O_3 in the NPB composite coating contributed to an increase in the surface's resistance to corrosion by functioning as a barrier against corrosive ions. According to the results suggested by the scanning electron micrographs (SEM), the deposition of nanoparticles within the coating appears to limit the growth of pores, which can lead to an enhancement in corrosion confrontation and a drop in current.

Figure 22.4 shows the impedance spectroscopy plots of NPB, NPB-Si, NPB-Al, and NPB-Si-Al hybrid nanocomposite formed by electroless coating on AZ91D. It can be seen that the variation of nanoparticles in the coating appears to affect the impedance plot. The impedance plot is a graphical representation of how a material affects the flow of electrons. The higher the impedance, the more the material resists the flow of electrons. It is interesting to note that when nanoparticles are added to the coating (NPB-Si, NPB-Al), there is a decrease in impedance and lowest decrement found in NPB-Si-Al). Which is ascribed to better corrosion coating, this may be due to the fact the hybrid layer not react with the surrounding materials.

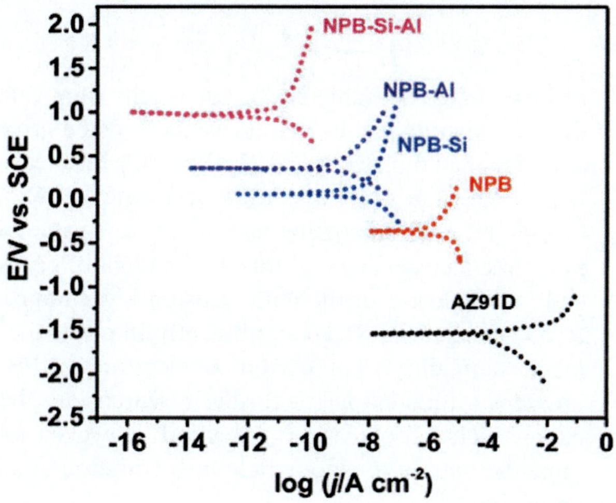

Fig. 22.3 Polarization plots of AZ91D coated with NPB, NPB-Si, NPB-Al, and NPB-Si-Al hybrid composites

Table 22.2 NPB, NPB-Si, NPB-Al, and NPB-Si-Al hybrid nanocomposites coatings electrochemical parameters

Coating	Corrosion potential (E_{cor}) V	Corrosion current density (j_{cor}) A.cm^{-2}	R_s Ω cm^2	R_c Ω cm^2	C_p μF
AZ91D alloy (without coating)	– 1.530	1.55×10^{-4}	4.9	2608	16.21
NPB	– 0.365	1.1×10^{-6}	8.1	5823	11.43
NPB-Si	0.056	7.99×10^{-7}	13.1	7358	10.21
NPB-Al	0.35	8.49×10^{-7}	14.6	9067	8.23
NPB-Al-Si	0.96	0.82×10^{-8}	12.6	10,299	6.32

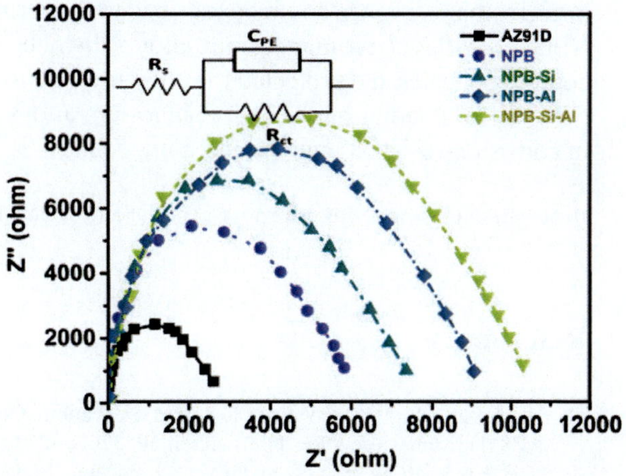

Fig. 22.4 Nyquist plots NPB, NPB-Si, NPB-Al, and NPB-Si-Al hybrid composites on AZ91D surface

The equivalent circuit model is used to fit the data and extract information about the coating (Fig. 22.4). The parameters extracted can be used to characterize the coating and its properties (Table 22.2). The results show that the addition of nanoparticles to the NPB coating results in improved corrosion protection performance and a denser structure than the electroless NPB coating, as measured by the charge transfer resistance (Rc) value and the constant capacitance (C_P), respectively. It was discovered that the NPB-Si-Al coating had the highest value of R_c and the lowest value of C_P; as a direct consequence of this, the amount of porosity in the material was reduced, and the thickness of the anticorrosion layer improved. It is evident that SiO_2 and Al_2O_3 nanoparticles can significantly improve the corrosion resistance of AZ91D magnesium alloy when used in an electroless NPB coating. This preliminary study provides a foundation for further research into the potential applications of these nanoparticles in corrosion protection. However, more research is needed to determine the optimum nanoparticle concentration and coating thickness for maximum protection.

22.4 Conclusion

The purpose of this preliminary study was to examine the impact of SiO_2 and Al_2O_3 nanoparticles on electroless Ni–P–B corrosion resistance enhancement on AZ91D alloy. The results of XRD and SEM analysis indicated that the incorporation of SiO_2 and Al_2O_3 nanoparticles into the AZ91D magnesium alloy surface resulted in a major enhancement in corrosion resistance. The results of polarization and EIS measurements indicated that the corrosion current density was decreased and the corrosion resistance was improved with the NPB–Si–Al coating. The improvement in corrosion resistance could be ascribed to the formation of a compact and adherent NPB–Si–Al layer on magnesium alloy substrate, which hindered the transport of corrosive species and protected the alloy from corroding. These findings suggest that NPB–Si–Al coating on AZ91D magnesium alloys may have potential applications in corrosion-resistant engineering components.

Declaration of Competing Interests The authors declare no competing financial interest.

References

1. H. Meifeng, L. Lei, W. Yating, T. Zhixin, H. Wenbin, Corrosion properties of surface-modified AZ91D magnesium alloy. Corros. Sci. **50**, 3267–3273 (2008)
2. K.H. Krishnan, S. John, K. Srinivasan, J. Praveen, M. Ganesan, P. Kavimani, An overall aspect of electroless Ni–P depositions—a review article. Metall. Mater. Trans. A. **37**, 1917–1926 (2006)
3. R.C. Agarwala, V. Agarwala, R. Sharma, Electroless Ni–P based nanocoating technology—a review. Synth. React. Inorg., Met.-Org., Nano-Met. Chem. **36**, 493–515 (2006)

4. E.M. Fayyad, A.M. Abdullah, M.K. Hassan, A.M. Mohamed, G. Jarjoura, Z. Farhat, Recent advances in electroless-plated Ni–P and its composites for erosion and corrosion applications: a review. Emergent Mater. **1**, 3–24 (2018)
5. B. Rezaee, E. Kermani, S. Ejlali, F. Biniyazan, H. Soleimanimehr, Nanoparticle concentration and heat treatment effects on microstructure and tribological behavior of the Ni–P nanocomposite coating. Adv. J. Sci. Eng. **2**, 71–78 (2021)
6. A. Varmazyar, S.R. Allahkaram, S. Mahdavi, Deposition, characterization and evaluation of monolayer and multilayer Ni, Ni–P and Ni–P–Nano ZnOp coatings. Trans. Indian Inst. Met. **71**, 1301–1309 (2018)
7. N. Ghavidel, S.R. Allahkaram, R. Naderi, M. Barzegar, H. Bakhshandeh, Corrosion and wear behavior of an electroless Ni–P/nano-SiC coating on AZ31 Mg alloy obtained through environmentally-friendly conversion coating. Surf. Coat. Technol. **382**, 125156 (2020)
8. S. Taşci, R.C. Özden, M. Anik, Corrosion and wear characteristics of electroless Ni–P, Ni–P–W and composite Ni–P–W/Al_2O_3 coatings on AZ91 sheet. Met. Mater. Int. **25**, 313–323 (2019)
9. V. Krishnakumar, R. Elansezhian, Dispersion stability of zinc oxide nanoparticles in an electroless bath with various surfactants. Mater. Today: Proc. (2021)
10. Q. Barati, S.M.M. Hadavi, Electroless Ni–B and composite coatings: a critical review on formation mechanism, properties, applications and future trends. Surf. Interfaces **21**, 100702 (2020)
11. H. Algul, M. Uysal, A. Alp, A comparative study on morphological, mechanical and tribological properties of electroless NiP, NiB and NiBP coatings. Appl. Surf. Sci. Adv. **4**, 100089 (2021)
12. P. Biswas, S. Samanta, A.R. Dixit, R.R. Sahoo, Investigation of mechanical and tribological properties of electroless Ni–PB ternary coatings on steel. Surf. Topogr. Metrol. Prop. **9**, 035011 (2021)
13. M. Lakavat, A. Bhaumik, S. Gandi, S.R. Parne, Electroless Ni-P-B coatings on magnesium alloy AZ91D: influence of nano Al_2O_3 on corrosion, wear, and hardness behaviour. Surf. Topogr. Metrol. Prop. **10**, 025021 (2022)
14. W. Wang, X. Jü, C. Xu, W. Zhang, N. Mitsuzaki, Z. Chen, Study on electroless plating Ni–WP ternary alloy with high tungsten from compound complexant bath. J. Mater. Eng. Perform. **29**, 8213–8220 (2020)
15. I.A. Shozib, A. Ahmad, A.M. Abdul-Rani, M. Beheshti, A review on the corrosion resistance of electroless Ni–P based composite coatings and electrochemical corrosion testing methods. Corros. Rev. **40**, 1–37 (2022)
16. V. Vitry, L. Bonin, Effect of temperature on ultrasound-assisted electroless nickel-boron plating. Ultrason. Sonochem. **56**, 327–336 (2019)
17. M. Islam, M.R. Azhar, N. Fredj, T.D. Burleigh, O.R. Oloyede, A.A. Almajid, S.I. Shah, Influence of SiO2 nanoparticles on hardness and corrosion resistance of electroless Ni–P coatings. Surf. Coat. Technol. **261**, 141–148 (2015)
18. Y. Wang, X. Shu, S. Wei, C. Liu, W. Gao, R. Shakoor, R. Kahraman, Duplex Ni–P–ZrO2/Ni–P electroless coating on stainless steel. J. Alloy. Compd. **630**, 189–194 (2015)
19. F. Cao, G.-L. Song, A. Atrens, Corrosion and passivation of magnesium alloys. Corros. Sci. **111**, 835–845 (2016)
20. Y. Song, D. Shan, E. Han, High corrosion resistance of electroless composite plating coatings on AZ91D magnesium alloys. Electrochim. Acta **53**, 2135–2143 (2008)
21. H. Hornberger, S. Virtanen, A.R. Boccaccini, Biomedical coatings on magnesium alloys–a review. Acta Biomater. **8**, 2442–2455 (2012)
22. Z. Bangwei, X. Haowen, Effect of alloying elements on the amorphous formation and corrosion resistance of electroless Ni–P based alloys. Mater. Sci. Eng., A **281**, 286–291 (2000)
23. J. Chen, G. Zhao, K. Matsuda, Y. Zou, Microstructure evolution and corrosion resistance of Ni–Cu–P amorphous coating during crystallization process. Appl. Surf. Sci. **484**, 835–844 (2019)
24. A.P. Meshram, M.P. Kumar, C. Srivastava, Enhancement in the corrosion resistance behaviour of amorphous NiP coatings by incorporation of graphene. Diam. Relat. Mater. **105**, 107795 (2020)

25. G.T. Guddla, S. Gandi, S. Ambadipudi, B.R. Ravuri, Design of Ni-based bulk metallic glasses with improved mechanical and corrosion properties. Curr. Appl. Sci. Technol. 115–131 (2021)
26. M. Sabzi, S. Mersagh Dezfuli, Deposition of Al2O3 ceramic film on copper—based heterostructured coatings by aluminizing process: Study of the electrochemical responses and corrosion mechanism of the coating. Int. J. Appl. Ceramic Technol. **16**, 195–210 (2019)
27. A. Kumar, R. Kumar, P. Bijalwan, M. Dutta, A. Banerjee, T. Laha, Fe-based amorphous/nanocrystalline composite coating by plasma spraying: effect of heat input on morphology, phase evolution and mechanical properties. J. Alloys Compd. **771**, 827–837 (2019)
28. J. Lin, J.J. Moore, B. Mishra, M. Pinkas, W.D. Sproul, The structure and mechanical and tribological properties of TiBCN nanocomposite coatings. Acta Mater. **58**, 1554–1564 (2010)
29. C. Tan, H. Zhu, T. Kuang, J. Shi, H. Liu, Z. Liu, Laser cladding Al-based amorphous-nanocrystalline composite coatings on AZ80 magnesium alloy under water cooling condition. J. Alloys Compd. **690**, 108–115 (2017)

Chapter 23
Elimination of Heavy Metals From Effluent Waters Using Activated Sewage Sludge and IPN Network

Nageswara Rao Lakkimsetty, S. Karunya, G. Kavitha, Motilal Lakavat, and Mohammed Ismail Iqbal

Abstract The presence of heavy metals in wastewater creates problems associated because of high toxicity, and there is a need to develop effective methods for water treatment. This research was aimed to remove heavy metals from wastewater using sewage sludge and polymers were an efficient adsorbent material. Firstly, the experimental study was carried out by activating sewage sludge using $ZnCl_2$ solution to produce activated carbon. This is followed by synthesizing an interpenetrating polymer network (IPN) using of poly methyl methacrylate and epoxy resin. The produced activated carbon was mixed with the synthesized interpenetrating polymer network with different concentrations of activated carbon 10, 20, 30, and 40% to produce sludge/polymer IPN adsorbent material. Sludge/polymer IPN adsorbent material was tested for its ability to remove Fe^{2+}, Cu^{2+}, and Zn^{2+} ions from wastewater samples. The parameters such as contact time, initial concentration of metal ions, adsorbent dosage, and pH value were studied. The overall results indicate that the highest removing capacity was attained at optimum conditions (pH = 8.0, contact time = 3 hr, adsorbent quantity = 4 g/100 mL, and initial metal ions concentration = 25 mg/L). The chemical structure of activated carbon and sludge IPN adsorbent was investigated using Fourier transform infrared spectroscopy.

Keywords Heavy metals · Water treatment · Polymer network · FTIR · Activated sewage sludge

---

N. R. Lakkimsetty (✉) · S. Karunya · M. Lakavat
Department of Mechanical and Industrial Engineering, College of Engineering, National University of Science and Technology, Muscat, Sultanate of Oman
e-mail: lnrao1978@gmail.com

G. Kavitha
Department of Chemical Engineering, RVR and JC College of Engineering (A), Guntur, A.P, India

M. I. Iqbal
Faculty of Engineering Department, University of Technology and Applied Sciences, Nizwa, Oman

A. K. Sahu et al. (eds.), *Energy Storage and Conservation*, Springer Proceedings in Energy, https://doi.org/10.1007/978-981-99-2870-5_25

23.1 Introduction

The presence of heavy metals causes a pollution of environment where these metals cause a global warming and causing a baleful effect on human, animal and plant life. Therefore, wastewater treatment is very important and should be given more attention and needs to be taken care of. There are various methods or processes used to remove heavy metals from wastewater such as: membrane filtration, precipitation, evaporation, and adsorption. All these methods are expensive because of producing a large quantity of sludge and this led to another treatment, while adsorption is a low cost, high efficiency, and flexible design process [1, 2]. The present study includes the removal of Fe^{2+}, Cu^{2+}, and Zn^{2+} ions from synthetic wastewater using sludge/polymer as low cost and more efficient adsorbent material. The combination of sewage sludge and polymer to produce a highly efficient and low-cost adsorbent could be applied in many wastewater treatment plants. The key objectives of this study are to optimize the condition for the removal of heavy metals using vinyl polymers, study various parameters for effective removal of heavy metals and characterization of developed polymer materials [3–5].

23.2 Methodology and Experimentation

23.2.1 Preparation of Adsorbent

Sewage sludge was collected from disposal area in Haya Company then dried in open air for 1 day followed by further drying in a dryer oven at 70 °C. The dryer sludge was then milled into small particles to start the activation. Activation process of sludge was carried out to activate the surface through the following steps, the milled sludge was first socked in $ZnCl_2$ and mixed it properly then left it for 24 hrs. The socked sludge washed with hydrochloric acid then with distilled water for several times till reach pH: 6 or more. The activated sludge was then dried again in the dryer at 70 °C to remove all moistures, after drying it was burned in a muffle furnace at 260 °C for 1 hr, finally it was powdered to produce fine activated carbon particles [6–8].

23.2.2 Synthesize of Interpenetrating Polymer Network

The bulk polymerization was carried out by taking 10 ml of methyl methacrylate monomer in a graduated cylinder then transferred it into round bottom flask. Followed by adding 0.17 g of the initiator benzoyl peroxide to the monomer. The water bath was heated till reach 70 °C, the round bottom flask with a magnetic bit was connected with the condenser and placed in the water under stirring and heating. The synthesized poly methyl methacrylate was then added to a certain amount of epoxy resin (A) without adding epoxy hardener (B) and mixed well. Different concentrations of activated carbon (10, 20, 30, and 40%) were then added to the mixture and mixed well, epoxy hardener (B) was finally added to the mixture and mixed well. Finally, the mixture of sludge/polymer IPN adsorbent material was poured in a silicon plate and left to dry in open air for 24 h [9, 10].

23.2.3 Preparing Heavy Metals Solutions

2.725 g of ferrous sulfate salt ($FeSO_4$), 2.5 g of copper sulfate salt ($CuSO_4$), and 2.465 g of zinc sulfate salt ($ZnSO_4$) were dissolved in 1000 ml of distilled water in volumetric flask, then diluted to 100 ml. To prepare the desired concentrations, a certain amount was taken from the diluted solution and transferred it into 50 ml capacity of graduated cylinder and filled it with distilled water up to the line. These steps were repeated to prepare stock solutions with four different concentrations (10 mg/L, 15 mg/L, 20 mg/L, and 25 mg/L).

23.2.4 Experiment Set up Using Batch Processing Method

The experiment had been carried out by taking 25 ml of metal ion solution and 2–3 ml of NaOH to adjust the pH of the solution were added to 100 ml capacity of conical flask. A desired amount of sludge/polymer IPN adsorbent material was added to the solutions in separated conical flask. The conical flasks were placed in water bath shaker at room temperature and middle speed for desired time and this procedure were repeated at different operating conditions (pH, contact time, adsorbent quantity, and initial metals ion concentration).

23.2.5 Analyze the Activated Sludge

Activated sludge had been characterized using Fourier transform infrared spectrometer. The FTIR spectra of activated carbon had been analyzed to know the chemical structure of sludge.

23.2.6 Analysis for Sludge/Polymer Adsorbent and Treated Water Samples

Final concentrations of Fe^{2+}, Cu^{2+}, and Zn^{2+} ions in the solutions had been measured after treatment using atomic adsorption spectrometer (AAS) in order to calculate the % removal of metal ions. Fourier transform infrared spectrometer (FTIR) was used to study the chemical structure of sludge/polymer adsorbent material.

23.3 Results and Discussions

23.3.1 Analysis of Treated Water Samples

Removal of Fe^{2+}, Cu^{2+}, and Zn^{2+} ions from synthetic wastewater was studied using batch method by adding a particular amount of adsorbent in prepared solutions containing iron, copper, and zinc ions. The experiments were carried out at room temperature and at the optimum pH which is 8.0. The mixture of synthetic wastewater and adsorbent was mixed in water bath at medium speed for a particular time. After selected time, the solutions were filtered using filter paper to separate them from adsorbent. The samples of heavy metals wastewater were then analyzed using atomic adsorption spectrometer (AAS) to measure the final concentration of metal ions in the solutions in order the determine the % removal. Different parameters that affect the removal efficiency of metal ions were studied in this experiment, which are.

23.3.2 Effect of Contact Time in Removal of Fe^{2+}, Cu^{2+}, and Zn^{2+} Ions

The effect of contact time was studied by preparing four samples of solutions containing 25 mg/L of Fe^{2+}, Cu^{2+}, and Zn^{2+} ions in conical flasks separately. The pH was adjusted in all four samples as 8. After that 4.0 g of sludge/polymer IPN adsorbent which contains 40% of activated carbon were added in all flasks, the samples were kept in water bath shaker for appointed time at room temperature. The samples were taken after 1, 2, 3, and 4 hr from the shaker as it required. The samples were then filtered by atomic adsorption spectrometer (Tables 23.1, 23.2 and 23.3).

From Fig. 23.1, it was noticed that the maximum adsorption of sludge/ polymer IPN was at first 3 h where the % removal of Fe^{2+}, Cu^{2+}, and Zn^{2+} ions reached 42.0%, 48.0%, and 48.0%, respectively. The removal rate was rapidly increased at the beginning and this is due the availability of large number of active sites in the adsorbent material. It was also observed after 3 h. The percentage of removal was decreased where the sludge/polymer IPN adsorbent started to reach equilibrium due

Table 23.1 Effect of contact time in removing Fe^{2+} ion (adsorbent: 4.0 g/100 ml; pH:8)

Serial number	Contact time, (hr)	Final concentration of Fe^{2+}, (mg/L)	% Removal
1	1	14.876	40.5
2	2	14.750	41.000
3	3	14.420	42.0
4	4	15.450	38.7

Table 23.2 Effect of contact time in removing Cu^{2+} ion (adsprbent:4.0 g/100 ml; pH:8)

Serial number	Contact time, (hr)	Final concentration of Cu^{2+}, (mg/L)	% Removal
1	1	14.75	38.0
2	2	14.785	42.0
3	3	14.485	48.0
4	4	15.550	42.0

Table 23.3 Effect of contact time in removing Zn^{2+} ion (adsprbent:4.0 g/100 ml; pH:8)

Serial number	Contact time, (hr)	Final concentration of Zn^{2+}, (mg/L)	% Removal
1	1	13.650	45.0
2	2	12.975	48.0
3	3	12.955	48.0
4	4	13.674	45.0

to deposition of metal ions on the surface of the adsorbent resulting in the reporting of these active sites [1].

Fig. 23.1 Effect of contact time for removing metal ions

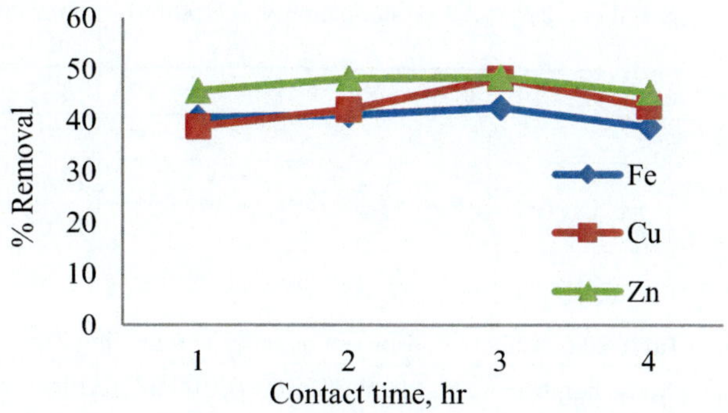

23.3.3 *Effect of Adsorbent Quantity in Removal of Fe^{2+}, Cu^{2+}, and Zn^{2+} Ions*

The study of adsorbent quantity was performed by preparing four samples of solutions containing 25 mg/L of Fe^{2+}, Cu^{2+}, and Zn^{2+} ions in conical flasks separately. The prepared NaOH solution was added to the samples to adjust the pH at 8. The selected amounts of sludge/polymer IPN adsorbent were added to the flasks, the samples were kept in water bath shaker for 3 hr duration at room temperature. After 3 hrs, all samples were taken from the shaker, filtered then analyzed using atomic adsorption spectrometer (Tables 23.4, 23.5 and 23.6).

Figure 23.2 shown the quantity of adsorbent material play an important role in removal of heavy metals. It was observed that the % of removal of Fe^{2+}, Cu^{2+}, and Zn^{2+} ions increased with an increase in adsorbent quantity where it reached 47.0%, 50.0%, and 51.0%, respectively. The reason for increase in % removal was because the addition of sludge/polymer IPN adsorbent led to an increase in the surface area and active site so these metal ions got more chance to stick on the adsorbent material [1].

Table 23.4 Effect of adsorbent quantity in removing Fe^{2+} ion (contact time: 3 h; pH:8)

Serial number	Adsorbent quantity, (g/100 ml)	Final concentration of Fe^{2+}, (mg/L)	% Removal
1	1	15.0	39.00
2	2	15.089	39.0
3	3	14.7	41.0
4	4	13.025	47.0

Table 23.5 Effect of adsorbent quantity in removing Cu^{2+} ion (contact time: 3 h; pH:8)

Serial number	Adsorbent quantity, (g/100 ml)	Final concentration of Cu^{2+}, (mg/L)	% Removal
1	1	13.95	44.0
2	2	12.9	48.0
3	3	12.7	49.00
4	4	12.5	50.0

Table 23.6 Effect of adsorbent quantity in removing Zn^{2+} ion (contact time: 3 h; pH:8)

Serial number	Adsorbent quantity, (g/100 ml)	Final concentration of Zn^{2+}, (mg/L)	% Removal
1	1	12.5	49.5
2	2	12.5	50.0
3	3	12.45	50.0
4	4	12.4	51.0

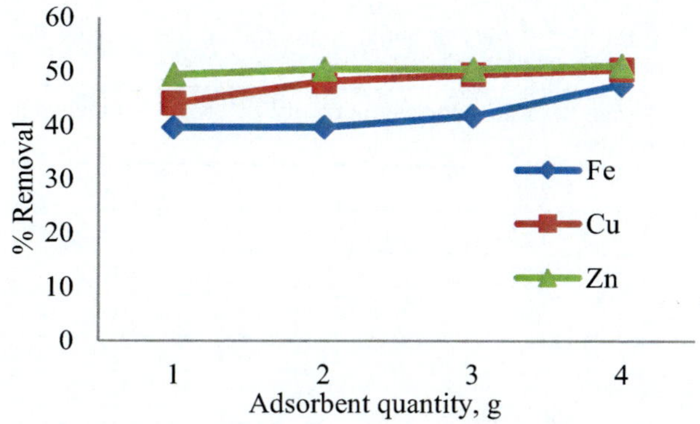

Fig. 23.2 Effect of adsorbent quantity in removing metal ions

23.3.4 Effect of Initial Concentration in Removal of Fe^{2+}, Cu^{2+}, and Zn^{2+} Ions

The effect of initial concentrations of Fe^{2+}, Cu^{2+}, and Zn^{2+} ion % removal was studied by preparing samples containing different concentrations (10, 15, 20, and 25 mg/L) of Fe^{2+}, Cu^{2+}, and Zn^{2+} ions in conical flasks separately. The experiment was investigated at room temperature and at pH = 8. Four grams of sludge/polymer IPN adsorbent were added to each sample then the samples were placed in water bath shaker for the optimum time, which was 3 h. After 3 h of treatment, all samples were taken from the shaker, followed by filtration using filter paper. After the completion of filtrations, the treated samples were analyzed using atomic adsorption spectrometer to know the final concentrations of Fe^{2+}, Cu^{2+}, and Zn^{2+} (Fig. 23.3, Tables 23.7, 23.8 and 23.9).

The adsorption capability of sludge/polymer adsorbent IPN for Fe^{2+} and Zn^{2+} was increased with increasing initial concentrations, while the adsorption capacity for Cu^{2+} ion was first increased till reach 15 mg/L of initial concentration then it remained constant even with increasing initial Cu^{2+} ion concentration. The maximum % of removal was at 25 ppm where it reached 53.0%, 52.0%, and 52.0% for Fe^{2+},

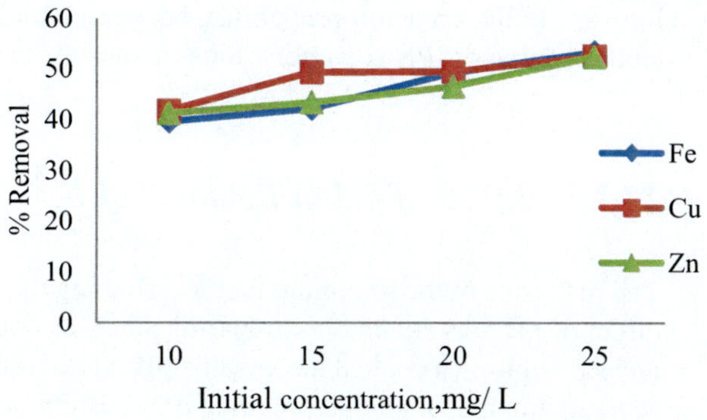

Fig. 23.3 Effect of initial concentration in removing metal ions

Table 23.7 Effect of initial concentration in removing Fe^{2+} ion (adsorbent: 4 g/100 ml; contact time: 3 h; pH:8)

Serial number	Initial concentration of Fe^{2+}, (mg/L)	Final concentration of Fe^{2+}, (mg/L)	% Removal
1	10	6.0	39.0
2	15	8.5	42.0
3	20	10.2	49.0
4	25	11.7	53.0

Table 23.8 Effect of initial concentration in removing Cu^{2+} ion (adsorbent: 4 g/100 ml; contact time: 3 h; pH:8)

Serial number	Initial concentration of Cu^{2+}, (mg/L)	Final concentration of Cu^{2+}, (mg/L)	% Removal
1	10	5.9	41.0
2	15	7.5	49.0
3	20	10.2	49.0
4	25	11.9	52.0

Table 23.9 Effect of initial concentration in removing Zn^{2+} ion (adsorbent: 4 g/100 ml; contact time: 3 h; pH:8)

Serial number	Initial concentration of Zn^{2+}, (mg/L)	Final concentration of Zn^{2+}, (mg/L)	% Removal
1	10	5.9	41.0
2	15	8.5	43.0
3	20	10.65	46.0
4	25	11.9	52.0

Cu^{2+}, and Zn^{2+} ions, respectively. The reason for this increase is that, the higher the initial metal ions concentration, the more concentration gradient which is due to increase in the collision probability between metal ions and the active sites on the sludge/ polymer IPN adsorbent, thus increasing % of adsorption [8].

23.3.5 *Effect of pH in Removal of Fe^{2+}, Cu^{2+}, and Zn^{2+} Ions*

The pH is one of the important factors affecting the adsorption of the metal ions. The effect of pH was studied by preparing samples containing 25 mg/ L of Fe^{2+}, Cu^{2+}, and Zn^{2+} ions in a conical flask with a pH range from 5 to 8 and at room temperature. A fixed amount of sludge/polymer IPN adsorbents were added to all samples and kept in water bath shaker for 3 h. After 3 hrs, the samples were taken from the shaker, then filtered using filter paper. After filtration process, the samples were analyzed by atomic adsorption spectrometer to obtain the final concentrations of Fe^{2+}, Cu^{2+}, and Zn^{2+} ions (Fig. 23.4, Tables 23.10, 23.11 and 23.12).

The obtained results indicate that the increase in pH leads to increase in % removal of Fe^{2+}, Cu^{2+}, and Zn^{2+} ions. The highest % of removal was obtained at pH 8 where it reached 42.0, 51.0, and 52.0% Fe^{2+}, Cu^{2+}, and Zn^{2+} ions. The absorption rate is increasing with increasing pH and this is due to the fact that at high pH levels, the

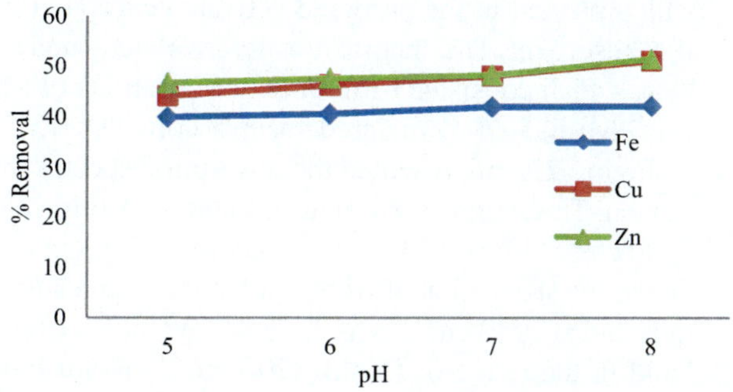

Fig. 23.4 Effect of pH value in removing metal ions

Table 23.10 Effect of pH in removing Fe^{2+} ion (adsorbent: 4 g/100 ml; contact time: 3 h)

Serial number	pH	Final concentration of Fe^{2+}, (mg/L)	% Removal
1	5	15.020	39.0
2	6	14.863	40.0
3	7	14.528	42.0
4	8	14.472	42.0

Table 23.11 Effect of pH in removing Cu^{2+} ion (adsorbent: 4 g/100 ml; contact time: 3 h)

Serial number	pH	Final concentration of Cu^{2+}, (mg/L)	% Removal
1	5	13.95	44.0
2	6	13.40	46.0
3	7	12.95	48.0
4	8	12.25	51.0

Table 23.12 Effect of pH in removing Zn^{2+} ion (adsorbent: 4 g/100 ml; contact time: 3 h)

Serial number	pH	Final concentration of Zn^{2+}, (mg/L)	% Removal
1	5	13.324	46.0
2	6	13.097	47.0
3	7	12.926	48.0
4	8	12.133	52.0

carbonate groups in the absorbent material increased the attraction of heavy metals due to the increase in negative charges on surface of the absorbent [5].

23.3.6 Sludge/polymer IPN Adsorbent Characterization

In this study, Fourier-transform infrared spectroscopy was used to locate the bonds which present in the prepared activated carbon and in the prepared sludge/polymer IPN adsorbent. The chemical structure of activated carbon and sludge/polymer adsorbent, which consisted different concentrations of activated carbon (10%, and 40%), was investigated by infrared spectroscopy [11, 12].

Figure 23.5 represented the absorption spectra in the infrared region for activated carbon. The sample showed four main absorption bands at (2800–3000 cm^{-1}, 2000–2200 cm^{-1}, 1400–1600 cm^{-1}, and 800–1300 cm^{-1}). The band at 2150 cm^{-1} formed due to the absorption of water molecules as a result of stretching mode of O–H while the band at 2950 cm^{-1} is assigned to the interaction of C–H with carbon surface. The band in the range of (1500–1700 cm^{-1}) is attributed to the hydrogen-bonded OH group of alcohols and phenols. The band at 1000 cm^{-1} showed a sharp absorption, which was assigned to either Si–O or C–O stretching vibrations.

Figure 23.6 represented the absorption spectra in the infrared region for prepared sludge/polymer IPN sample containing 10% of activated carbon. The sample showed three main absorption bands at (3100–3600 cm^{-1}, 2800–3000 cm^{-1} and 800–1700 cm^{-1}). The band at 3100 cm^{-1} corresponds to O–H starching vibration. The band in the range of 2800–3000 cm^{-1} attributed to C-O group of carbonyls. The

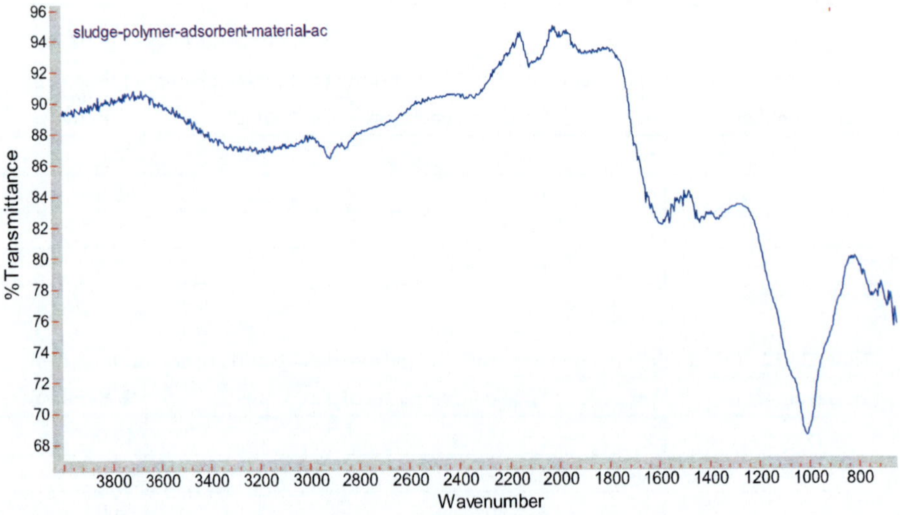

Fig. 23.5 Infrared absorption spectroscopy for activated carbon

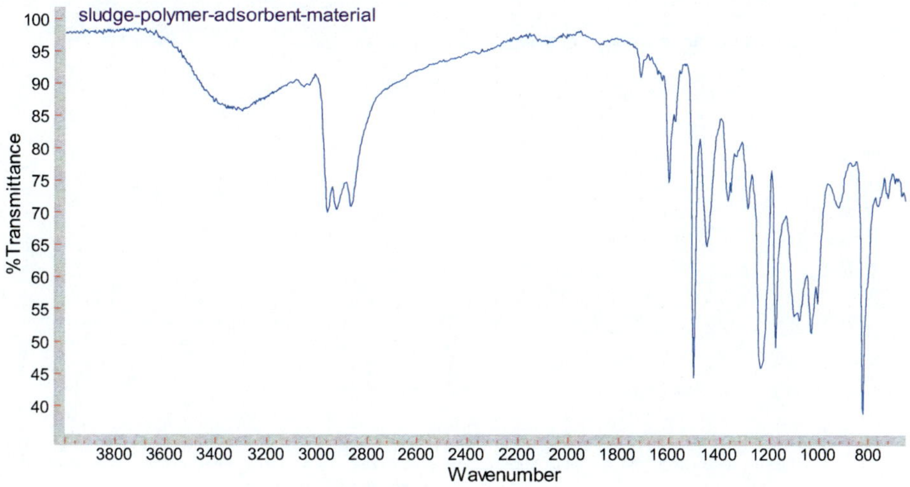

Fig. 23.6 Infrared absorption spectroscopy for polymer IPN contains 10% activated carbon

band at 1700 cm^{-1} is assigned to the interaction of C-H with carbon surface while the band at 1600 cm^{-1} attributed to C–O stretching and -OH deformation [13].

Figure 23.7 represents the absorption spectra in the infrared region for prepared sludge/polymer IPN sample containing 40% of activated carbon. The sample showed three main absorption bands at (3100–3600 cm^{-1}, 2800–3000 cm^{-1}, and 800–1700 cm^{-1}). The band at 3100 cm^{-1} corresponds to O–H starching vibration. The band in the range of 2800–3000 cm^{-1} attributed to C–O group of carbonyls. The band at 1700 cm^{-1} is assigned to the interaction of C–H with carbon surface while the band at 1600 cm^{-1} attributed to C–O stretching and –OH deformation [14, 15].

23.4 Conclusions

The large amount of the sewage sludge, which is produced during the wastewater treatment process, has been utilized in this work. Interpenetrating polymer network (IPN) was synthesized by combining poly methyl methacrylate and epoxy resin then added to the activated carbon, which was attained from chemical activation of sewage sludge to produce sludge/polymer IPN adsorbent. The prepared sludge/polymer IPN adsorbent material was tested to remove Fe^{2+}, Cu^{2+}, and Zn^{2+} ions and showed that the sludge/polymer IPN adsorbent has the ability to adsorb heavy metals from wastewater samples. The overall best conditions for removing metal ions were obtained at contact time of 3 h., pH 8, 4 g of adsorbent quantity, and initial metal ions concentration of 25 mg/L where the adsorption capacity of removing Fe^{2+}, Cu^{2+}, and Zn^{2+} ions was up to 53.484%, 52.560%, and 52.468% respectively. The functional groups in activated carbon and sludge/polymer IPN adsorbent have been identified using FTIR.

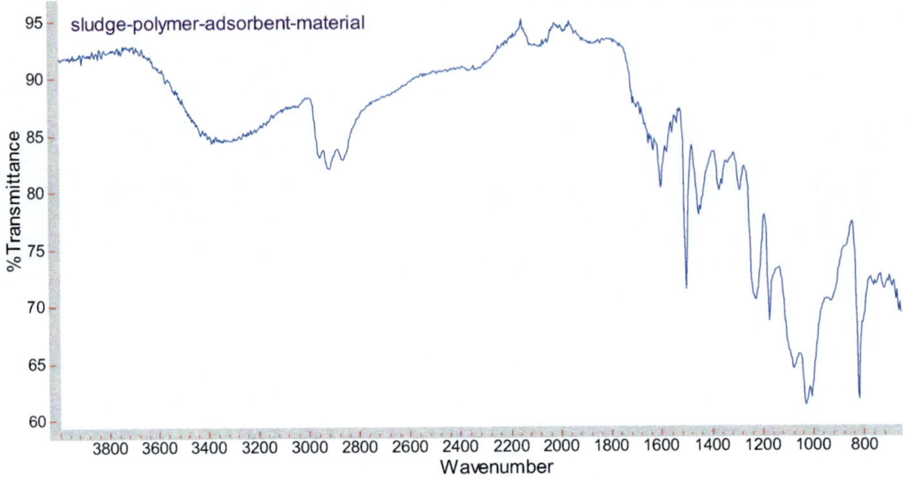

Fig. 23.7 Infrared absorption spectroscopy for polymer IPN contains 40% activated carbon

References

1. M. Abdel-Aziz, Removal of Fe^{++} from wastewater using sludge-polymer hybrid adsorbents. Middle Pomeranian Sci. Soc. Environ. Protect. **18**, 28–45 (2016)
2. S. Aslan, Adsorption of heavy metals onto waste tea. Eur. Sci. J. 111–117 (2016)
3. G. Kavitha, N.R. Lakkimsetty, N. Chittibabu, M.V. Rao, and F. Shaik, Bioprocessing optimization for efficient simultaneous removal of lead and chromium by Gracilaria corticata seaweed biomass. In AIP Conference Proceedings 2690, 020050 (2023) https://doi.org/10.1063/5.0119484
4. S. Fathya, S. Karunya, N.R. Lakkimsetty, S. Feroz, Treatment of petroleum wastewater using synthesized hematite (α-Fe_2O_3) photocatalyst and optimization with response surface methodology. Int. J. Environ. Anal. Chem. 1–20 (2020)
5. Shamsa Al Saadi, S. Feroz and Nageswara Rao, L., (2015), Treatment of Industrial Wastewater by Solar Nano Photocatalysis. Int. J. Chem. Tech. Res. **8**(7):177–182
6. N.S. AL-Reyamia, S. Feroz, N.R. Lakkimsetty, Environmental impact on usage of treated effluents from ISTP for irrigation. J. Water Process Eng. **36**, 101363 (2020)
7. N.R. Lakkimsetty, S. Karunya, S. Feroz, L. Motilal, P. Saidireddy, G. Suman, Synthesis, characterization and application of polymer composite materials in wastewater treatment. Mater. Today: Proc. **59**(3), 1726–1734 (2022). https://doi.org/10.1016/j.matpr.2022.04.050
8. X. Wang, X. Zhang, Adsorption and desorption of nickel (II) ions from aqueous solution by a lignocellulose/montmorillonite nanocomposite. PMC J. **10** (2015)
9. N.R. Lakkimsetty, J.R. Al Said, S. Feroz, Removal of zinc and copper from aqueous solution by using modified sugarcane bagasse as a low-cost adsorbent. Int. J. Mech. Prod. Eng. Res. Dev. **10** (2020)
10. O.H. Al-Abri, N.R. Lakkimsetty, S. Feroz, Pretreatment of oil produced water using low cost adsorbents. Int. J. Mech. Prod. Eng. Res. Dev. **10**(3), 2249–6890 (2020)
11. N.R. Lakkimsetty, G. Prabhakar, M.V. Rao, S. Feroz, Experimental study on removal of zinc metal ions from aqueous solution using *Ulvalobata sp.* As an adsorbent. Int. J. Innov. Res. Sci. Technol. **2** (2016)
12. N.R. Lakkimsetty, G. Prabhakar, S. Feroz, Biosorption of chromium from aqueous solution by using *Citrus reticulata*: equilibrium and kinetic studies. J. Chem. Pharm. Res. **7**(6), 117–127 (2015)

13. N.R. Lakkimsetty, G. Prabhakar, S. Feroz, Asian J. Biochem. Pharm. Res. **3** (2013)
14. R. Venkateswara Rao, S. Janusz Kozinski, N.R. Lakkimsetty, Biosorption of lead ions from aqueous solution using ficus benghalensis L. J. Eng. Hindawi Publishing Corporation, Article ID 167518. (2013)
15. N.R. Lakkimsetty, S. Feroz, G. Prabhakar, Biosorption of chromium ions from aqueous solution by using Citrus reticulata: Equilibrium and kinetic studies. Caled. J. Eng. **8** (2012)

Chapter 24
Thai: A Global Energy Solution in Extracting Heavy Oil

Mohammed Ismail Iqbal, K J Sabareesaan, and Sarwat Jahan

Abstract At the time when the global crude oil is depleting at an alarming rate, the world is now banking on unconventional resources to fuel the developing economies. Solution to the rising demand for fuel lies in the economic development of technologies that can help us produce unconventional energy resources such as tar sands or oil sands, shale gas, coal bed methane in the most efficient and cost-effective manner. One such prolific source of crude oil is the Oil Sands; its accumulation around the world is proven to be around 249.67 billion barrels (39,694 MMSCM). As the crude found in oil sands is heavy in nature; the major difficulty faced is the extraction of crude. Researchers have applied various in-situ processes like cyclic steam stimulation (CCS), and Steam assisted gravity drainage (SAGD) but THAI process has proved to give far more satisfactory results when compared with other processes. The THAI process produces 50% less greenhouse gases and lesser footprint than the other processes, which are very hazardous, and has high investment cost. Petrobank, the developer of this technology, claims that this process has the sweep efficiency of around 60–70%. The paper focuses on the available technology for THAI process and comparison among the various well patterns for most effective extraction of crude; its implementation, development, advantages and current status and modifications, which could be done to improve the quality of production and reduction of the cost of the process. Finally, it will discuss the introduction of a catalyst along with the THAI process, which could enhance or upgrade the quality of crude produced.

M. I. Iqbal (✉) · K. J. Sabareesaan
Faculty of Engineering Department, University of Technology and Applied Sciences, Nizwa, Oman
e-mail: mdiqbal1988@googlemail.com

S. Jahan
Department of Businsess and Accountancy, Lincoln University College, Selangor, Malaysia

A. K. Sahu et al. (eds.), *Energy Storage and Conservation*, Springer Proceedings in Energy, https://doi.org/10.1007/978-981-99-2870-5_26

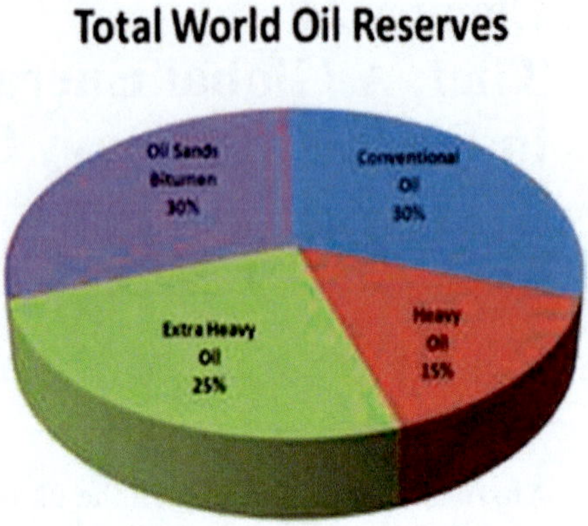

Fig. 24.1 Total world oil reserves by IEA

24.1 Heavy Crude Oil Reserves in World

The crude oil having API gravity less than 22 (sp. gr. 0.92) and viscosity ranging from 100 to 10,000 centipoises is referred to as the heavy oils or bituminous oil. According to International Energy Agency (IEA), the total world crude oil resource is 9–13 bbls of which 2/3 of the total oil resources are heavy oil and bitumen (Fig. 24.1) [1].

Today the total heavy crude oil production stands approximately equal to 5 million bbls per day whereas India produces only 30,000 bbls per day of heavy oil with API gravity ranging from 13 to 17.

24.2 Problem Statement

Industries are increasingly focusing on the unconventional hydrocarbon recovery due to continuous increase in energy demand. As a result, new methods and technologies are being developed to meet the demand. The heavy oil recovery includes a number of processes of which Steam-based thermal recovery is being extensively used. But it has been found that these processes are not efficient enough and can only produce 30–50% OOIP [2]. But due to the advent of technology and introduction of new techniques like THAI (Toe to heel Air Injection), we can be able to produce up to 80% OOIP [3].

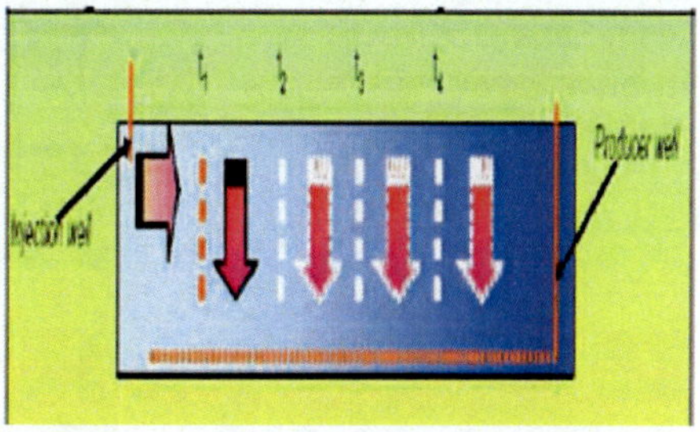

Fig. 24.2 Short distance displacement [6]

24.3 Toe to Heel Air Injection Process

THAI process is a combustion method for recovery of heavy oil, which combines an injector well and a horizontal production well. In this process, a combustion front is created that travels along the horizontal production well thus allowing the heated oil to drain towards the production well due to gravity [4].

In spite of the inherent benefits of the Insitu combustion processes, limited control on flood front and poor sweep efficiencies have resulted in limited success of this process. In conventional horizontal floods, gravity segregation between the hot combustion gases and cold heavy reservoir oil further reduces the sweep efficiency to an undesirable value. Hence, gravity-stabilization of insitu combustion processes is necessary.

The main objective of the toe to heel air injection is to prevent gravity segregation of the injected and the displaced fluid and to provide a near upright displacement front that eventually results in higher sweep efficiencies. It is an integrated horizontal well process and operates on "**Short Distance Displacement**" **(SDD)** [5]. Due to the SDD feature, THAI is categorized as the new method for the extraction of heavy oil. The striking feature of THAI is the near complete elimination of the gravity over-ride results in an efficiency which is two-four times higher than steam in energy cost and lower emission loading. Migration of injected air to combustion zone occurs due to diffusion of oxygen into the crude ahead of combustion front (Fig. 24.2). The balance between the oxygen flux required burning the fuel ahead and the removal of mobilized fluid and combustion products is the key requirement for stability and growth of THAI process (Fig. 24.3) [6].

24.4 Ignition and Air Requirement

In a forward combustion process, the injection of air causes oxidation of the crude oil at the reservoir temperature and results in slow temperature rise. The rise in temperature causes the increase in rate of oxidation of crude. Depending upon the crude composition the speed of oxidation is decided, i.e. asphaltic crude provides a good medium for propagation of oxidation [8, 9]. Normally, it takes 1 or 2 days

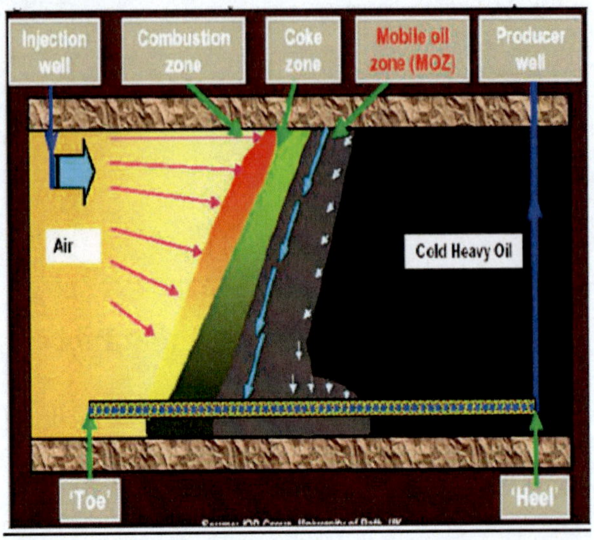

Fig. 24.3 Toe to heel air injection process schematic [7]

to reach the ignition limit, so the injected air can be pre-heated in order to reduce the ignition time. Sometimes pre-heating is required to reduce the effect of reverse combustion process. The below study has been taken from the reservoir simulation data obtained by SCHLUMBERGER while performing THAI simulation. During the process, the average porosity was kept as 25%, vertical permeability as 70 md and horizontal permeability as 700 md.

Reaction 1: Heavy oil fraction combustion

$$C_{32} - C_{47} + 45.17O_2?37.46N_2 - CO_X + 29.71\text{ water}$$

Reaction 2: Medium oil fraction combustion

$$C_{20} + 25.08O_2?20.80N_2 - CO_X + 16.51\text{ water}$$

Reaction 3: Light oil fraction combustion

$$C_6 - C_{12} + 11.66O_2?9.76N_2 - CO_X + 7.67\text{ water}$$

Based upon the reactions, energy factors were obtained, which control reaction initiation and development [9]. Following is shown in Table 24.1.

Two types of mixtures of air were proposed for making a combustion front:

Table 24.1 Showing activation energy and reaction rate of all the hydrocarbon components [10]

Reaction	Reaction rate (lbmol/day)	Activation energy constant	Reaction enthalpy (BTU/lbmol)
Reaction 1	$3.4054E^{+10}$	32785	8142400
Reaction 2	$0.2816E^{+10}$	32785	4521600
Reactio 3	$0.4035E^{+10}$	32785	2102400

24.5 Enriched Air Injection (50% NO_2 and 50% O_2)

The linear increase in cumulative air and increase of average reservoir temperature and reservoir enthalpy shows good air injectivity and sustained combustion front [11]. The water is mostly injected as the heat scavenger to recover heat from the burned out regions. Figure 24.4 represents the production profile during the enriched air injection. Nearly 58% of the OOIP is recovered until gas breakthrough and ultimate recovery being 74% of OOIP (Original Oil in Place). Excellent reservoir volumetric sweep and near vertical displacement front are observed.

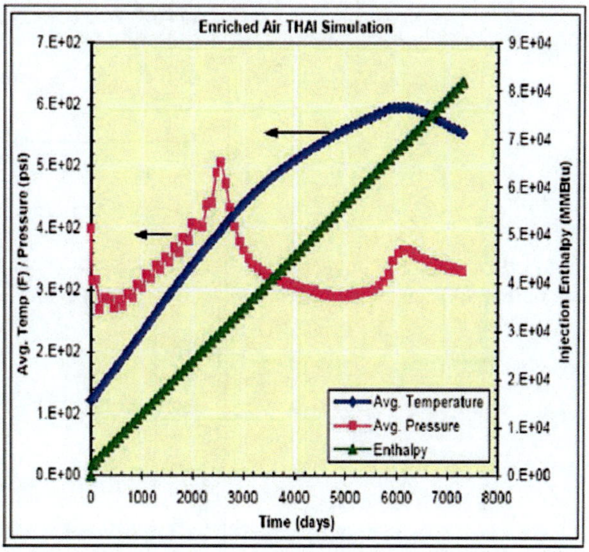

Fig. 24.4 Average pressure, temperature and injected enthalpy for enriched air

24.5.1 Normal Gas and Flue Gas Injection (79% N_2 and 21% O_2)

Excellent reservoir sweep having near vertical displacement front is also observed in normal air and fuel air injection (Figs. 24.5 and 24.6). However, lower availability of oxygen concentrations results in comparatively low fireflood front thereby resulting in lower amount of oil production [12, 13]. As the time progresses, the combustion front starts losing temperature, which is associated with the oxygen depletion in the reservoir (Fig. 24.7). Based upon the study, two scenarios were developed:

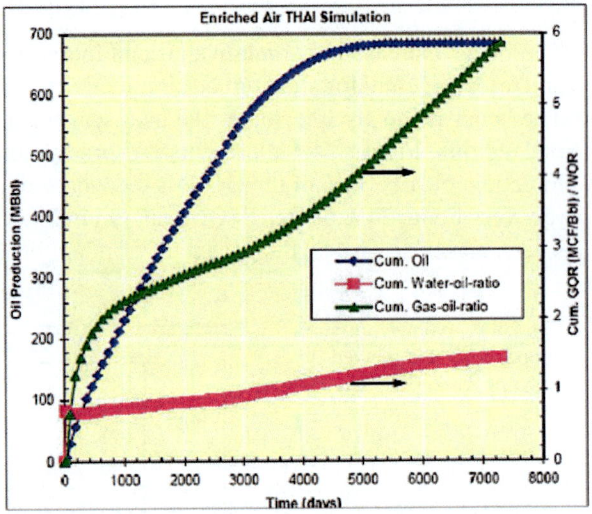

Fig. 24.5 Cumulative water–oil, gas–oil ratio and oil production

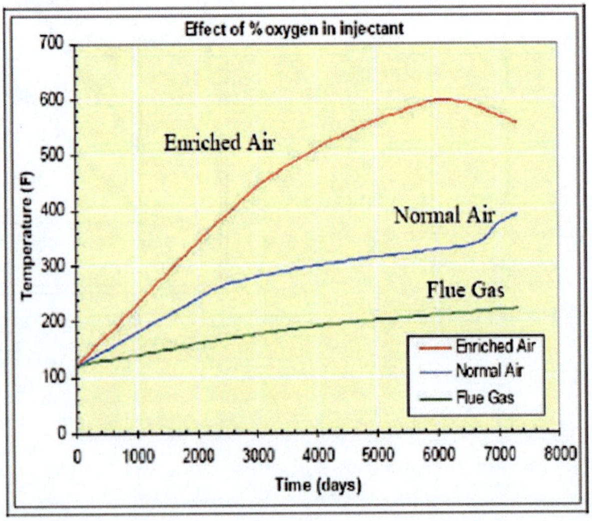

Fig. 24.6 Effect of oxygen concentration on temperature

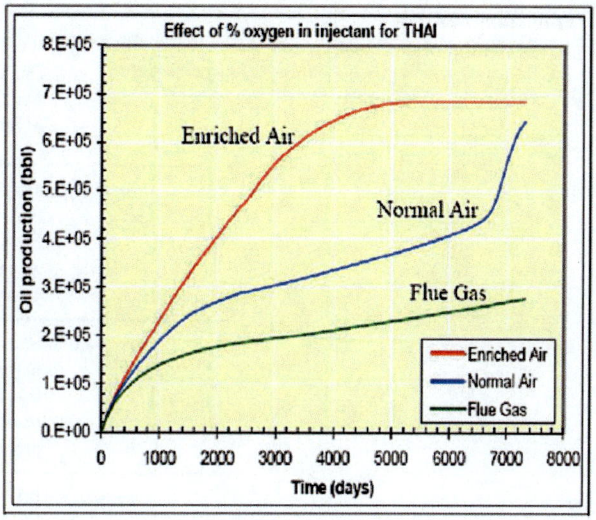

Fig. 24.7 Effect of oxygen concentration on oil production

1. Rate of injection of air is doubled or
2. Increasing the concentration of oxygen in the injection air.

24.6 Well Array for Thai Process

In this case, four patterns combining horizontal and vertical wells were tested by ECLIPSE (software for reservoir simulation by Schlumberger). The control parameter for this simulation was done by keeping a constant injection volume for all the cases at 5MMscfd, while each of the production well was maintained for a constant bottom hole pressure (Fig. 24.8).

It was concluded that using horizontal well as the injector well helps to create a better combustion front. It was concluded that for the patterns given above, the horizontal injector and horizontal production accounted for best recovery case as it delivered higher temperatures and improved sweep efficiency across the array. Additionally, it was given that the heat transfer effect through tubing promoted better quality of crude with low viscosity since the surrounding was kept hot.

24.6.1 Effects of heterogeneity on Thai Process Development

Three scenarios were considered to study the effect of combustion over the varying range of porosity and permeability (Fig. 24.9).

1. Homogenous scenario.
2. A fully heterogeneous scenario populated with randomly generated properties.
3. Stochastic property distribution.

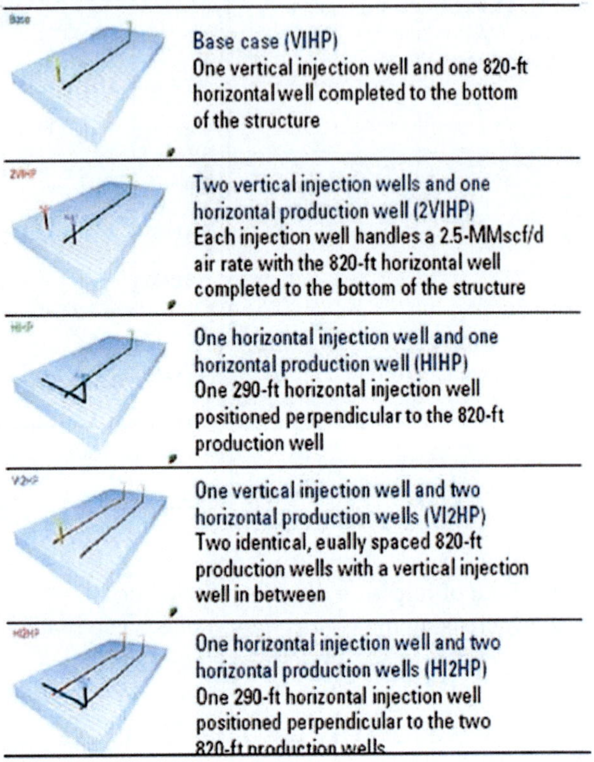

Fig. 24.8 Various combination of well array

For these conditions, the average porosity was kept at 25% and 750 md of permeability [14].

The results obtained after 5 years of simulation done on ECLIPSE simulation software are given in Table 24.2.

Therefore from the data's obtained we can say that gravity forces and the pressure drop caused by production well outcomes the effect of heterogeneity. Though a decline is observed between the homogenous media and the stochastic/random media, but this decline in production is far better than the other thermal and extraction processes.

24.7 Energy Returns on Investment (Eroi) Using Thai

The EROI analysis is based on Whitesands THAI performance as provided by the project permit documents. The figure below shows the EROI. The energy input is around 94,000,625 GJ/day, which consists of use of diluents naphtha; fuel gas, connected electrical requirement, bitumen and produced gas. The energy out is equal to 8,500,774 GJ/day, which includes saleable products, electrical exports, losses and

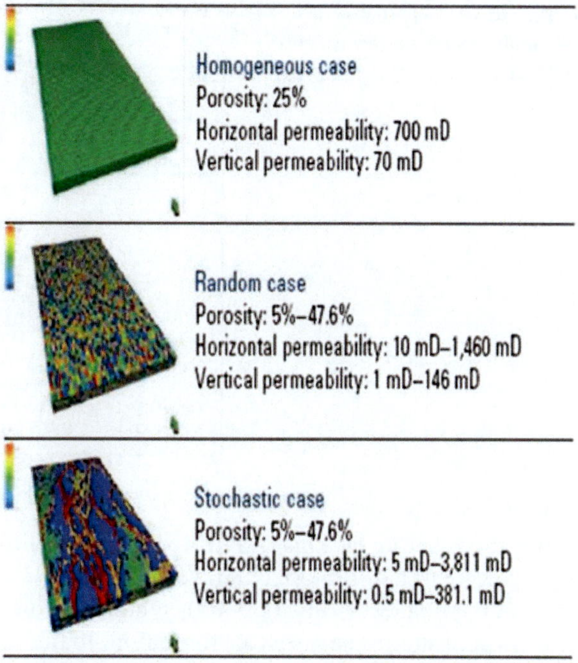

Fig. 24.9 Various heterogeneous systems developed by eclipse

Table 24.2 Cumulative production data after 5 years of production

Case	Np	Np below base case (%)
Homogeneous	159,890	0.00
Stochastic	156,316	−2.24
Random	152,560	−4.58

uses[15]. The energy balance as calculated by Petrobank is 90.6% implying that for every 10 units of energy taken from the reservoir one unit is used up in the process (Fig. 24.10).

24.8 Case Study of "Down Hole Upgrading of Wolf Lake Oil Using Thai"

There are many extensive tests of THAI and CAPRI that have taken place. The current case qualifies the upgrading nature of extent during an experiment dry and wet phases of THAI and CAPRI.

Multiple 3-D in-situ combustion tests have been done in a low pressure stainless steel cell, measuring 0.6 m * 0.4 m * 0.1 m equipped having ports for air injection and production Heavy Wolf Lake crude oil (10.5° API) was mixed with homogenous

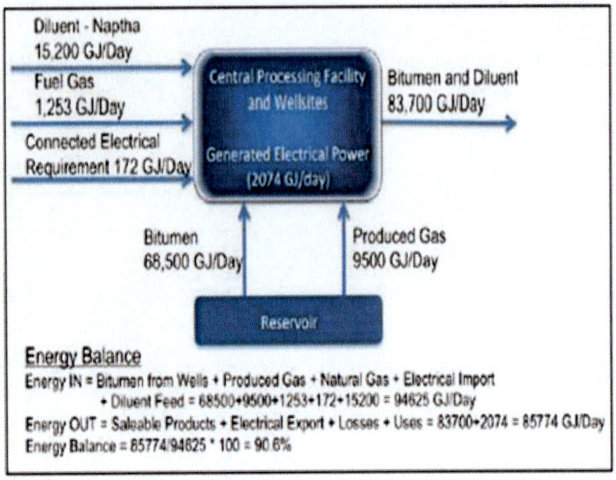

Fig. 24.10 Calculation of energy returns on investment of Thai

wet sands before packing. A Ni-Mo hydrogen sulfurization HDS catalyst was packed in down part half of horizontal well.

In the first test (Run 2000–07), synthetic brine was used in order to stimulate the connate water in the reservoir formation. In the second test (Run 2001–01), a tracer was used to investigate the mechanism of up gradation oil by CAPRI.

24.8.1 Conclusions of Test Run 2000–07

1. High oil recovery is achieved up to 87% of OOIP.
2. A stable high temperature combustion front propagated to the Horizontal Production Well.
3. Before the combustion front reached the Horizontal production well, the API increase was 2°. The effect of catalyst is clearly noted as the API increased from 14° to 24° as combustion front reached HP.
4. Viscosity of produced oil is 10~20 cp whereas oil in reservoir has 24,400 cp.

24.8.2 Conclusions of Test Run 2000–01

1. In this case, the recovery is 82% of OOIP.
2. The combustion front has temperature 500~600 °C.
3. The oil produced is upgraded to 23° API (Table 24.3).

Table 24.3 Results of test run 2000–07/2000–01

Run	2000–07	2001–01
Recovery method	Primary	
Combustion mode	Dry and Wet	
Overall period (hr)	12.25	12
Pre-ignition period (hr)	2.25	2.25
Air injection period (hr)	10.0	9.75
Dry phase (hr)	7.0	6.75
Wet phase (hr)	3.0	3.0
Normal water injection (hr)	1.5	3
WOR (m^3/1000 S m^3)	0.3	0.3
Air injection rate (1/min)	10	
Peak temperature (°C)	500–600	
Produced gas composition (% Average) CO_2 CO O_2	14.56 4.88 3.25	15.38 5.49 3.88
CO/(CO+CO_2)	0.251	0.263
H/C ratio	0.05	0.0
O_2 utilization (%)	84.5	81.5
Furn consumption (*%OOIP*)	7.7	9.6
Oil recovery (*%OOIP*)	87.1	82.0
Residual oil (*%OOIP*)	2.9	5.39
AOR (Sm^3/m^3)	1130	1070
Combustion front velocity (m/hr)	0.05	0.05

24.9 Benefits of Thai Process

1. The THAI process has environmental benefits over the other thermal processes like SAGD. The THAI process requires minimum water and hence lesser use of natural gas. The THAI uses nominal volume of steam for the beginning of process and no longer requires after air injection has started. The THAI projects have smaller capital investment and have lower operating cost than SAGD.
2. SAGD is a "net water user" whereas THAI process produces water which is in fact clean enough for industrial use or immediate reinjection.
3. Carbon dioxide emission is relatively very less in THAI process.
4. Higher potential resource recovery (projected 70–80% based on laboratory tests and numerical simulation).
5. Ability to operate in lower quality reservoirs compared with other steam-based recovery methods.
6. Ability to produce partially upgraded bitumen reduced diluents requirements for transportation.

24.10 Conclusion

The environmental polluting factors are very less as compared to the other extraction methods, and the results of the simulation of use of horizontal injection and two horizontal production wells can enhance the recovery of the bitumen from the oil sands. Additionally, the use of the CAPRI catalyst in the horizontal production well can also increase the API index of the oil saving the millions of dollars in upgrading the oil. Thus, THAI along with CAPRI is the best method suggested for the extraction of the bitumen from the oil sands.

The other advantage is the ease of the land reclamation, which is not possible in the case of the surface mining, and there is no construction of tailing ponds which further another advantage of the THAI on surface mining. The use of the water in the THAI process is minimum than any other thermal methods. Hence, THAI gives the perfect ideal solution for the extraction of the bitumen.

References

1. M. Ismail Iqbal, O. Isaac, I. Al, S. Khuthbuddin, A. Ameen, Hazard identification and risk assessment with controls (Hirac) in oil industry-A proposed approach. Mater. Today Proc. **44**(January), 4898–4902 (2020). https://doi.org/10.1016/j.matpr.2020.11.800
2. K. Harish, K. Srivvastava, Study on effect of meditation on individual's performance at workplace. Int. J. Manag. (IJM), **12**(1), 379–389 (2021). https://doi.org/10.34218/IJM.12.1.2021.033
3. M.I. Iqbal, O. Issac, C. Author, Study and analysis of accident causation theory for improving safety performance in oil and gas. Mater. Today **11**(12), 2181–2191 (2020). https://doi.org/10.34218/IJARET.11.12.2020.206
4. D.K. Gupta, M.I. Iqbal, Reservoir simulation to diagnose the causes of reduced well production efficiency in coal bed methane. J. Earth Sci. Climatic Change 6(01), 1–10 (2015). https://doi.org/10.4172/2157-7617.1000248
5. M. Ismail, I. Alrajawy, O. Isaac, A. Ameen, Study the impact of safety awareness program (SAP) as moderating variable for reduction of accidents in oil and gas industry–a proposed framework. Int. J. Manage. Human Sci. (IJMHS), **5**(1) (2021). https://ejournal.lucp.net/index.php/ijmhs/article/view/1331
6. T. Kumar Dora, K. Vamsi Krishna, M. Ismail Iqbal et al., An experimental analysis on nanoparticles role in drilling fluids. Mater. Today Proc. https://doi.org/10.1016/j.matpr.2022.06.001
7. T. Kumar Dora, M. Ismail Iqbal, K. Vamsi Krishna et al., Application of low temperature fracturing fluid system in oil reservoir–a breakthrough approach. Mater. Today Proc. https://doi.org/10.1016/j.matpr.2022.06.141
8. M.I. Iqbal, Stress analysis in drill string and selection during oil well drilling with. **20**(6), 6165–6176 (2022). https://doi.org/10.14704/nq.2022.20.6.NQ22622
9. M.I. Iqbal, S.R. Mahapatra, I. Alrajawy, A. Ameen, Sand consolidation by chemical treatment using indigenously developed chemicals. Psychol. Educ. **57**(8), 3748–3751 (2020). https://doi.org/10.17762/pae.v58i3.4581
10. C.V. Deustch, J.A. McLennan, Guide to SAGD reservoir characterization using geostatics
11. S. Jahan, I. Alrajawy, S. Abid Ali, P. Martin, Analysing the impact of Globalisation on business performance: case of UAE food industry. Int. J. Qual. Res. **15**(2), 665–678 (2021). https://doi.org/10.24874/IJQR15.02-19

12. S. Jahan, S. Ali, T. Hussaini, Impact of internet usage on the academic performance of undergraduate students in Karachi. Univers. J. Educ. Res. **9**(3), 579–585 (2021)
13. S. Jahan, I. Alrajawy, Exploring the relationship between corporate social responsibility (CSR) and organisational profitability-a study of the UAE manufacturing industry. Int. J. Psychosoc. Rehabil. **24**(03) (2020). https://doi.org/10.37200/IJPR/V2413/PR2021129
14. M. Krishanamacharyulu, N.C. Rajyalakshmi, O. Isaac, K.S. Harish, M.I. Iqbal, K. Srivvastava, Study on effect of meditation on individual's performance at workplace. Int. J. Manag. (IJM) **12**(1), 379–389 (2021). https://doi.org/10.34218/IJM.12.1.2021.033
15. M.I. Iqbal, D.K. Gupta, Reservoir simulation to diagnose the causes of reduced well production efficiency in coal bed methane. J. Earth Sci. Climatic Change **6**(1), 1 (2015)

Chapter 25
Factors Influencing the Consumer Purchase Behavior of Electric Vehicles in United Arab Emirates

Sarwat Jahan, Sidra Abid, Parul Martin, Zafarullah Khan, and Mohammed Ismail Iqbal

Abstract As of 2021, the cumulative number of electric vehicles worldwide increased 43% from 2020, exceeding 10 million. In this study, the researcher surveyed and analyzed important electric vehicle purchase considerations for electric vehicle owners. The researchers aim to discover a potential method for supplying electric vehicles through this study. It further aims to develop personalized marketing ideas for corporations by empirically examining the elements influencing consumers' purchases of electric vehicles and predicting market demand for electric cars. The investigators have determined the market situation of electric vehicles by a review of the relevant literature and analyzed prior studies on the factors influencing their purchasing intent. Through empirical investigations, differences in electric car buying determinants by gender, age, and degree of performance relevance are evaluated. This research has the target audience of all these people in UAE who drive cars and have their own vehicle to collect relevant data. The data have been collected from 100 car owners across the different Emirates of UAE by using convenience sampling technique. The SPSS statistics program has been utilized for this purpose. Factors influencing the consumer purchase intention for electric vehicles are based on distance, charge time, new technology, degree of autonomous driving development, design, price, charging infrastructure, maintenance, and repair. In addition, the essential parameters have been determined, and a study of the average difference has been undertaken according to gender, age, and performance relevance.

S. Jahan (✉)
School of Business, City University Ajman, Ajman, United Arab Emirates
e-mail: sarwat_usman@hotmail.com

S. Abid
Univesiti Sains Malaysia, Main Campus, Penang, Malaysia

P. Martin
City University Ajman, Ajman, UAE

Z. Khan
Department of Business and Accountancy, Lincoln University College, Petaling Jaya, Malaysia

M. I. Iqbal
University of Technology and Applied Sciences, Nizwa, Oman

A. K. Sahu et al. (eds.), *Energy Storage and Conservation*, Springer Proceedings in Energy, https://doi.org/10.1007/978-981-99-2870-5_27

Keywords Electric vehicles in UAE · Purchase Behavior · Charging time · Autonomous driving

25.1 Introduction

As the planet's climate continues to shift for the worse as a result of human-caused pollution, protecting the planet's natural resources has risen to the forefront of global consciousness. Market size and sales of eco-friendly cars like electric vehicles are predicted to continue to develop as the significance of people's view of eco-friendly has increased and eco-friendly legislation are becoming a trend. By 2020, there will have been over 10 million electric cars sold throughout the globe, a rise of 43% annually. As COVID-19 expanded, the number of new automobiles registered fell by 16% from the previous year, while the percentage of electric vehicles among all new cars rose by 70%. This is attributed to the development of electric car supply policy, the implementation of new incentives connected to electric vehicles to overcome the COVID-19 economy, the growth in the number of models of electric vehicles, and the decrease in the price of batteries. In 2030, there will be 145 million EVs in use throughout the globe, a 30% increase from today's numbers [1]. By 2020, countries all over the world are expected to have increased their efforts to enforce carbon emission standards or zero-carbon emission vehicles, and 20 of these countries have already announced policies prohibiting the sale of internal combustion engine vehicles and mandating the sale of carbon-free vehicles [1].

Greenhouse gas emissions are predicted to decrease dramatically as the industry shifts toward more environmentally friendly vehicles like electric automobiles, which produce much less pollutants than conventional combustion engine cars fueled by gasoline or diesel [2]. Despite the increasing interest in electric cars, there is a lack of data on the elements that influence buyers' decisions and how happy they are with their purchases. This research aims to shed light on the variables that influence customers' decisions to buy electric cars, with the hope of identifying actionable insights that can be used to accelerate the adoption of this technology.

25.2 Theoretical Backgrounds

25.2.1 An Explanation of What is Electric Vehicle

An electric vehicle is one that is propelled by an electric motor using electricity generated by either fossil fuels or renewable energy sources. Since an electric car's propulsion comes from electricity stored in a battery and converted into mechanical motion by a motor, the vehicle itself doesn't consume any fossil fuels. There is no longer a need for internal combustion engines in cars; as such, the electric vehicles powering electric vehicles serve as the equivalent of the engines found in traditional vehicles, and the storage batteries they utilize as fuel for those vehicles.

Hybrid electric vehicles (HEV), plug-in hybrid electric vehicles (PHEV), battery electric vehicles (BEV), and fuel cell electric vehicles (FCEV) are all subcategories of electric vehicles. HEVs are a convenient way for those used to conventional automobiles to try out environmentally friendly vehicles. Hybrid electric vehicles (HEVs) are vehicles that store electrical energy in a battery inside the vehicle's body via engine drive and regenerative braking, and then use an electric motor to boost fuel economy and mileage. A plug-in hybrid electric vehicle (PHEV) is a kind of electric car that falls between a full hybrid and a basic electric vehicle. The battery in a PHEV is enormous, and the vehicle is equipped with a charging port. Pure electric power, emission-free driving, and other eco-friendly benefits are all within reach with the PHEV since it combines the best of electric and hybrid vehicles. The blended power technique also has the potential to extend the vehicle's range. The BEV is propelled entirely by an electric motor and not by a gasoline or diesel engine. A BEV's only power source is an electric motor and battery, therefore it produces no emissions and makes almost little noise. A battery electric vehicle (BEV) may receive power from an external charger and use that power to propel the vehicle. BEVs have several benefits, but the largest is that they may be utilized as eco-friendly cars and cost considerably less to run than cars that use gasoline or diesel [3].

25.2.2 The Response of Consumers to Electric Vehicles

One should expect more enthusiasm for electric vehicles among the young and the well-educated. Furthermore, the impression of oil prices, greenhouse gas emission reductions, and eco-friendliness while driving electric cars is positively correlated with environmental awareness and comprehension. Electric car buyers have voiced several apprehensions, including those related to range and scepticism about emerging technology [3]. It has been suggested that customers' unfavorable views of electric cars might be mitigated by providing them with more information about the mileage of electric vehicles and enhancing the interface design of electric vehicles [4]. By using the planned theory of behavior (TPB), Moons and De Pelsmacker (2012) investigated what elements are associated with a preference for EVs. This led researchers to conclude that individuals' subjective norms, together with their feelings and impressions of EVs, play a role in shaping their intentions to purchase and utilize such cars. Furthermore, respondents who reported a strong ambition to utilize EVs also reported feeling introspective about their driving conduct while using EVs [5]. It is crucial to study how to enhance the efficiency of electric car charging systems [6], since charging time seems to be the second most significant issue for customers.

25.2.3 Factors Affecting the Purchase of Electric Vehicles

Electric vehicle adoption is slowed by factors such as the high cost of entry, short range, lack of charging stations, and inaccurate information available to customers who have little experience driving EVs. Consumers in United Arab Emirates were asked what elements were most influential in their decision to buy an electric car, and the results showed that reduced gasoline prices and governmental subsidies were the most significant economic incentives. In addition, concern for the environment has been an important driving force in recent years. Maximum range was the most significant consideration for 45% of buyers, followed by performance (24%), vehicle price (17%), design (9%) and governmental subsidies (5%). State subsidies were found to have risen to the number one position (45%) from the number two position (25%), which seems to have improved customers' outlook on and propensity to buy electric cars. However, the lack of charging infrastructure is cited by the vast majority of those who are hesitant to purchase electric vehicles (82%), which can be interpreted as a sign that consumers are still wary of the convenience of electric vehicles despite the fact that they have to travel further to reach a charging station. Plus, 94% of people stated they would consider buying an electric car, which is a 3% increase from the previous year [7].

As a result of a lack of investment in charging stations, the availability of EVs in United Arab Emirates is a major issue. Since electric cars' range per charge is currently inferior to that of conventional gasoline-powered automobiles, more fast-charging infrastructure needs to be put in place. The limited range of electric cars is a drawback that might be ameliorated by expanding access to quick charging stations. However, the fast-charging infrastructure is woefully inadequate when weighed against the electric car adoption rate when considering the time needed for charging [8].

The elements that impact consumers' choices for electric cars have been researched for decades and are broken down into four categories: socioeconomic, psychological, mobility, and social [9]. The purchase price, operation cost, battery leasing cost, and fuel cost are all examples of economic features. Charging availability and charging station density are examples of infrastructural features [10]. As people become more eco-conscious, researchers are looking at how attributes like lower carbon dioxide emissions and lower pollution levels influence consumer behavior [10]. Studies have also been done lately to build marketing strategies by segmenting the market, such as determining "who will buy electric cars" to help bring them to the general public, and evaluating "characteristics of early adopters" to help identify potential customers [11, 12]. Authors have also examined the impact of real world driver experience on the adoption of EVs. Some authors claim, "individual preferences change significantly after a real experience with an electric vehicle" and concerns regarding commonly held misconceptions such as "driving range, top speed, fuel cost, battery life and charging in city centres and train stations" [13].While driver experience proved to be an effective method for changing perceptions about EVs, some authors have pointed to the impact that symbolic meanings have on the adoption of EVs. Semiotics are important in two ways: first, cars do not simply provide mobility, but symbolize certain ideas. Second, the ideas that cars symbolize relate to a consumer's identity [14, 15]. Researchers have employed multidimensional perspectives of value to understand consumers' opinions toward adopting EVs in UAE [16] and Indonesia [17], among other countries in the world [18]; however, fewer studies have focused on the UAE. Furthermore, EVs not only provide basic transportation, but are also part of a broader solution for addressing the increase in CO2 emissions. An investigation of the effects of relevant values on consumers' opinions toward adopting EVs is valid for predicting the purchase of "environmentally friendly" light-duty road transport [19].

The proposed hypotheses for this study were established on the basis of Theory of Planned Behavior. Being a major energy consumer, United Arab Emirates is diligently looking forward to reduce its energy requirements and greenhouse emissions. Electric vehicles (EVs), a sustainable form of automobile transportation, can reduce the country's dependence on gasoline while greatly reducing its carbon footprints [20]. The study uses a TPB model in order to predict adoption intention of 100 customers towards the purchase of EVs and the hypothesis are as follows:

H1 The purchasing factors of electric vehicle is influenced by the gender.
H2 The purchasing factors of electric vehicle is influenced by the age.
H3 The purchasing factors of electric vehicle is influenced by the income level.
H4 The purchasing factors of electric vehicle is influenced by the perceived level of performance given by electric vehicles.

25.3 Research Methodology

The present investigation is exploratory and descriptive and was approached in two stages. In the first stage, a bibliographic search was carried out in order to determine a state of the art and antecedents of similar studies of problem. The secondary research was used to explain underlying theories of buying behavior for electric vehicles and its extensions in previous studies [21]. In the second stage, with a simple cross-section design, the researcher sought to explain the phenomenon under study by conducting a survey questionnaire as data collection instrument and convenience sampling to recruit 100 car owners in the different restaurants of UAE. Second stage of the study involves researcher's intention of conducting a primary research by using quantitative research approach. This research has the target audience of all these people in UAE who drive cars and have their own vehicle to collect relevant data. The SPSS statistics program has been utilized for this purpose. The data collection instrument for the study chosen was a questionnaire for conducting a survey. Questionnaire and convenience sampling are commonly used primary data collection processes in past studies.

This study set the following factors influencing the purchase of electric vehicles. A questionnaire was prepared on the Likert scale (1 point to 5 points) for the important factors influencing the purchase of electric vehicles. After providing an explanation of the goal of the research to one hundred different persons who were driving vehicles as of January 2022, a survey was carried out. It was accomplished by the respondents immediately filling out their replies on the questionnaire. The statistical application IBM SPSS for Win 22.0 was used in order to do an analysis on the data that was gathered for this investigation.

25.4 Data Analysis and Findings

Regarding the demographic and social features of the people who participated in the research, this study looks at their gender, age, and the typical monthly family income. In terms of their monthly income, 9 people (9%) earned less than two million won, 17 people (17%) earned between two million won and three million won, 35 people (35%) earned between three million won, and 40 people (40%) earned more than five million won. There were 59 males (or 59%) and 41 females (or 41%). Among the participants, 3% were in their 20s, 21% were in their 30s, 42% were in their 40s, 23% were in their 50s, and 11% were in their 60s.

25.4.1 Analysis of Electric Car Buying Factors

On the Likert scale, the scores for all of the elements related to buy motivation ranged from 3.56 points out of 5 to 4.39 points out of 5 at the highest point. When looking at each specific aspect of purchase motivation, the average value of the significance of mileage is 4.07 points (standard deviation = 0.795), the significance of charging time is 4.35 points (standard deviation = 0.769), the significance of automobile new technology is 3.88 points (standard deviation = 0.769), and the significance of the degree of development of autonomous driving is 3.75 points (standard deviation = 0.978). 3.56 points were awarded for the design (standard deviation = 0.903), 4.31 points were awarded for the charging infrastructure (standard deviation = 0.716), 4.39 points were awarded for the ease of maintenance and repairs (standard deviation = 0.764), and 4.18 points were awarded for the incentive (standard deviation = 0.672).

25.4.2 Gender-Based Analysis of Electric Vehicle Purchase Factors

An independent sample T-test was performed to investigate the gender gap in the variance associated with the decision to purchase an electric car. The relevance of

Table 25.1 Analysis of the difference in importance values of electric vehicle purchase factors according to gender

Factors affecting the purchase of electric vehicles	Male (N = 59)		Female (N = 41)		t	p	Difference of mean	S.E. of difference
	M	SD	M	SD				
Mileage	4.14	0.798	3.98	0.790	0.992	0.324	0.160	0.161
Charging time	4.24	0.773	4.51	0.597	−2.004	0.048	−0.275	0.144
New technologies	3.80	0.783	4.00	0.742	−1.318	0.191	−0.203	0.156
Autonomous driving	3.61	1.000	3.95	0.921	−1.758	0.082	−0.341	0.197
Design	3.68	0.860	3.39	0.945	1.553	0.124	0.288	0.185
Price	4.24	0.751	4.41	0.706	−1.203	0.232	−0.177	0.147
Infrastructure	4.29	0.744	4.37	0.623	−0.566	0.573	−0.078	0.137
The ease of maintenance and repair	4.25	0.822	4.59	0.631	−2.275	0.025	−0.331	0.153
The incentives	4.20	0.761	4.15	0.527	0.443	0.659	0.057	0.129

charging time and the phase of maintenance and repair varied considerably at the 5% level between male and female buyers of electric cars. This was one of the variables that influenced the purchase of electric vehicles. At the 10% threshold of relevance, it was discovered that people of different genders place different weightings on the significance of autonomous driving. It was discovered that the charging time was of greater significance to women (M = 4.51) than it was to males (M = 4.24) (t = 2.004, p0.05). It was discovered that women (M = 4.59) viewed maintenance and repair to be more essential than males (M = 4.25) (t = −2.275, p0.05). This was based on the results of a survey. It was discovered that women (M = 3.95) place a higher value on autonomous driving than males (M = 3.61) do (t = −1.758, p0.10) (see Table 25.1).

25.4.3 *Comparison of Electric Vehicle Purchase Factors by Age*

We utilized a T-test for independent samples to look at the typical premium for purchasing. Based on the median value of 45.54, two age groups were created for further examination. The relevance of elements such as mileage, charging time, price, infrastructure, the phase of maintenance and repair, and the incentives varied greatly according to age as one of the factors that influenced the purchase of electric cars. It was discovered that older people placed a higher level of significance on all relevant aspects than younger people did. It was found that those in the older age bracket had a greater average value of significance than those in the younger age bracket. Factors including range (M = 4.31 versus M = 3.84), charging time (M = 4.61 versus M = 4.10), and price (M = 4.51 versus M = 4.12), as well as charging infrastructure (M = 4.63 versus M = 4.16), maintenance and repair (M = 4.35 versus M = 4.02), and incentives (M = 4.35 versus M = 4.02), were analyzed to evaluate their relative relevance (see Table 25.2).

25.4.4 *Performance-Based Analysis of Electric Vehicle Purchase Factors*

The first group consisted of those who placed a high priority on the automobile's performance, while the second group prioritized other aspects of the purchase, such as the environment, the brand, the design, and the price. The independent sample T-test was then used to examine the potential significance of the mean difference in purchasing motivation. It was discovered that the population that places a higher priority on performance places a higher priority on charging time, autonomous driving, pricing, and the phase of maintenance and repair than the population that does not. The group that placed a greater priority on performance had mean values that were much higher than the other group, which did not. High priority was placed on charge time (M = 4.47 versus M = 4.20), the degree of autonomous driving advancement (M = 3.91 versus M = 3.56), price (M = 4.44 versus M = 4.16), and the ease of maintenance and repair (M = 4.53 versus M = 4.22) (see Table 25.3).

25.5 Conclusion and Discussion

Given the limited number of studies in the UAE on the attitudes of consumers regrading EVs, the current research contributes to the advancement of knowledge on sustainable transport by gaining an understanding of the factors that influence university students' perceptions, preferences, and valuation regarding the adoption of EVs in the UAE. The relevance of each aspect was validated by the findings of this research, which investigated the variables that influence the decision to acquire electric cars. In addition, a t-test was conducted to determine which characteristics vary according to the extent to which age, gender, and performance are significant. The ease of maintenance and repair (M = 4.39) is the primary feature that encourages individuals to invest in electric vehicles. After charging infrastructure (M = 4.32), price (M = 4.31), incentive (M = 4.18), mileage (M = 4.07), car new technology (M = 3.88), and the degree of self-driving development (M = 3.75) and design (M = 3.56), charging time (M = 4.35) was determined to be a significant consideration in buying electric vehicles.

Table 25.2 Analysis of the difference in importance values of electric vehicle purchase factors according to age

Factors affecting the purchase of electric vehicles	Young (N = 51)		Old (N = 49)		t	p	Difference of mean	S.E. of difference
	M	SD	M	SD				
Mileage	3.84	0.784	4.31	0.742	−3.034	0.003	−0.463	0.153
Charging time	4.10	0.806	4.61	0.492	−3.866	0.000	−0.514	0.133
New technologies	3.80	0.800	3.96	0.735	−1.011	0.315	−0.155	0.154
Autonomous driving	3.71	1.045	3.80	0.912	−0.459	0.647	−0.090	0.196
Design	3.59	0.829	3.53	0.981	0.317	0.752	0.058	0.182
Price	4.12	0.840	4.51	0.545	−2.783	0.007	−0.393	0.141
Infrastructure	4.20	0.775	4.45	0.580	−1.853	0.067	−0.253	0.137
The ease of maintenance and repair	4.16	0.857	4.63	0.566	−3.287	0.001	−0.476	0.145
The incentives	4.02	0.735	4.35	0.561	−2.510	0.014	−0.327	0.130

Table 25.3 Analysis of the difference in importance values of electric vehicle purchase factors according to the importance of performance

Factors affecting the purchase of electric vehicles	Performance-oriented group (N = 55)		Other factor-oriented group (N = 45)		t	p	Difference of mean	S.E. of difference
	M	SD	M	SD				
Mileage	4.15	0.678	3.98	0.917	1.020	0.311	0.168	0.160
Charging time	4.47	0.604	4.20	0.815	1.865	0.066	0.273	0.142
New technologies	3.91	0.752	3.84	0.796	0.414	0.680	0.065	0.156
Autonomous driving	3.91	0.967	3.56	0.967	1.819	0.072	0.354	0.194
Design	3.56	0.856	3.56	0.967	0.044	0.965	0.008	0.185
Price	4.44	0.601	4.16	0.852	1.864	0.066	0.281	0.151
Infrastructure	4.35	0.673	4.29	0.727	0.400	0.690	0.057	0.141
The ease of maintenance and repair	4.53	0.663	4.22	0.850	1.968	0.052	0.305	0.155
The incentives	4.18	0.611	4.18	0.747	0.029	0.977	0.004	0.139

Next, the average difference according to gender was verified in the significance of charging time, the level of autonomous driving advancement, and the ease of maintenance and repair when examining the results of the difference in the relevance of buying motivation depending on gender. Third, based on the findings of the age-related variations in the significance of buying motivation, the average difference according to gender was confirmed in the following categories: range, charging time, cost, charging infrastructure, repairability, and incentives. The senior population placed a greater emphasis on the significance of every crucial aspect compared to the younger population. Because of this, it will be required to raise the advantages of mileage, charging time, affordability, infrastructure, simplicity of maintenance and repair, as well as incentives, in order to entice older people (those who are at least 46 years old) to acquire electric cars.

The results of the comparison of the importance of purchasing motivation between the group that prioritizes performance when selecting a car and the group that does not indicate that there is an average difference in charging time, the degree of autonomous driving development, price, and the ease with which maintenance and repair can be performed. As a result of this, it was shown beyond a reasonable doubt that in order to convince those who place a high value on performance to buy electric cars, it is required to place an emphasis on the superiority of the charging time, autonomous driving, pricing, and simplicity of maintenance.

25.5.1 Managerial Implications

The current research contributes to the existing academic literature in several ways. Issues such as environmental concerns, trendiness, operation costs, drivability, other people's opinions and government incentives are among the most important factors that affect potential adopter attitudes with no significant difference between male and female responses. This is in line with other studies that examine the importance of a variety of incentives that promote the adoption of EVs, 'real world' driver experience, and the symbolic value that vehicles have regarding potential adopter attitudes. Consumers believe that objective information is trustworthy and that electric automobiles have a positive impact on environmental preservation, as shown by the past findings. Presently, electric car manufacturers mostly employ information broadcast across networks, which can be easily ignored, as a means of promoting their products. Electric vehicle manufacturers are encouraged to increase consumer interest in and demand for their products by encouraging a green and environmentally sensitive way of life.

Given these findings, it's evident that customers both believe that objective information is reliable and think that electric automobiles have a positive impact on environmental preservation. Electric car producers rely mostly on potentially ignored network-based advertisements to promote their products at now. It has been suggested that electric car manufacturers advertise the benefits of going green and safeguarding the environment to increase consumer interest in and demand for EVs.

Consumers don't notice any major differences in the way EVs and traditional cars function, so they treat them similarly. Conversely, electric cars, a new kind of eco-friendly technology product, have cutting-edge innovations and technologies like voice systems and automated parking systems, and have driving operations that are optimized in contrast to conventional automobiles. Due to this, it has been proposed that EV makers provide more possibilities for buyers to interact with the cars. As a result, consumers' perspectives would shift, the electric car industry would expand, and customers' knowledge of the industry would increase.

Consumers believe the accessibility of charging facilities for electric vehicles will influence their decision to buy. In order to make charging electric cars easier, it has been proposed that the government first use a prototype layout of charging piles in key cities as a model, and then attract investment from relevant industries via subsidies. The goal of doing so is to facilitate the process of charging electric automobiles.

Consumers feel that the price of batteries as well as their lifespan will influence their decision to acquire batteries. Because of this, it has been recommended that makers of electric vehicles should embrace improved methods for battery servicing, such as battery leasing, while simultaneously increasing their investment in the research and development of battery technology. By establishing the concept of a car recycling economy, manufacturers may reduce the cost of batteries via the recycling, disassembly, and reuse of trash and scrap autos and their components, so contributing to the automobile industry's long-term viability and positive growth. This can include the recycling of power batteries and the disassembly of automobiles.

25.5.2 Future Research Directions

Firstly, future researchers are encouraged to employ alternative research methods from this study while studying electric automobiles in order to promote the widespread adoption of these vehicles. As the title suggests, this research focuses only on battery-powered automobiles. Secondly, it is suggested that future studies examine the possibility that various energy vehicles based on distinct guiding principles are associated to diverse impacts on consumer demand. Third, although some attention is paid to technology associated with electric vehicles, the study's primary emphasis is on customer demand. Researchers of the future should focus on bridging the gap between the automotive industry and customers, both from a technical and an industrial standpoint, while studying electric cars. Lastly, because of the study's time and money constraints, this survey was only sent to residents in the seven Emirates United Arab Emirates. The study's findings may not be universally accepted due to regional variances in United Arab Emirates with any other region, i.e. Africa and Europe because of the changes in infrastructure, income level and the concern of environment by the car owners. Future scholars may look at the state of affairs in many different geographical locations, i.e. Middle East as a whole to serve as pointers for policymakers and car companies in their efforts to promote electric automobiles.

References

1. S.R. Kim, Ch.H. Kim, Report and prospect of IEA's global electric vehicle market. World Energy Market Insight **21**(14), 1–22 (2021). https://www.iea.org/reports/global-ev-outlook-2021
2. A.M. Lewis, J.C. Kelly, G.A. Keoleian, Evaluating the life cycle greenhouse gas emissions from a lightweight plug-in hybrid electric vehicle in a regional context, in *2012 IEEE International Symposium on Sustainable Systems and Technology (ISSST),* pp. 1–6 (2012). DOI:https://doi.org/10.1109/ISSST.2012.6403806
3. S. Jahan, I. Alrajawy, S. Abid Ali, P. Martin, Analysing the impact of Globalisation on business performance: case of UAE food industry. Int. J. Qual. Res. **15**(2), 665–678 (2021). https://doi.org/10.24874/IJQR15.02-19
4. R.M. Krasuse, S.R. Carley, B.W. Lane, J.D. Graham, Perception and reality: public knowledge of plug-in electric vehicles in 21 U.S. cities. Energy Policy **62**, 433–440 (2013). https://doi.org/10.1016/j.enpol.2013.09.018
5. T. Franke, I. Neumann, F. Bühler, P. Cocron, J.F. Krems, Experiencing range in an electric vehicle: understanding psychological barriers. Appl. Psychol. **61**(3), 368–391 (2012). https://doi.org/10.1111/j.1464-0597.2011.00474.x
6. I. Moons, P. De Pelsmacker, Emotions as determinants of electric car usage intention. J. Market. Manag. **28**(3–4), 195–237 (2012). https://doi.org/10.1080/0267257X.2012.659007
7. S. Ro, B.S. Shin, K.J. Lee, Y.H. Ki, H.S. Ahn, Development of a smart grid monitoring system with anti-Islanding function for electric vehicle charging. J. Instit. Internet Broadcast. Commun. **12**(5), 31–37 (2012). https://doi.org/10.7236/JIWIT.2012.12.5.31
8. SISUN News, [Motorgram] What makes you buy an electric car? 'Environmental protection.' The reason is also increasing (2019). http://www.sisunnews.co.kr/news/articleView.html?idxno=103548

9. S. Jahan, P. Martin, Exploring the value of your website analytics. EPRA Int. J. Econ. Bus. Rev. 66–69 (2019). https://doi.org/10.36713/epra2999
10. J.S. Shin, Energy cost reduction main course and government policy effects on electric car stock. Master's Thesis, Korea University (2017)
11. F. Liao, E. Molin, B. Wee, Consumer preferences for electric vehicles: a literature review. Transp. Rev. **37**(3), 252–275 (2017). https://doi.org/10.1080/01441647.2016.1230794
12. C. Rudolph, How may incentives for electric cars affect purchase decisions? Transp. Policy **53**(C), 113–120 (2016). https://doi.org/10.1016/j.tranpol.2016.07.014
13. S. Jahan, A. Bhaumik, S. Abid, K. Tiwari, Consumer behaviour towards mobile marketing in UAE restaurants: application of technology acceptance model. Front. Artif. Intell. Appl. (2022). https://doi.org/10.3233/faia220390
14. R. Bennett, R. Kottasz, S. Shaw, Factors potentially affecting the successful promotion of electric vehicles. J. Soc. Mark. **6**(1), 62–82 (2016). https://doi.org/10.1108/JSOCM-08-2015-0059
15. Y.J. Baek, Analysis on consumer preference of electric vehicles in South Korea. Master's Thesis, Yonsei University (2017)
16. M. Ismail, I. Alrajawy, O. Isaac, A. Ameen, Study the impact of safety awareness program (SAP) as moderating variable for reduction of accidents in oil and gas industry–a proposed framework. Int. J. Manage. Human Sci. (IJMHS) **5**(1) (2021). https://ejournal.lucp.net/index.php/ijmhs/article/view/1331
17. M.I. Iqbal, Stress analysis in drill string and selection during oil well drilling with. **20**(6), 6165–6176 (2022). https://doi.org/10.14704/nq.2022.20.6.NQ22622
18. S. Jahan, S. Hijazi, W. Mahdaoui, The role of teaching styles in driving quality of student learning and achievement: evidence from universities in UAE. J. Contemp. Issues Bus. Gov. **27**(4), 878–897 (2022). https://www.cibgp.com/issue_156_318_volume+27%2C+Issue+5%2C+Winter++and+Spring+2021.html. Accessed 11 Sept 2022
19. Y.M. Kwon, A study on the promotion of electric vehicle distribution: focusing on the analysis of satisfaction factors of electric vehicle users. Master's Thesis, Korea University (2018)
20. S. Jahan, I. Alrajawy, Exploring the relationship between corporate social responsibility (CSR) and organisational profitability-a study of the UAE manufacturing industry. Int. J. Psychosoc. Rehabil. **24**(03) (2020). https://doi.org/10.37200/IJPR/V2413/PR2021129
21. M.I. Iqbal, S.R. Mahapatra, I. Alrajawy, A. Ameen, Sand consolidation by chemical treatment using indigenously developed chemicals. Psychol. Educ. **57**(8), 3748–3751 (2020). https://doi.org/10.17762/pae.v58i3.4581

Chapter 26
15-Level MultiLevel Inverter with NLMS-based IVCIMD for THD Reduction

Check for updates

Karimulla Peerla Shaik, Karimulla Syed, Mahaboob Shareef Syed, and Mohammed Ismail Iqbala

Abstract The Diode Bridge Rectifiers (DBR) play a major role in the Induction Motor Drive (IMD) and the usage of DBRs constitutes a high total harmonics distortion in the input currents. In this project, the universal bridge configuration is used for the induction motor drive, which aims to reduce the total harmonic distortion in the input current of IMD. A reduced switch Indirect Vector Controlled Induction Motor Drive (IVCIMD) with a 54-pulse AC-DC converter which is fed by a Multiple Level Inverter (MLI) is proposed for applications of medium power units. This eventually reduces the Total Harmonic Distortion (THD) at the input end and the supply current is mostly obtained as pure sinusoidal. An 18-pulse AC-DC converter is mainly used at the secondary winding of each of the three transformers that are isolated from each other. Thereby, the phase shifting is implemented between the transformers, which constitute a 54-pule AC-DC converter. Performance of IVCIMD is improved by incorporating a 15-level MLI with a binary standard cascading configuration with a universal bridge using IGBT switch model at the motor end. The proposed IVCIMD is featured with an efficient switching which works with a fundamental frequency-based Nearest Level Modulation Strategy (NLMS) that reports a very low switching loss. The proposed IVCIMD aims to provide an efficient signal input (sinusoidal) using a universal bridge configuration of MLI, and the simulations for different operating

K. P. Shaik (✉)
Malla Reddy College of Engineering and Technology, Maisammaguda, Telangana, India
e-mail: karimullapeerlashaik@gmail.com

K. Syed
Faculty of Engineering Department, University of Technology and Applied Sciences-Shinas, Shinas, Oman

M. S. Syed
Vignan's Lara Institute of Technology and Science, Guntur, Andhra Pradesh, India

M. I. Iqbala
Faculty of Engineering Department, University of Technology and Applied Sciences, Nizwa, Oman

A. K. Sahu et al. (eds.), *Energy Storage and Conservation*, Springer Proceedings in Energy, https://doi.org/10.1007/978-981-99-2870-5_28

conditions are analyzed using the Simulink/MATLAB for a medium power application which reported a reduced harmonic distortions and improved input signal waveforms.

Keywords IVCIMD · Multilevel inverter · NLMS · THD · Universal bridge

26.1 Introduction

The voltage at the stator for a medium power IMD is usually about FewKV and the voltage of the grid which is supplied to the IMDs as input will range over an 11 to 33kV. In such conditions for traditional medium power, IMD will be incorporated with a step-down transformer. The stepped down voltage is given to a module of 3-ph diode rectifier bridge (DBR) which converts the AC-DC supply, and a 3Ph VSI is used to drive the IM. DBR unit eventually leads to a low performance due to inducing the harmonics in the system, which is not permissible by the top international standards. To get rid of this problem, this paper aims to go for an approach of multiple isolated transformer technique instead of traditional step-down transformer and also the DBR unit is replaced by the universal bridge, which has a synchronous switching operation. Thus, the harmonics produced in the system will be reduced and also, they can be efficiently canceled by the internal universal bridges. Thus, ensuring the input current to have less THD which is drawn from the multiple winding transformers [1–7].

Also, besides the drawbacks of traditional IMD on the side of grid, the drive also undergoes various issues with respect to the motor terminal. As it will generate a large THD in voltage and currents of the stator winding, there will be a stress at the IGBT switch and the insulation deteriorates due to a large motor winding. The switching of IGBT switches will be high in the DBR, which result in the large switching losses. Hence, the synchronous switching in the universal bridge will be a better solution for improving the switching losses and hence the performance also improves. The Multiple Level Inverters (MLIs) will have less current THDs and consume less voltage, also has 2-level VSI.

There can be other MLI models [12] for different DC-AC/AC-DC applications. The main three configurations for MLIs are used for VFIMDs that are diode, cascaded and fly capacitor. However, all these configurations have major contributions on the drive side only, whereas the supply end is just briefed. In this work, to achieve a phase voltage with a 15-level, a binary tree technique is used, which requires only 36 IGBT switches. Where this count increases to 84 in a cascaded configuration. Modular transformer provides phase shift in primary and secondary windings of the transformer. Thus, the highly dominant harmonics can be eliminated on the drive side of the IMD. The proposed IVCIMD is highly suitable for the medium power industrial applications with variable speeds in the IMD [8–11]. The proposed IVCIMD is having a much better performance due to the elimination of dominant harmonics in the drive when compared to the existing IMD model.

26.2 Proposed IVCIMD Design

IVCIMD proposed system is shown in Fig. 26.1. Supply end of IVCIMD has 54-pulse AC-DC converter and motor end has a 15-level cascaded MLI with a binary base (1:2:4). The supply is 415 V, 50 Hz AC, which drives 7.5Kw (10hp), 415 V, 50 Hz IMD.

A. **15Level Cascaded MLI**

Every phase is distributed with three 1Ph VSIs, which are connected in cascade. The H-bridge drives the load of 7.5kW, which is mathematically termed as

$$m_i = \hat{V}_{aol}/(E4V_{dc}/\pi) \tag{26.1}$$

Here mi: modulation index; Vao1: fundamental phase voltage; Vdc: DClink voltage

Fundamental voltage peak value is defined as,

$$\hat{V}_{aol} = m_i E4V_{dc}/\pi = (0.7x7x4V_{dc})/\pi = 6.68V_{dc}$$

The rms value of fundamental voltage (V01) is defined as,

$$V_{aol} = \hat{V}_{aol}/\surd 2 = 6.68V_{dc}/\surd 2 = 4.72V_{dc}$$

B. **Design of 54-Pulse AC-DC Converter**

Transformers T1, T2, T3 are designed to be a delta to obtain 18-pulse converter. Due to the incorporation of the isolated transformers, current drawn by the proposed IVCIMD will have 54 pulses in current grid. The phase shifts of the transformers are calculated by

$$N_{sec.} = N_{dc}/N_t,\ \alpha_{sec.} = \left(360/N_{pulse}\right)/N_t \text{ and } \alpha_{prim.} = \alpha_{sec.}/N_t \tag{26.2}$$

Here Nsec, Ndc, Nt and Npulse are the secondary windings of transformers.

In this paper, the multiple winding transformer which is isolated from each other will have a phase shift of +20°, 0° and −20°.

The voltages of the winding Vsw and Vlw are calculated as shown below,

$$V_{LL}/\mathrm{Sin}120 = V_{Sw}/\mathrm{Sin}\alpha = V_{Lw}/\mathrm{Sin}(60 - \alpha) \tag{26.3}$$

Figure 26.2 represents the phase shifting angles of multi winding transformers, in the winding voltages of Vps and Vpl are calculated from the delta configuration, which is illustrated in Fig. 26.3 and is defined as,

$$V_{an}/\sin 166.6 = V_{pS}/\sin 6.7 \tag{26.4}$$

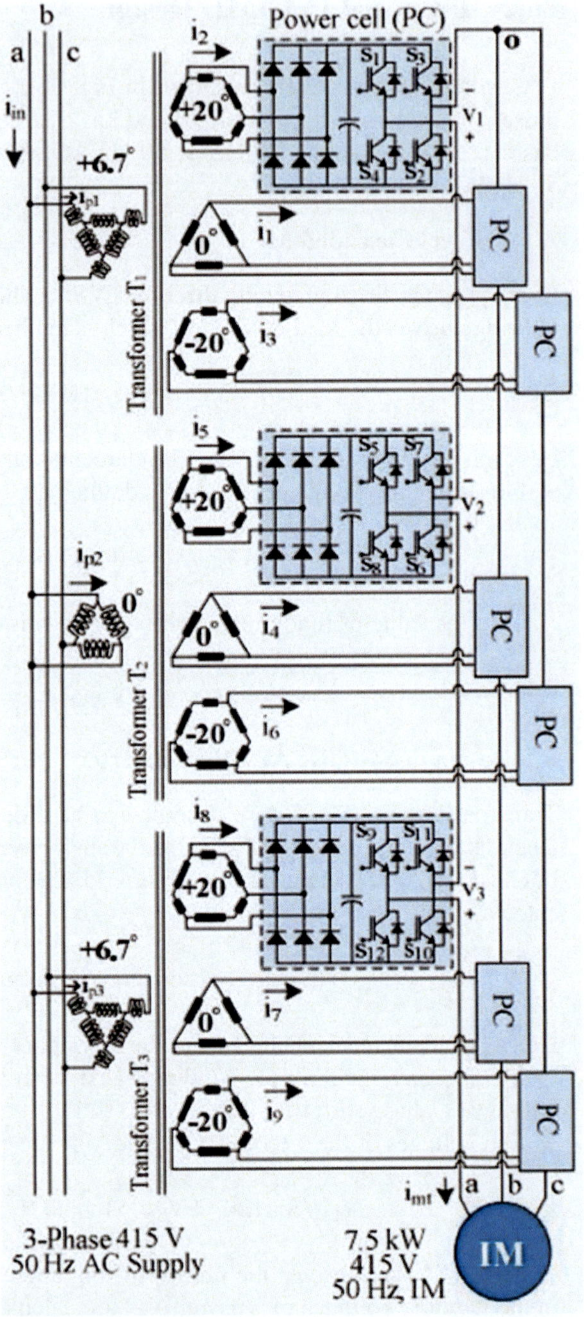

Fig. 26.1 Power cell configuration

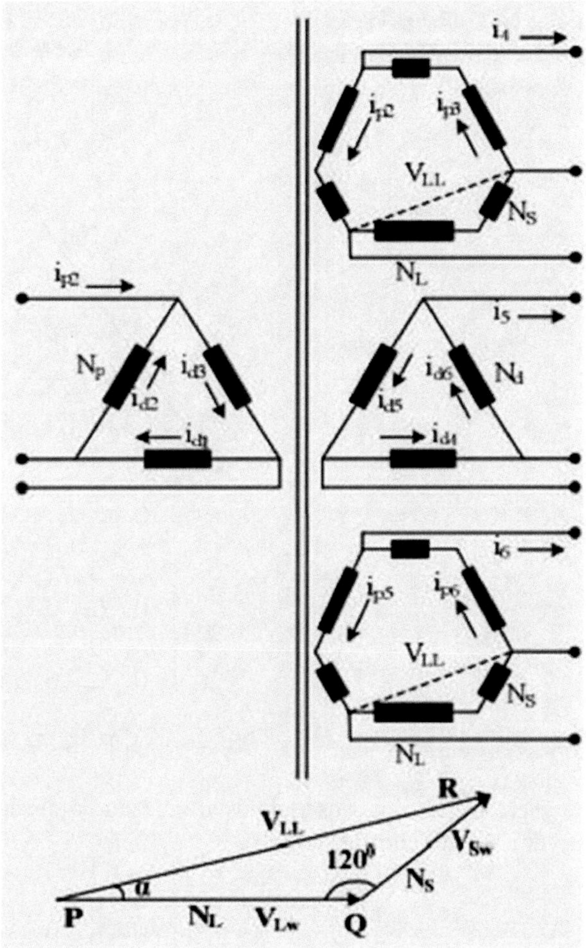

Fig. 26.2 Delta style winding

26.3 Dynamic Modelling of IM

The IM is expressed in the volt-ampere equation, which is termed in time differential operator and is defined as,

$$p[i] = [L]^{-1}\{[v] - [R][i] - \omega_r[G][i]\} \tag{26.5}$$

$$v = \left[v_{ds}\ v_{qs}\ v_{dr}\ v_{qr}\right]^T \tag{26.6}$$

$$i = \left[i_{ds}\ i_{qs}\ i_{dr}\ i_{qr}\right]^T \tag{26.7}$$

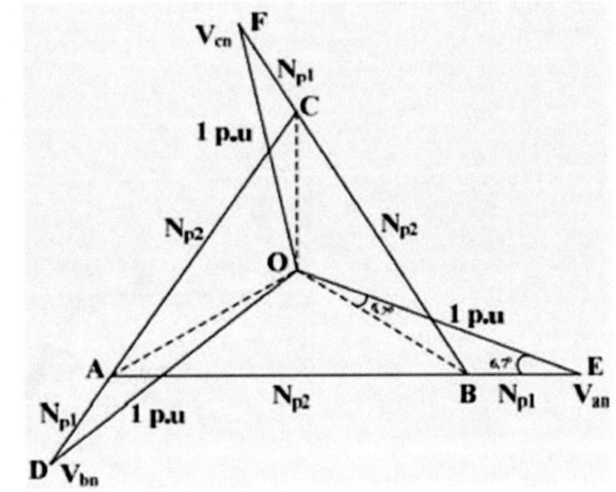

Fig. 26.3 Triangle calculation of the vectors

$$L = \begin{bmatrix} L_s & 0 & L_m & 0 \\ 0 & L_s & 0 & L_m \\ L_m & 0 & L_r & 0 \\ 0 & L_m & 0 & L_r \end{bmatrix} \tag{26.8}$$

$$R = \mathrm{diag}\left[R_s \; R_s \; R_r \; R_r \right] \tag{26.9}$$

where the ids, iqs, vqs and vds are the stator currents and voltages. idr, iqr, vqr and vdr are rotor currents.

$$G = \begin{bmatrix} 0 & 0 & 0 & 0 \\ 0 & 0 & 0 & 0 \\ 0 & L_m & 0 & M_m \\ L_m & 0 & L_m & 0 \end{bmatrix} \tag{26.10}$$

and voltages in d-q axis.

Rotor speed derivative is:

$$p\omega_r = (P/2)(T_e - T_L)(1/J) \tag{26.11}$$

where Te—Developed Electromagnetic Torque; TL—Load Torque, P—Number of pole and J-Moment of inertia.

$$T_e = (3/2)\,(P/2) L_m \left(i_{qs} i_{ds} - i_{ds} i_{qr} \right) \tag{26.12}$$

26.4 Proposed IVCIMD Control Algorithms

In proposed model, two control techniques are considered, namely, Indirect Vector Control which produces three modulation signals that are fed as inputs for nearest modulation strategy dependent by which an IM is controlled and Modified NLMS.

A. **Indirect Vector Control (IVC)**

The IVC of an IM is represented in the Fig. 26.2 and the algorithm checks for the output value to be valid. The speed sensed from the IM is compared to the reference speed model and the error is calculated, this is given as input to PI-controller and O/P torque is obtained from controller unit. The reference torque is defined as

$$T^*_{ref}(k) = T^*_{ref}(k-1) + k_p\{s_c(\mathrm{k}) - s_c(\mathrm{k}-1)\} + k_i s_c(\mathrm{k}) \tag{26.13}$$

The above equation, the gains of PI controller are kp and ki; depict error are se(k), se(k−1), reference torques are T^*_{ref} (k), T^*_{ref} (k−1).

$$i^*_{ref\cdot d} = \mathrm{I_{mag}} \text{if } \omega_\mathrm{r} < \omega_\mathrm{b} \text{ and } i^*_{ref\cdot d}(k)\,(\omega_\mathrm{b}/\,\omega_\mathrm{r})\,\mathrm{I_{mag}} \text{if } \omega_\mathrm{r} > \omega_\mathrm{b} \tag{26.14}$$

where Imag: Magnetizing current, wr: sensed speed, wb: rated speed, respectively.

$$T^*_{ref}(k) = \frac{3}{2}\frac{P}{c}\mathrm{L_m I_{mag}}\, i^*_{ref.q},\, i^*_{ref.q} = \frac{2}{3}\frac{2}{P}\cdot\frac{T^*_{ref}}{i^*_{ref.d}L_m} \tag{26.15}$$

The currents in both the phases are sensed and thereby the current at the third phase is evaluated by using ic=(−ia−ib). They are then converted to dq, which is compared to reference values and the error is given as input to the PI controllers. The complete system model is shown in Fig. 26.4. The output is determined by the following mathematical expressions:

$$V^*_d = V^*_d(k-1) + k_{pd}\{er_d(k) - er_d(k-1)\} + k_{id}er_d(k) \tag{26.16}$$

$$V^*_d = V^*_q(k-1) + k_{pq}\left\{er_d(k) - er_{qd}(k-1)\right\} + k_{iq}er_q(k-1) \tag{26.17}$$

$$\begin{bmatrix} v_{ref\_a} \\ v_{ref\_b} \\ v_{ref\_c} \end{bmatrix} = \begin{bmatrix} Cos\theta_r & -sin\theta_r \\ Cos(\theta_r + \frac{2\pi}{3}) & -sin(\theta_r - \frac{2\pi}{3}) \\ Cos(\theta_r + \frac{2\pi}{3}) & -sin(\theta_r + \frac{2\pi}{3}) \end{bmatrix} \cdot \begin{bmatrix} v^*_d \\ v^*_q \end{bmatrix} \tag{26.18}$$

where

$$\theta_\mathrm{r} = \int \omega_\mathrm{e}\mathrm{dt} = \int (\omega_{\mathrm{r}+}\omega_{\mathrm{sl}})\,\mathrm{dt} = (\theta_{\mathrm{r}\,+}\theta_{\mathrm{sl}}) \tag{26.19}$$

The slip speed is finally estimated as

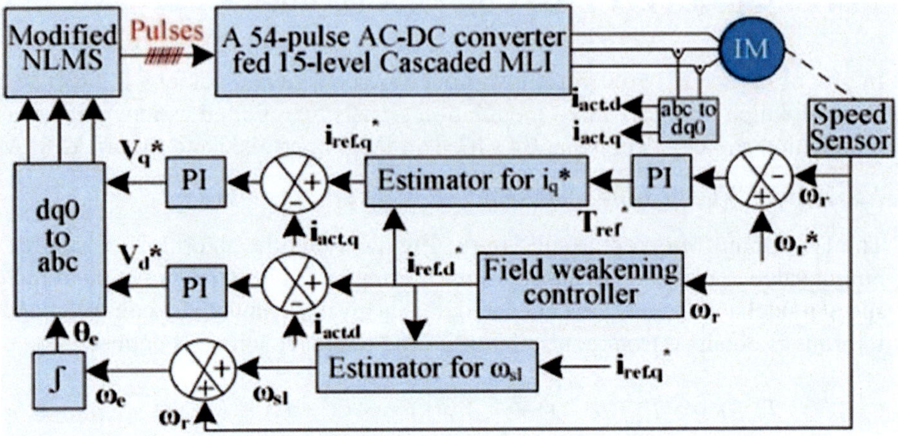

Fig. 26.4 Graphical representation of IVC

$$\omega_{\text{sl}}(\text{t}) = \frac{R_r}{L_r}\frac{i^*_{ref.q}(k)}{i^*_{ref.d}(k)} \tag{26.20}$$

B. **Modified NLMS**

There is a mandatory switching of power cell of each transformer leg for 4 times, 2 times and this happens for 1/4 of every Ph-voltage cycle. This mandatory switching can be achieved by the drive only if it is independent of the frequency, which is incorporated in the proposed IVCIMD. As the motor speed is directly proportional to the frequency of the signals, which are obtained from the IVC as shown in Fig. 26.5.

In this paper, modifications are made to NLMS, which calculates required fixed levels and also finds quotient of ratio of instantaneous value of modulating signal and predefined step, required level can be found for that particular duration of time. The predefined step is defined as

$$\text{ES}_p = 1 \text{ and } E = (N-1)/2 \tag{26.21}$$

where E: 15-level-Inverter number of steps, Sp: fixed steps, N: number of levels. The sequence for the predefined switching activity is shown in Table 26.1.

Algorithm of modified NLMS is shown in Fig. 26.3. The switching depends on the values of L as illustrated in the Fig. 26.3.

Fig. 26.5 Algorithm for modified NLMS

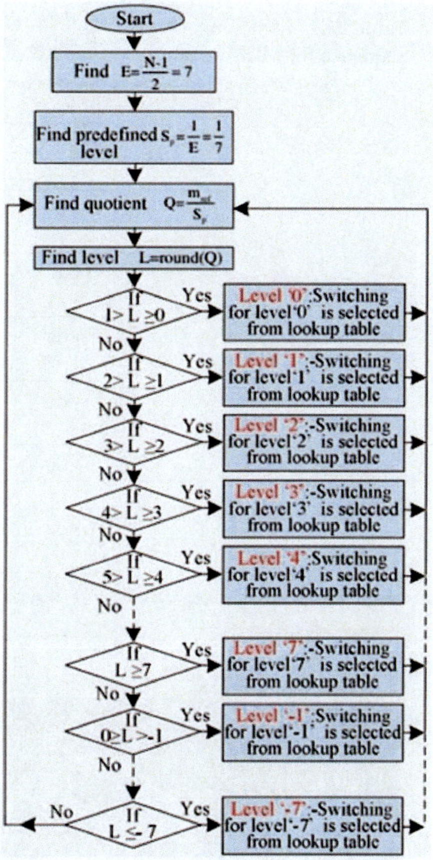

26.5 Simulation Results

Proposed IVCIMD simulation results are obtained by designing the models in Simulink/MATLAB and performance parameters of IM are obtained. Simulations of proposed-IVCIMD are shown in figures illustrated below.

In this paper, an IVCIMD with 54pulse AC-DC converter which is fed by MLI is proposed and been developed. IVC in proposed model senses motor speed and 3modulating signals are produced. The modified NLMS is eventually fed by these modulating signals.

Figure 26.6 represents the phase current and voltage containing distortions at supply end whereas Figure 26.7 represents value of THD of 33.18% at supply end.

Figure 26.8 represents phase current and voltage at the drive end whereas Figure 26.9 represents value of THD at the drive end, which is reduced to 3.49% by using proposed IVCIMD and NLM controlling techniques.

Table 26.1 Switching Sequence for MLI

Positive levels		Negative levels	
Level	Switching combination	Level	Switching combination
0	S1S3, S5S7, S9S11	−Vdc	S3S4, S5S7, S9S11
Vdc	S1S2, S5S7, S9S11	−2Vdc	S1S3, S7S8, S9S11
2Vdc	S1S3, S5S6, S9S11	−3Vdc	S3S4, S7S8, S9S11
3Vdc	S1S2, S5S6, S9S11	−4Vdc	S1S3, S5S7, S11S12
4Vdc	S1S3, S5S7, S9S10	−5Vdc	S3S4, S5S7, S11S12
5Vdc	S1S2, S5S7, S9S10	−6Vdc	S1S3, S7S8, S11S12
6Vdc	S1S3, S5S6, S9S10	−7Vdc	S3S4, S7S8, S11S12
7Vdc	S1S2, S5S6, S9S10		

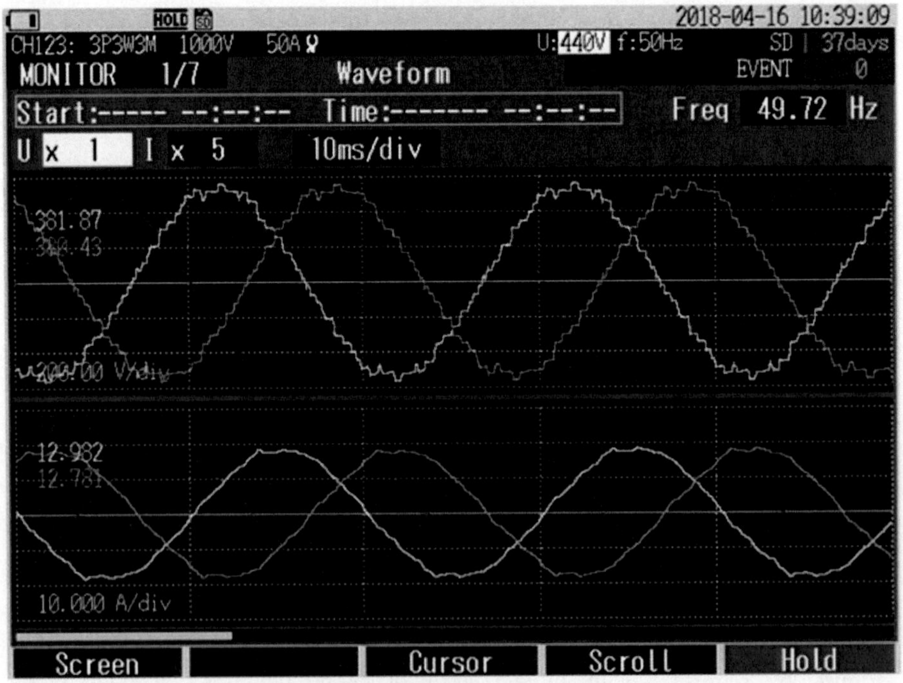

Fig. 26.6 Phase voltage and currents at supply side

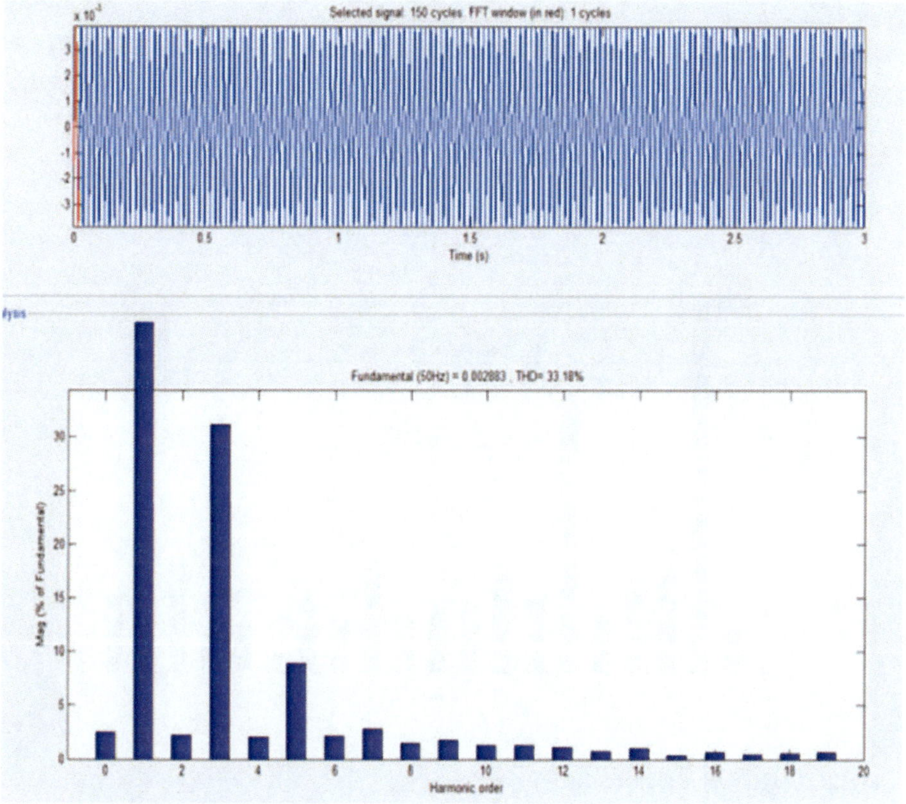

Fig. 26.7 THD value at supply side

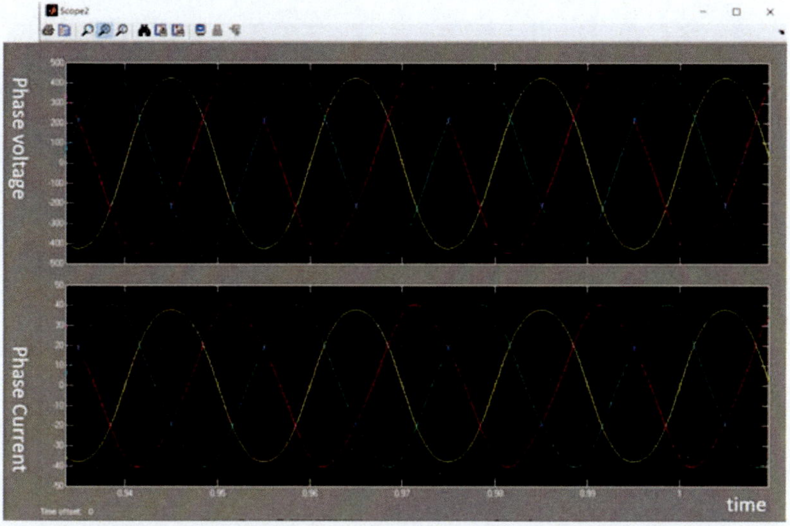

Fig. 26.8 Phase voltage and currents at drive side

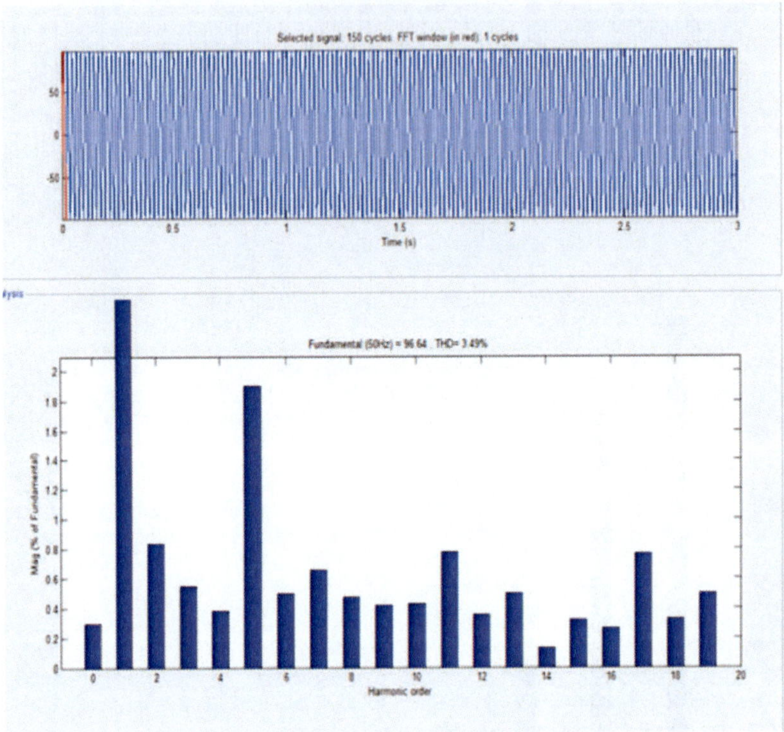

Fig. 26.9 THD value at drive side

26.6 Conclusion

An IVCIMD with a 54-pulse-AC-DC converter which is fed by a multiple level inverter (MLI) is proposed in this paper. This AC-DC converter has nearly shaped the input signals to pure sinusoidal by eliminating dominant harmonics which reduced THD to minimum. Also, universal bridge with configuration of IGBT is used as 15-level-Cascaded MLI to improve the performance at drive side of proposed drive. The modified NLMS has reduced the switching losses by the predefined switching levels. Thereby ensuring a better efficiency of proposed drive for medium power applications.

References

1. S. Singh, B. Singh, Optimized passive filter design using modified particle swarm optimization algorithm for a 12-pulse converter-fed LCI–synchronous motor drive. IEEE Trans. Indust. Appl. **50**(4), 2681–2689 (2014)
2. J. N. Forestieri, M. Farasat, A. M. Trzynadlowski, Indirect real and reactive power control of induction motor drives. IEEE J. Emerg. Selected Topics Power Electron. Early Access
3. M. Kang, B.O. Woo, P. Enjeti, I.J. Pitel, Auto connected-electronic transformer-based multi pulse rectifiers for utility interface of power electronic systems. IEEE Trans. Indust. Appl. **35**(3), 646–656 (1999)
4. D.A. Paice, *Power Electronic Converter Harmonics Multi-Pulse Methods for Clean Power* (IEEE Press, New York, 1996)
5. P. Kant, B. Singh, A. Chandra, K. Alhaddad, Twenty pulse AC-DC converter fed 3-level inverter based vector controlled induction motor drive, in *IECON 2017-43rd Annual Conference of the IEEE Industrial Electronics Society* (Beijing, 2017), pp. 2225–2230
6. S. Choi, B.S. Lee, P.N. Enjeti, New 24-pulse diode rectifier systems for utility interface of high-power AC motor drives. IEEE Trans. Indust. Appl. **33**(2), 531–541 (1997)
7. B. Singh, S. Gairola, B. N. Singh, A. Chandra, K. Al-Haddad, Multipulse AC–DC converters for improving power quality: a review. IEEE Trans. Power Electrons. **23**, 260–281 (2008)
8. P. Kant, B. Singh, A multi-pulse AC-DC converter fed 5-level NPC inverter based VCIMD, in IEEMA Engineer Infinite Conference (eTechNxT). New Delhi, India, pp 1–6 (2018)
9. B.K.Bose, *Modern Power Electronics and AC Drives* (Prentice Hall PTR, Chaps. 2 & 8)
10. N.A. Azeez, A. Dey, K. Mathew, J. Mathew, K. Gopakumar, M.P. Kazmierkowski, A Medium-voltage inverter-fed IM drive using multilevel 12-sided polygonal vectors, with nearly constant switching frequency current hysteresis controller. IEEE Trans Indust. Electrons. **61**(4), 1700–1709 (2014)
11. Q.A. Le, D.C. Lee, Elimination of common-mode voltages based on modified SVPWM in five-level ANPC inverters. IEEE Trans. Power Electrons Early Access

Chapter 27
Novel Approach for Performance Analysis of Photovoltaic and Wind Hybrid Energy System

Shabbier Ahmed Sydu, Karimulla Peerla Shaik, Mahaboob Shareef Syed, and Karimulla Syed

Abstract A solar photovoltaic (PV) is a standout among the most encouraging sustainable power source resources that proselytes solar energy into power with condition neighborly way. In any case, it has low productivity and high relative expenses. With a specific end goal to conquer these downsides, a grid-connected PV energy system ought to be required to fulfill the load request. The PV system is thought to be a most encouraging innovation, due to its appropriateness in conveyed age. In dispersed age applications, the PV system works in two distinct modes: grid connected mode and island mode. In the grid-connected mode, most extreme power is removed from the PV system to supply greatest accessible power into the grid. The wind and solar PV system are connected to the regular load through DC-DC Boost converter. In the remaining solitary mode, the converter needs to keep up consistent voltage and recurrence paying little heed to load unbalance or nature of the present which can be exceedingly misshaped, if the load is non-straight. The point of this paper is to study, outline and execute investigation of hybrid system comprising of wind and solar PV system connected to grid.

Keywords Electrical Power and Energy Systems Photovoltaic (PV) system · Wind System · T-Source Inverter · PV capacity · MPPT Control · Z-Source Inverter · Pulse Width Modulation (PWM) · Total Harmonic Distortion (THD) · DC-DC Boost converter

---

S. A. Sydu
Department of Engineering, E&E Section, University of Technology and Applied Sciences Al-Musannah, Al-Musannah, Sultanate of Oman

K. P. Shaik
Department of EEE, Malla Reddy College of Engineering and Technology, Hyderabad, India

M. S. Syed (✉)
Vignan's Lara Institute of Technology and Science, Guntur, India
e-mail: alif.shareef@gmail.com

K. Syed
University of Technology and Applied Sciences Shinas, Shinas, Sultanate of Oman

A. K. Sahu et al. (eds.), *Energy Storage and Conservation*, Springer Proceedings in Energy, https://doi.org/10.1007/978-981-99-2870-5_31

27.1 Introduction

The need for electricity could quickly rise due to the development and industrialization of the global population in the immediate future. This rise in the demand for energy needs electricity to increase its output [1]. Latest reports have projected that global net power production would rise by 24.4 trillion kilowatt hours (up 41%) and by 33.3 trillion kilowatt hours (up 92.5%), from 17.3 trillion kilowatt hours in 2005, in 2015. Today, fossil fuels provide a substantial share of energy, in particular coal, owing to its low costs [2]. The intensified usage of fossil fuels, however, contributes to most of the environment pollution and greenhouse gas emissions, the primary cause of the global warming. Emissions from CO_2 and mercury, for example, are estimated to rise, respectively, by 35% and 8% by 2020, as a consequence of the projected development of two power generations. In reality, two major problems for developed countries are possible exhaustion of fossil fuel supplies and the unpredictable oil price [3]. Renewable energy options should be used in the energy mix to address the issues associated with power production from fossil fuels. The light obtained from the sun is one of the green energies that can be used for this goal. Photovoltaic processes can turn this light into renewable electricity. The use of electricity production photovoltaic (PV) systems began in the 1970s and currently is fast increasing globally [4]. Many companies genuinely consider these programs to have a promising future. The European Photovoltaic Industry Association (EPIA), for instance, anticipates a worldwide gross combined photovoltaic potential of 200 GW by 2020 and 800 GW by 2030, as seen in Fig. 27.1. A new survey of over 2.26 GW of photovoltaic potential in the IEA reporting country was also submitted by the International Energy Agency (IEA) in the year 2007 [5]. This indicates that overall installed ability of these countries is rising by more than 50% year on year and reaches about 7.8 GW [6].

In three major fields of use, PV systems are typically used: (1) satellite applications, which provide support for solar arrays to satellites; (2) off-grid applications, which use solar arrays to power remote electricity grid-excluded loads; and (3) grid-connected, or network-connected, energy-supplying systems for the local and electricity grid regions. The PV market is currently controlled by grid-connected PV systems, especially in Europe, Japan, and the US in particular [7–10]. In 1990, but, as seen in Fig. 27.2, this figure was growing to over 90% by the end of 2007, for example, just 27% of cumulatively deployed PV power was linked to the grid. On the façades and roofs of buildings on the shades of parking areas, grid-connected PV systems may be built. You may also be mounted as power stations aimed at pumping all the power into the grid [11–15].

These systems also face a significant challenge in terms of the growing usage of PV systems because of their high cost of capital, as demonstrated by the energy they generate at a price per KWh. The use of recent developments to build low cost photovoltaic cells and offer rewards for consumer use to construct these systems will address this challenge. The fact that the increasing implementation of grid-based PV systems, particularly large systems in the megawatt order, will trigger some operating

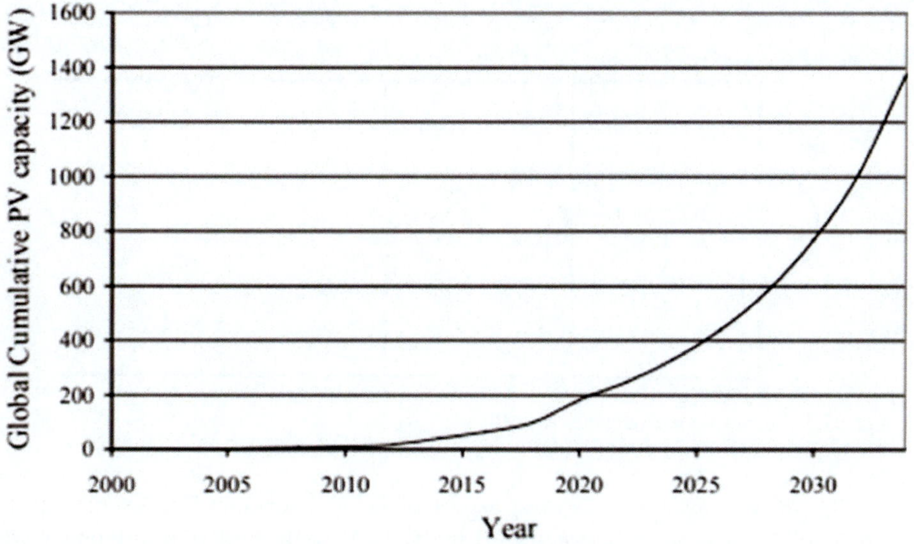

Fig. 27.1 Expected global cumulative PV capacity based on EPIA data

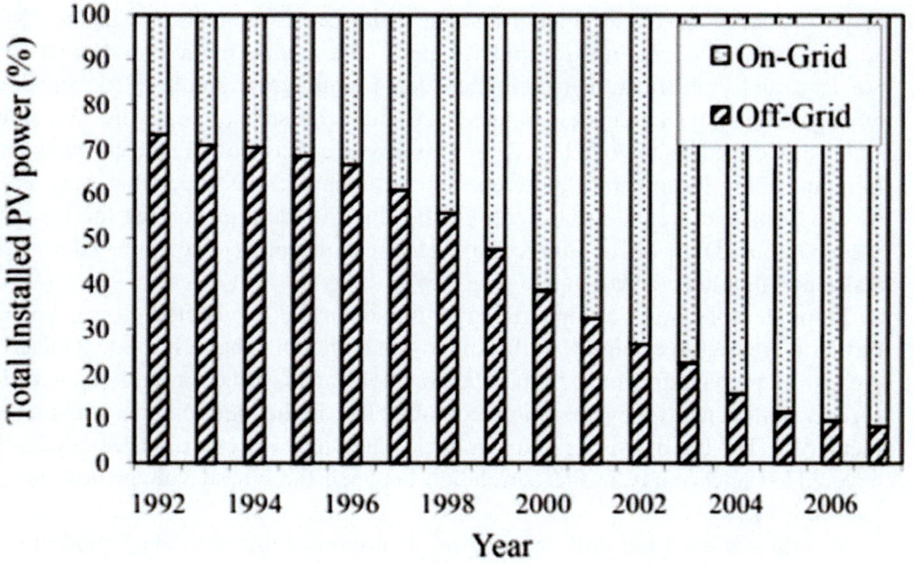

Fig. 27.2 Percentages of on-grid and off-grid PV power in the IEA-reporting countries

problems in the electric grid is another main issue faced by the large number of PV systems [16]. The primary objective of this research study is an accurate evaluation of the impacts of deploying photovoltaic systems on grid reliability and the identification of ways to minimize the operational challenges created by their construction [17].

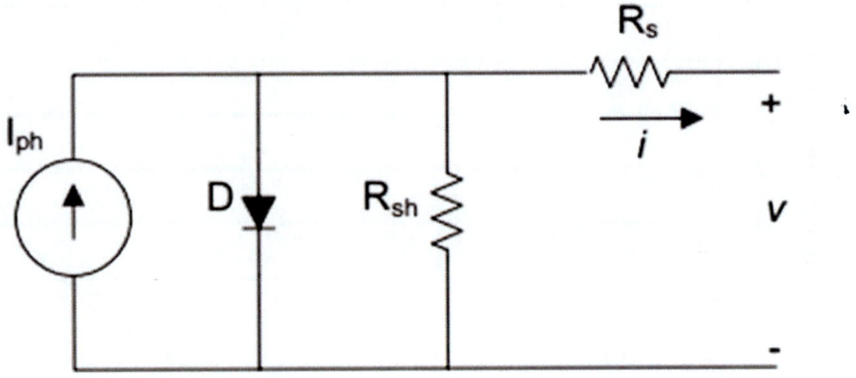

Fig. 27.3 Single-diode model of a PV cell

The solar array is the costliest aspect of the PV scheme. The average cost of photovoltaic modules is between 4.00 and 5.00/W, so the bulk of development work in this field focuses on the creation of low-cost solar cells that are effective and appropriate. Progress in this area would have a major influence on the wide usage of photovoltaic systems [18]. Many PV device efficiency studies include the use of the PV-array irradiance model and the atmospheric temperature in the PMPP corresponding PV-array maximum DC output. The models in the literature differ in precision and sophistication and are thus ideal for numerous studies [19]. One of the most common physical models for electrical characteristics of a specific PV cell is a single diode model shown in Fig. 27.3. This model consists of: (1) the current source Iph, describing the p-n link between the cell, a shunt resistance, Rsh, accounting for the leakage currents of the light, (2) the shunt diode representing the PV cell's p–n junction, (3) the drifting force of the electro-hole pairs (ideally, the value of this resistance should be zero).

Through replacing the single diode with two parallel-connected diodes, you can further increase the accuracy of the single-diode model. The first diode depicts the present of propagation in the quasi-neutral field of the function and has a 1. The second diode reflects the generation recombination in the junction area with an ideal factor of 2. The big downside of this model is that there are two diode equations that have added uncertainty in the correlation between the output voltage and the cell current [20].

Several studies have concentrated on determining the individual diode model parameters from the PV cell datasheets and on the effects of radiation and temperature on these parameters. Additional research suggests the use of new models to best reflect cell characteristics in PV cells. Some of these models increase the model's efficiency by the use of comprehensive physical processes in PV. In addition, soft computing approaches are used to model the output of PV cells, utilizing patterns of I-V curves at particular working conditions, under various operating conditions through training the PV models. The efficiency of PV cells was recently suggested with a mathematical model utilizing polynomial. This model is beneficial if the maximum power point is to be found in real time. The model has no physical significance, however, and its precision primarily depends on the measured cell data available.

Physical PV models typically include cell-level knowledge and thus are helpful in studying the PV device specifics like maximum power monitoring algorithms and partial shading impacts. These models are not appropriate, however, in the presence of PV systems to determine the efficiency of the electricity network. The key explanation for this is that these studies include the measurement, over long periods, of the power produced from the PV device under various conditions. Simpler versions are also commonly favored since the measurement quantity is decreased considerably. For e.g., the single-diode model may be optimized if the shunt resistance is indefinitely high or if shunt resistance and series resistances are omitted. Consequently, in the key model equation, the voltage and current are decoupled. The irradiance and temperature at all times are directly related to the maximum power produced from a PV device by other simpler models. These models are generally used to test the efficiency of the electric grid in the presence of a PV device. However, they should be validated with one of the full physical models in order to verify its accuracy before using those models in the study.

27.1.1 Wind Energy System

First of all, wind power was used for sailing Nile ships about 5000 years ago. In the 1700s and 1800s, the Europeans used it to pump water and food. The first electrical generating windmill was built in the USA in 1890. In 1979, on Howard Knob Mountain near Boon, a grid-linked wind turbine generator with a power of 2 MW. On Berger Hill in Orkney, Scotland, a 3-MW turbine was commissioned in 1988. In lighting buildings that are at distant locations and not linked with the grid, the energy produced by the wind is used. Today, wind power generation stations are accessible in limited sizes and can be attached to the power grids in standalone systems and larger power stations [21]. The world's wind power was about 39,294 MW in 2003, and India's wind power was 1550 MW in 2003.

27.1.2 System Configuration of Wind Energy

In Fig. 27.4 below, you can see the schematic of the wind energy device.

This machine comprises a wind turbine, which converts the cinematic energy of wind into spinning motion and is fitted with a gear box to balance the turbine speed and the power transformer, a generator that transforms mechanical energy into electric power, a rectification device which converts the AC voltage into DC [22].

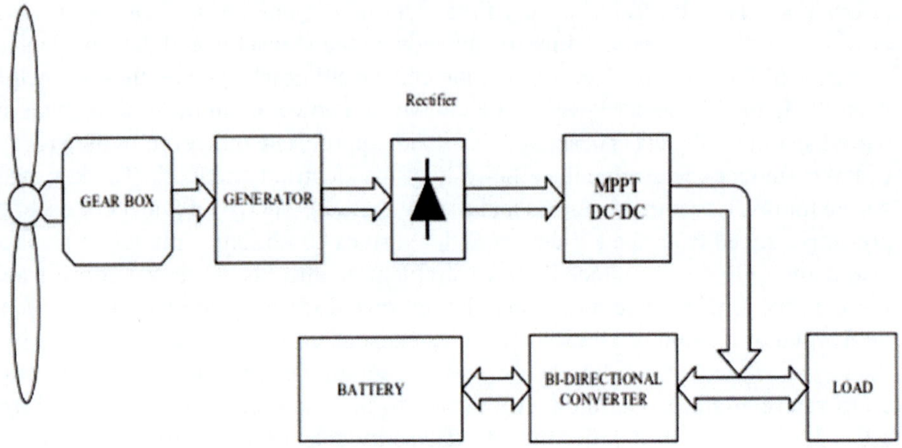

Fig. 27.4 Overall Block diagram of wind energy system

27.2 Proposed Methodologies for T-Source Inverter

The key aim of the research proposed is to reflect on this subject and establish a way of managing it. The following works were done for this purpose. The T-Source inverter is designed with a high voltage gain, lower typical leakage current and less passive part number, seen in Fig. 27.5 for transformers with a less grid linked PV device. With multiple shooting techniques, the proposed device is evaluated. Comparisons are made between basic boost control, maximum boosting control, maximum constant boost control, changed space vector PWM and maximum boosting of modified PWM space vector methods centered on voltage gain, stress voltage, input power ribs, popular mode voltage and common mode leakage currents to avoid improved shooting through systems for the T-Source inverter based grid linked PV device. The stability and performance of the device proposed were determined.

A new MPPT algorithm is built with the right interruption size to obtain the full usable power from the PV module, in order to increase Transformer's efficiency less grid-connected PV device. The findings are contrasted with traditional P&O and DP P&O algorithms for grid-linked PV system perturbation and observation. The proposed system's dynamic model is built using state space techniques. From the state-space model, both the input and output disruptions are controlled by the DC connection controller and the AC grid current controller. The actual and reactive power of the grid is regulated using the PI controller in the synchronous revolving reference frame by means of the voltage-oriented control in the grid-side converter. In order to predict the stability limit, the output of the controller suggested is analyzed separately. Finally, a 1 kW test model has been developed for the T-source Solar Energy Conversion Device, which is contrasted with the current system.

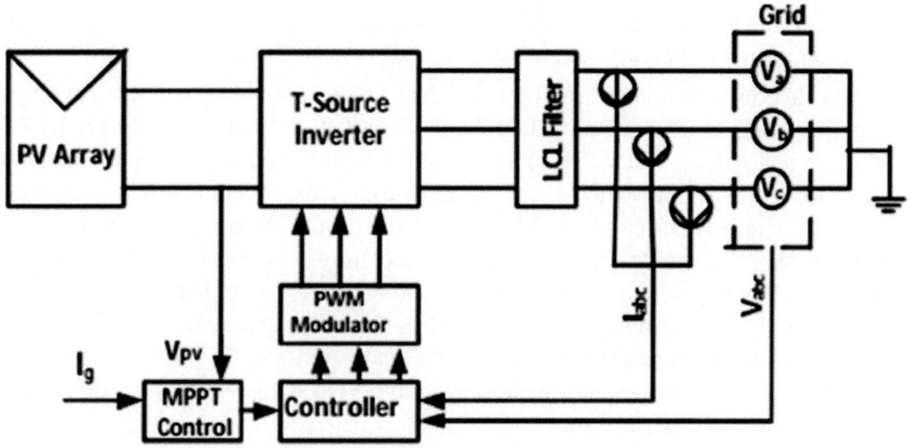

Fig. 27.5 Block diagram of TSI-based grid-connected PV system

27.3 Building of a Proposed System

Parameters for PV device performance analysis the performance parameters used during this study are those developed by the IEA. The specifications for the performance analysis of the grid PV system have been established. The total device output, solar resource, system errors overall, power ratio, PV scanner performance, and PV system efficiency are described in many parameters. There are a variety of parameters used for the description. For this analysis, energy and efficiency parameters have been determined using documented monitoring data.

The key goal of the study suggested in this thesis is to increase the penetration rate of electricity network systems, PV systems. This can be accomplished by precise quantification and study of the effect on electrical network efficiency of wide network-connected photovoltaic systems installed. The creation of a modern and smart approach is implemented to accomplish this goal. The approach accurately uses the data available to deliver precise practical reports on electricity network output without overestimating or underestimating the effects of the PV device. In order to calculate of the performance power profile of the PV device, the technique uses the historical environmental data obtained over many years. The approach often takes into consideration the real electrical network details. In order to detect future functional challenges, an interface between the output power of the PV device and the electrical network components should then simulate.

When the operational problems resulting from the implementation of PV systems, especially due to power fluctuations, have been established, numerous techniques are analyzed in depth to mitigate these problems. Such techniques require construction, usage of dumping loads and service below the full power point for energy storage systems. Their economic dimensions are discussed in the analysis of mitigation strategies. The economic dimension is important for photovoltaic systems because of their high cost, which is expressed in the price of their electricity.

Hybrid power should be used for reducing the need for energy storage. The impact of power likelihood inadequacy (DPSP), reasonably high-power generation (REPG), and energy-to-charge ratio (ELR), fraction of PV and wind power and PV and wind power covers were investigated against device dimension and efficiency. The technological viability of the photovoltaic wind hybrid device was tested in the unique demand range. The Life Cycle Cost (LCC) technique for economic assessment has been developed and simulated using the model for the autonomous photovoltaic systems, independent wind systems and PV wind hybrids. Detailed study of the cost analysis of the power supply for grid line expansion for a PV wind hybrid device. The optimal mix of solar photovoltaic wind hybrid systems varies from 0.70 to 0.75 of solar power to charge ratio.

The PV-wind hybrid device returns the lowest unit's costs in competition with separate solar and wind systems, retaining the same amount of DPSP. The sniveled energy costs of the PV-wind hybrid device are often smaller for all load specifications than those of the single solar PV or wind system. For rural electrification, the PV-Wind Hybrid Alternative is theoretically viable.

27.3.1 Wind Power System and Track Its Maximum Power Point, Implement Hybrid System

The ever-increasing need for renewable energy sources has necessitated a high decline of fossil fuels worldwide. The promising solution to satisfy the continuously growing demand for electricity is photovoltaic and wind power. These systems are highly recommended for remote applications since they are more appropriate. These systems produce diesel generators that are typically designed for brief intervals in order to satisfy the maximum demand for electricity when the need is met by a shortage of sufficient energy. In order to minimize the need for the diesel engine, a battery bank may be used. The photovoltaic device is not used to shield batteries from deep releases; however, the inclusion of a wind system will protect batteries from deep discharges and thereby prolong their existence. An advanced green energy system integrates solar and wind energy production to satisfy the load requirement of a given site with ample insolation and wind speed.

A model was built in the present paper, modeling, evaluating and optimizing the hybrid solar PV-wind device, taking due account of the different parameters appropriate to build the standalone wind/solar and hybrid system (wind + PV) with battery back-up. The different design specifications involve details on load, wind machine model, solar panel, battery and site requirements such as latitude, longitude, wind speed and solar insolation. The parameters considered are: REPG or life-cycle expense, or imposed energy cost or life cycle cost units to maximize the size of the hybrid device for a given DPSP.

The T-Source inverter is a single-phase power converter consisting of a combined pulse width modulation (PWM) scheme in a linked inductor and a condenser in the network to provide a buck-boost capacity. Mathematical modeling and interpretation in this chapter was addressed to forecast the efficiency of the TSI. Even for various firing methods, voltage gains, voltage strains, inductor current torsion and typical mode voltage, the boost factor is computed and simulated using MATLAB/Simulink.

27.4 MATLAB Simulation

The suggested Z-source converter is a first PWM control technique. In this method, it is compared with a carrier signal to generate the PWM pulse that the positive or negative value of the DC and the reference signal is shown in Fig. 27.6, demonstrates the scheme of basic boost power. Positive and negative values are generated by the standard PWM system as pulses are created when crossing the reference signal. Here, the inverter switches to shoot in status as the waveform is wider than the positive side, Vp, or smaller than the negative line VN. It also functions as a standard PWM courier.

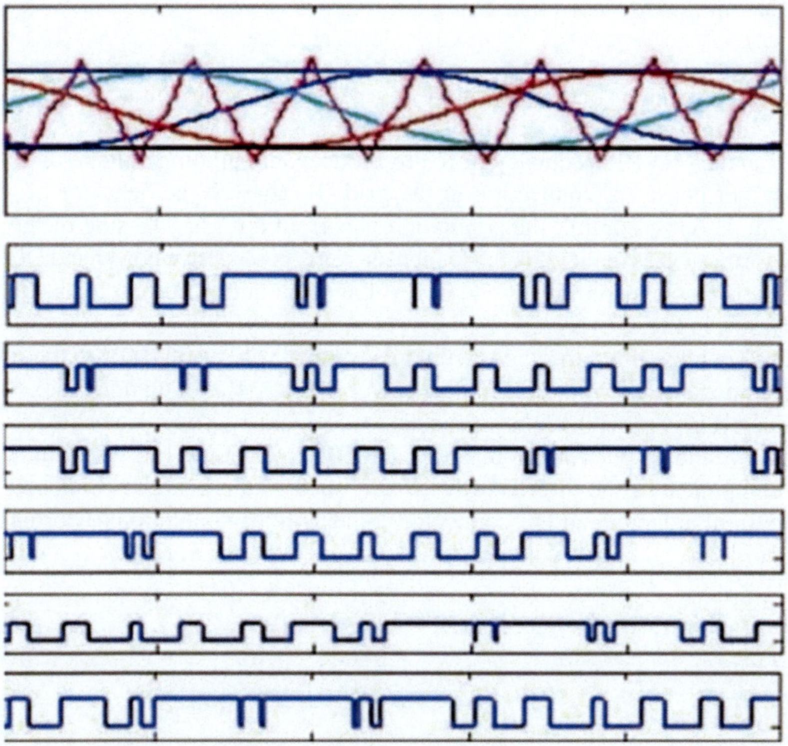

Fig. 27.6 Schematic of simple boost control

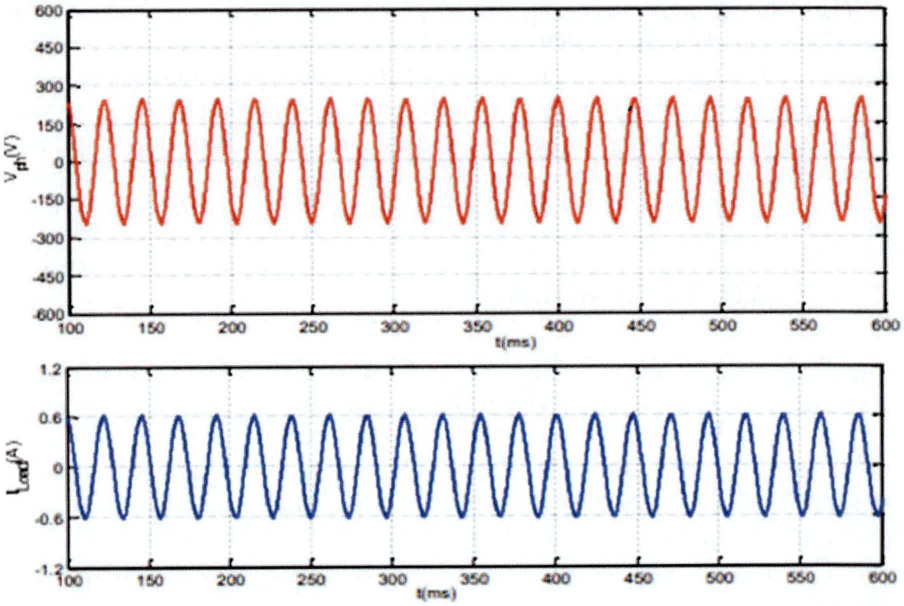

Fig. 27.7 Simulation results of TSI with simple boost control: **a** DC side parameter Vdc, IL, Vc & Vp V. **b**Output side parameter Vph & IL

The simulation is performed on the inverter with a photovoltaic output voltage of 100 V, an Impedance Network induced 5mH, capacity 1000μF, an Index Modulation = 0.7 and an ID-RP frequency of 0.3 to produce an outgoing voltage of 230 V and 50 Hz per phase for connection to the grid. To analyze the behavior of TSI with the simple boost control, the simulation is carried on. The DC side of the inverter waveform, such as input voltage, condenser voltage, inductive power and DC connect voltage is displayed in Fig. 27.7a. The voltage and output current load side inverter are shown in Fig. 27.7b

The simulation is carried out with a PV output voltage of 100V to the inverter, a 5mH impedance network inductance, power is 1000μF, modulation index=0.7 and a service cycle is set at 0.42, boost factor is 6,338 and 10 kHz switchable frequencies to achieve the output voltage of 230V and 50Hz per point. Fig. 27.8a indicates the DC component of the inverter wave shape, such as input stress, condenser tension and current inductor and DC connector voltage, in the simulation performance. The inverter load and output voltage of the load side are seen in Fig. 27.8.

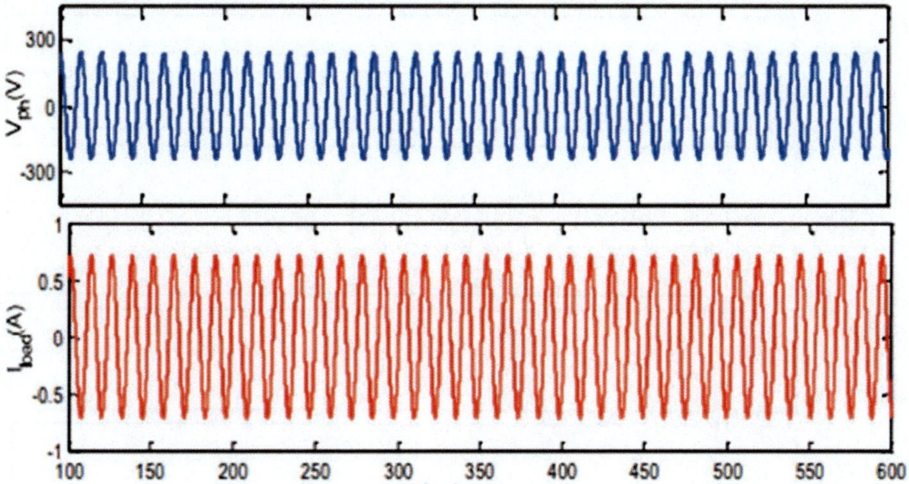

Fig. 27.8 TSI simulation results' inverter load and output voltage of the load side

27.4.1 Modified Space Vector Pulse Width Modulation (MSVPWM)

PW scheme space vector was more commonly used because of its large modulation index and low current harmonics. In Impedance Source Space Vector, PWM is mildly changed to monitor the fire and the space vector PWM (MSVPWM) is modified Space Vector. It also has an extra T0 shot to raise the DC inverter link voltage and the time span T1, T2 and TZ. The shoot through states is spread equally to and step through T0/26 within zero voltage cycles (TZ). Inside shooting duration, the zero voltage intervals should be shortened and the active states T1 and T2 are unchanged.

The simulation is conducted with PV output voltage feed at the inverter 100 V for the study of TSI principles with full boost power, the impedance network has a 5 mH capacity of 1,000 μF, the service cycle modulation index = 0.315, a boost factor of 2.72 and 10 kHz switching frequency to generate an output voltage of 230 V and 50 Hz per step to link to the inverter. The simulation effects are seen in the DC side, as input voltage, condenser voltage, inductor current and DC relation voltage, as seen in Fig. 27.9a. Output currents on the load side inverter are shown in Fig. 27.9b.

The application of high voltage tension on control devices induces temperature stress. By correct regulation of duty by shoot, the switching voltage tension in T-source inverter is minimized. This improves the TSI's durability. T-source inverters and Z-source inverters (ZSI) are presented in Fig. 27.10 for tension and voltage rise comparison. On gain G1, the voltage of TSI is smaller than ZSI, so the shot over time is spread similarly nil in TSI.

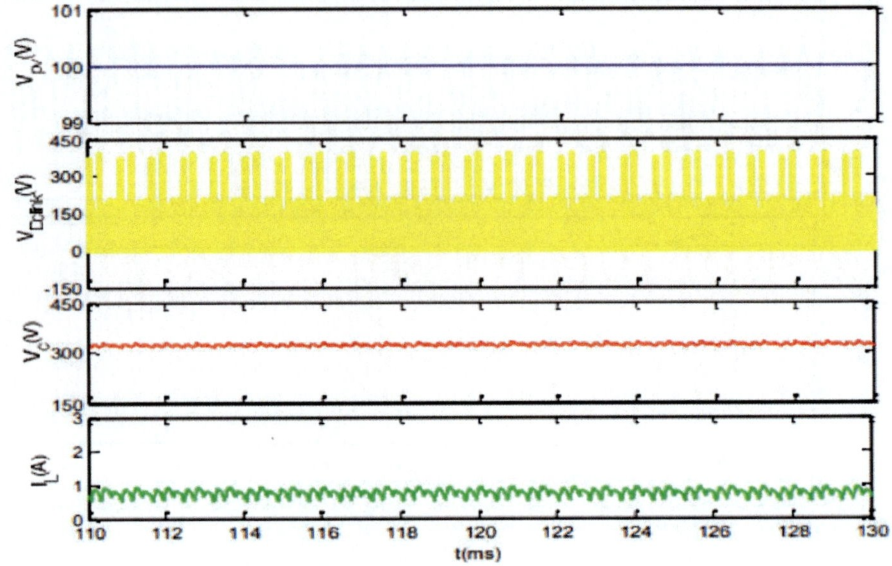

(a)

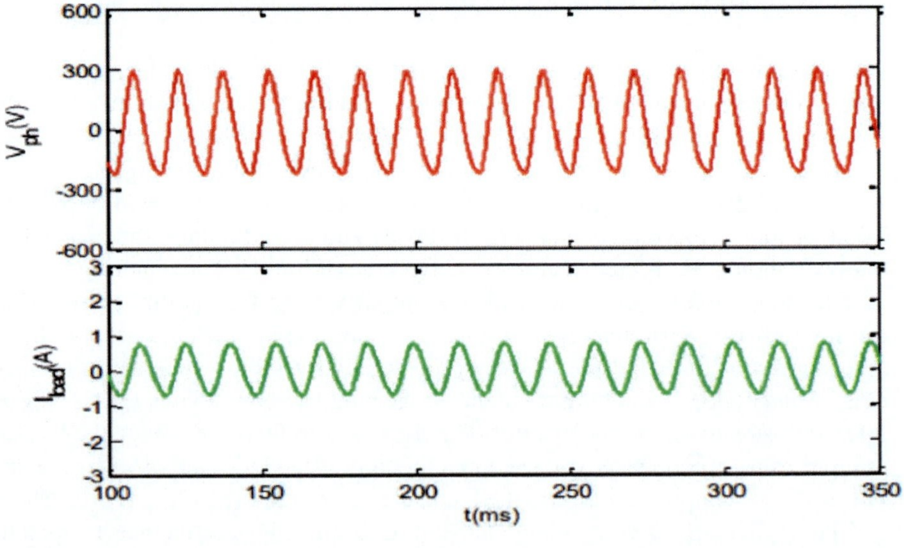

(b)

Fig. 27.9 TSI simulation effects for the **a** DC side parameter. **b** Performance side of the boost control

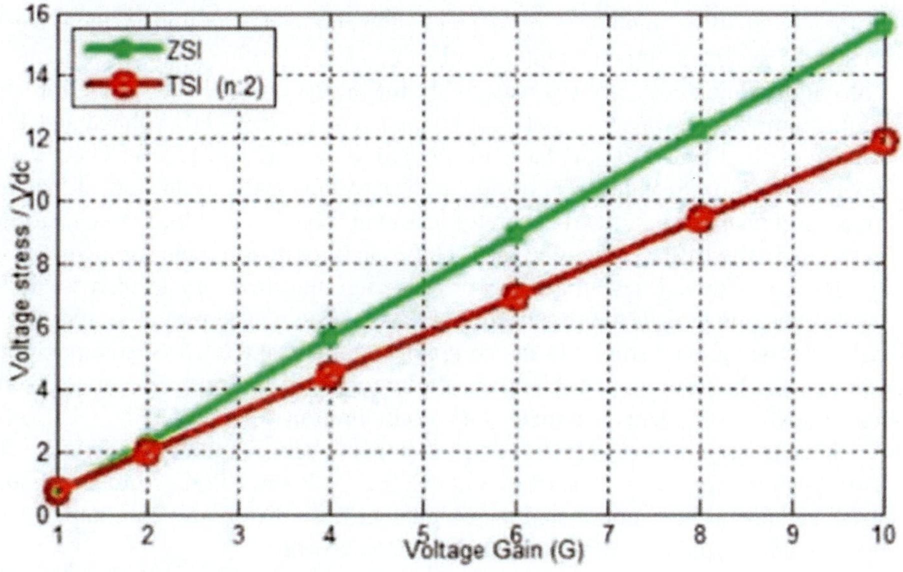

Fig. 27.10 Voltage stress and voltage gain comparison of TSI and ZSI

27.5 Conclusion

In the presence of PV systems, a variety of methods may be used to test the electrical network efficiency. There are no details on the impacts of the PV device power variations through the deterministic and probabilistic techniques. On the other side, this role can be accomplished by using temporal details inside the study with the method focused on chronological simulations. However, none of the studies found here can be used using a long background of data (in recent years historical data) while maintaining the time details of the variations in photovoltaic capacity. In order to provide a detailed measurement of the output of the method, the value of using long-term historical data is to provide several potential trends produced by the PV system. In addition, the findings obtained from long historical data will lead to advising the device operator about the "when" and "how frequently" inappropriate network output. Accordingly, effective operating arrangements should be planned to mitigate issues resulting from the implementation of the PV device. These arrangements involve selecting the operational power component, storage specifications and operation below the MPP.

To conclude, an approach should be established which can resolve the previously stated disadvantages of the current approaches and encourage the analysis of various solutions to reduce the magnitude of operational problems in delivery networks induced by broad PV systems installing. By this approach conclude that the most effective approach has been identified by the study of the T-Source Inverter (TSI).

Finally, there are improvements in irradiation and load conditions to the PV grid-connected system's output evolution and its control algorithms. The total harmonic TSI output voltage deviation is reasonably low at maximum load. Furthermore, in both situations, the waveforms are not warped. The present THD voltage and THD voltage of the PVGCS on the basis of the TSI were 3.62 and 1.76%. The PV grid device linked to TSI induces sinusoidal power voltage and current against load and irradiation phase changes and generates less than 5% of overall harmonic distortion as set out in the IEEE 1547 guidelines. Thus, the filter requirement for the proposed system is minimized, resulting in error reduction and therefore an increase in the overall performance of the proposed TSI-based system. The proposed TSI and its controller had good disruptions and comparison monitoring in 188 in all operational conditions. Thus, the T-source inverter is the better alternative over a Z-source inverter for a transformer less grid-connected PV configuration.

The newest invention of interfacing instruments between the grid and utilities is using custom power devices such as STATCOM, DVR and UPQC in order to resolve stress and power interruption and increase the efficiency of power by compensating reactive and harmonic power generated or load absorbed.

References

1. H. Abu-Rub, A. Iqbal, S.M. Ahmed, F.Z. Peng, Y. Li, G. Baoming, Quasi-z-source inverter-based photovoltaic generation system with maximum power tracking control using ANFIS. IEEE Trans. Sustain. Energy. **4**(1), 11–20 (2013)
2. Ahmad Al, Constantinos Sourkounis, Variable step size P&O MPPT algorithm for PV systems. in *Proceedings of the 12th international conference on optimization of Electric and Electronic Equipment OPTIM* (2010), pp. 1097–1102
3. B. Subudhi, R. Pradhan, A comparative study on maximum power point tracking techniques for photovoltaic power systems. IEEE Transaction on Sustainable Energy, 4(1), 89–98 (2013).
4. B. Yang, W. Li, Y. Gu, W. Cui, X. He, Improved transformer less inverter with common-mode leakage current elimination for a photovoltaic grid-connected power system. IEEE Trans. Power Electron. **27**(2), 752–762 (2012)
5. F. Bradaschia, M.C. Cavalcanti, P.E.P. Ferraz, F.A.S. Neves, E.C. Dos Santos, J.H.G.M. Da Silva, Modulation for three-phase transformer less Z-source inverter to reduce leakage currents in photovoltaic systems. IEEE Trans. Industr. Electron. **58**(12), 5385–5395 (2011)
6. A. Bulawka, S. Krauthamer, R. Das, Power conditioning subsystems for photovoltaic central-station power plants: state-of-the-art and advanced technology. IEEE Trans. Energy Convers. **EC-l**(1), 47–53 (1987)
7. M. Castilla, J. Miret, A. Camacho, J. Matas, L.G. De Vicuna, Reduction of current harmonic distortion in three-phase grid-connected photovoltaic inverters via resonant current control. IEEE Trans. Industr. Electron. **60**(4), 1464–1472 (2013)
8. M.C. Cavalcanti, G.M.S. Azevedo, B.A. Amaral, K.C. De Oliveira, F.A.S. Neves, Z.D. Lins, Efficiency evaluation in grid connected photovoltaic energy conversion systems. in *Proceedings of the 36th IEEE Power Electronics Specialists Conference* (2005), pp 269–275
9. S.A. Sydu, Modeling and analysis of a dynamic voltage restorer (DVR) for robustness constraints with 3-level inverter and sinusoidal pulse width modulation (SPWM). J. Adv Res. Dyn. Control. Syst. **10**(07), 1972–1978 (2018)
10. A. Chatterjee, A. Keyhani, D. Kapoor, Identification of photovoltaic source models. IEEE Trans. Energy Convers. **26**(3), 883–889 (2011)

11. Y. Chen, K. Smedley, Three-phase boost-type grid-connected inverter. IEEE Trans. Power Electron. **23**(5), 2301–2309 (2008)
12. C-S. Chiu, T-S. Chiang, Y-T. Lee, Maximum power control of wind energy conversation systems via a T-S fuzzy model-based approach, in *International Conference on Fuzzy Systems*, vol. 6 (2010). pp. 18–23
13. M.A.G. De Brito, L. Galotto, L.P. Sampaio, G. De Azevedoe Melo, C.A. Canesin, Evaluation of the main MPPT techniques for photovoltaic applications. IEEE Trans. Industr. Electron. **60**(3), 1156–1166 (2013)
14. K.C.A. De Souza, M.R. De Castro, F. Antunes, A DC/AC converter for single-phase grid-connected photovoltaic systems, in *Proceedings of the 28th Annual IEEE Conference on Industrial Electronics Society IECON 02*, vol. 4 (2002), pp. 3268–3273
15. D. Sera, R. Teodorescu, J. Hantschel, M. Knoll, Optimized maximum power point tracker for fast-changing environmental conditions. IEEE Trans. Industr. Electron. **55**(7), 2629–2637 (2008)
16. O. Ellabban, J. Van Mierlo, P. Lataire, Experimental study of shoot-through control methods for Z-source inverter. Eur. Power Electron. (EPE) J. **21**(2), 18–29 (2011)
17. O. Ellabban, J. Van Mierlo, P. Lataire, A DSP-based dual-loop peak DC-link voltage control strategy of the Z-source inverter, IEEE Trans. Power Electron. **27**(.9), 4088–4097 (2012)
18. O. Ellabban, J. Van Mierlo, P. Lataire, Voltage mode and current mode control for a 30kw high-performance Z-source inverter, in *Proceedings of the IEEE Conference on Electrical Power & Energy* (2009), pp. 1-6
19. J.M. Espi Huerta, J. Castello-Moreno, J.R. Fischer, R. Garcia-Gil, A synchronous reference frame robust predictive current control for three-phase grid-connected inverters. IEEE Trans. Industr. Electron. **57**(3), 954–962 (2010)
20. S. A. Sydu, A novel method to enhance of power quality in distribution system using IDVR. Int. J. Recent. Technol. Eng. **8**(2S11), 1028–1034 (2019)
21. F. Liu, S. Duan, F. Liu, B. Liu, Y. Kang, Variable step size INC MPPT method for PV system. IEEE Trans. Industr. Electron. **55**(7), 2622–2628 (2008)
22. C.J. Gajanayake, D.M. Vilathgamuwa, P.C. Loh, Development of a comprehensive model and a multiloop controller for Z-source inverter DG systems. IEEE Trans. Industr. Electron. **54**(4), 2352–2359 (2007)